REFRACTORY CARBIDES

STUDIES IN SOVIET SCIENCE

PHYSICAL SCIENCES

A Continuation Order Plan is available for this series. A continuation order will bring
delivery of each new volume immediately upon publication. Volumes are billed only upon
actual shipment. For further information please contact the publisher.

STUDIES IN SOVIET SCIENCE

REFRACTORY CARBIDES

Edited by
Grigorii V. Samsonov
Institute of Problems in Materials Science
Academy of Sciences of the Ukrainian SSR
Kiev, USSR

Translated from Russian by
Noel B. Vaughan

CONSULTANTS BUREAU • NEW YORK AND LONDON

Library of Congress Cataloging in Publication Data
Main entry under title:

Refractory carbides.

   (Studies in Soviet science)
   Translation of Tugoplavkie karbidy.
   Includes bibliographical references.
   1. Transition metal carbides — Addresses, essays, lectures. 2. Refractory transition metal compounds — Addresses, essays, lectures. I. Samsonov, Grigoriĭ Valentinovich, ed. II. Series.
TA418.26.T8313   620.1'43   73-83892
ISBN 978-1-4684-8600-1   ISBN 978-1-4684-8598-1 (eBook)
DOI 10.1007/978-1-4684-8598-1

The original Russian text, published by Naukova Dumka Press in Kiev in 1970, has been corrected by the author for the present edition. This translation is published under an agreement with the Copyright Agency of the USSR (VAAP).

ТУГОПЛАВКИЕ КАРБИДЫ
Г. В. Самсонов
TUGOPLAVKIE KARBIDY
G. V. Samsonov

© 1974 Consultants Bureau, New York

Softcover reprint of the hardcover 1st edition 1974

A Division of Plenum Publishing Corporation
227 West 17th Street, New York, N.Y. 10011

United Kingdom edition published by Consultants Bureau, London
A Division of Plenum Publishing Company, Ltd.
4a Lower John Street, London W1R 3PD, England

Preface

The present stage of technological development makes new and ever more complex demands on materials that have to work under conditions of high temperature and pressure, in high vacuum, and in corrosive media. In consequence special importance is now attached to the refractory compounds of transition metals of groups IV to VI with such nonmetals as boron, carbon, silicon, and nitrogen. These compounds possess high melting points, great hardness, and high refractory and corrosion-resisting properties. The most widely used and important compounds of this type from a technological point of view are the carbides, which are already fairly widely used in various fields of technology.

The present collection of papers contains the results of recent investigations into methods of producing high-purity carbides and also components made of the carbides and their alloys. Great attention has been paid to the study of a wide range of properties of the carbides and of alloys based on them, viz., the electro- and thermophysical, thermodynamic, mechanical, and chemical properties, and also to the utilization of the carbides as wear- and abrasion-resistant materials.

In contrast to many previous publications dealing with carbides, the results presented in this collection relate to the properties of carbides having a definite phase composition, corresponding to a higher degree of purity. In some of the contributions the physical and chemical properties of the carbides are interpreted in terms of certain solid-state models and concepts concerning the types of chemical bonding in these compounds.

Contents

II. METHODS OF PRODUCING COMPONENTS OF CARBIDES AND ALLOYS BASED ON THEM

III. REFRACTORY-CARBIDE COATINGS

IV. PHASE DIAGRAMS OF SYSTEMS CONTAINING CARBON

V. PHYSICAL PROPERTIES OF CARBIDES

VI. THERMODYNAMIC PROPERTIES OF CARBIDES

VII. MECHANICAL AND ABRASIVE PROPERTIES OF CARBIDES

Some Problems in the Theory of the Properties of Carbides

G. V. Samsonov

Features of the electron structure of carbides of d-transition metals
are discussed in terms of the configuration-localization model. It is
shown that the factors which mainly determine the level of the properties of carbides are the statistical weight of the carbon atoms with
sp^3-configurations, the localization of electrons in the d^5-configura-
tion, and the concentration of excess nonlocalized electrons in the
carbide lattice.

From the practical point of view, carbides form the most important class of inorganic refractory compounds.

The nature of the properties of the carbide phases has not been
adequately studied, and this has made it difficult to systematize
their physical and chemical properties. The most widely adopted
method of approach to the analysis of the nature of the properties
of carbides consists of the use of various modifications of the
density-of-states model [36-40]. Unfortunately this model describes satisfactorily only the physical properties of carbide phases
(and even then only those with the simplest structures). Attempts
have been made to utilize other models to elucidate the nature of
the properties of carbides, in particular the configuration-localization model. The main principle of this [6, 13, 16-18, 43, 43] lies
in the fact that on the formation of the condensed state from the
isolated atoms the valence electrons separate into those localized
within the frameworks of the atoms and unlocalized electrons; the
localized part of the valence electrons forms a broad spectrum of

1

configurations in which the greatest statistical weights belong to
the configurations that are most stable in terms of energy, i.e.,
those to which the minimum store of free energy can be ascribed.
The nonlocalized valence electrons are in a statistical exchange
with the electrons that form stable configurations (i.e., the most
stable in terms of energy) and are responsible also for the internal
electron—electron interaction.

The following conclusions can be drawn from a consideration,
on the basis of the given model, of the special features of the electron structure of carbides of d-transition metals, the refractoriness of which is the reason for the very great practical interest
in them.

In the formation of scandium monocarbide, as a result of the
small degree of localization of the valence electrons within the
framework of its atoms (the statistical weight of atoms with stable
d-configurations — SWASC — is about 18% for metallic scandium
[19]), a large part of the valence electrons are transferred to the
nonlocalized state; although this leads to stabilization of the sp^3-
configurations of the carbon atoms,* it is at the same time responsible for the great store of energy in scandium carbide and
for the strong electron—electron interaction, which is the cause
of the great difficulties experienced in obtaining this carbide [20]
in the pure state. This phase is stabilized only in the presence of
acceptor impurities such as oxygen atoms which are capable of
capturing part of the nonlocalized electrons of scandium to form
stable s^2p^6-configurations ($s^2p^4 + 2e = s^2p^6$); consequently, it is
usually only the oxycarbides of scandium that are fairly easily
formed.

Turning now to titanium carbide TiC, we find that localization
of the valence electrons is higher at the titanium atoms (SWASC =
43%), which substantially reduces the energy stored in TiC as
compared with ScC, thereby rendering the formation of the carbide
phase relatively simple by the stabilization of the sp^3-configura-
tions of the carbon atoms and correspondingly producing high values of the melting point and hardness. (However, in the case of
TiC it must be recorded that oxycarbides are very easily formed in

---

* By stabilization of the sp^3-configurations is meant the shift, for example, of the
equilibrium $sp^3 \rightleftharpoons sp^2 + p$ to the left in the presence of additional free or readily
mobile electrons.

comparison with the carbide, and also that it is extremely difficult
to remove oxygen from it.) The high values are due to the formation of sp^3-configurations at a moderate proportion of nonlocalized
electrons. That this proportion is still moderate is due to the fact
that at the temperature of formation of TiC the titanium atoms assume a bcc lattice as a result of the conversion, with rise in temperature to the $\alpha \rightarrow \beta$ transformation, of the less stable, in terms
of energy, d^x -configuration ($x < 5$) to the more stable d^5-configura-
tions [21]. Hence the statistical weight of the d^5-configurations in
β -Ti should be not 43% as in α-Ti, but of the order of 60% (in
vanadium, which has a SWASC of 63%, a bcc lattice without any kind
of polymorphic transformation appears for the first time in the
series of d-transition metals).

It must again be emphasized that the high melting point and
hardness of TiC are due to the existence of sp^3-configurations,
stabilized by the nonlocalized electrons of the titanium atoms, but
"diluted" as regards energy by the considerably less stable d^5-
configurations of the titanium atoms. Extrapolation of the concentration dependence of the hardness of TiC to zero carbon content
is known to lead to the hardness value of titanium [7]. In the same
way it is to be expected that as titanium atoms are removed from
the TiC lattice the hardness should approximate to that of diamond,
if the sp^3-configurations are retained in the stabilized state. In
point of fact this does not occur, since as titanium atoms are removed from the carbide delocalization of the valence electrons
of carbon takes place with breakdown of the sp^3-configurations.
It is also known that as a result of delocalization the sp^3-config-
urations in diamond breakdown with rise in temperature with the
formation of sp^2-configurations, which determine the presence of
the plane networks in the structure of graphite and of the p-elec-
trons which are responsible for the conductivity of graphite. In
the absence of oxygen this process goes on actively at temperatures of the order of 2000°C [35] and gives rise to the transformation of diamond into graphite. At the same time TiC re-
tains its fcc lattice without any polymorphic transformations up to
the melting point (3000°C), i.e., the whole "significance" of TiC
lies in the fact that in its lattice breakdown of the sp^3-configura-
tions is to a considerable extent deferred to higher temperatures
than those at which breakdown of the sp^3-configurations occurs in
diamond, in whose lattice there are no stabilizers of the type of
titanium in TiC. Thus TiC represents a unique compromise as its

hardness, although less than that of diamond, is retained at temperatures which diamond cannot in general reach. The concepts of titanium as an electron donor in TiC are confirmed by the experimental results reported in [1, 8].

With the transition from TiC to ZrC and HfC, the sp^3-configurations of the carbon atoms are retained in a stable state as a result of the increase in SWASC to 52 and 55%, respectively (rising to 60% at the $\alpha \rightarrow \beta$ transformation), although the total proportion of nonlocalized electrons decreases somewhat. This reduction is due not to the greater values of the SWASC of the metals, since in general these values are the same for all metals of group IV at the $\alpha \rightarrow \beta$ transition, but to the transfer of part of the nonlocalized electrons to the deeper $4f$ and $5f$ states which are vacant in isolated atoms, and to their localization at these levels. This leads in particular to a fairly steep rise in the melting points of the series of carbides TiC – ZrC – HfC (3150, 3530, and 3890°C) as a result of the decrease in the total energy store, but not to any substantial change in hardness. The latter is determined mainly by the statistical weight of the sp^5-configurations, which remain almost constant for all the carbides of group IV metals. It may even be supposed that over the series TiC – ZrC – HfC stabilization of the sp^3-configurations decreases somewhat on account of localization of part of the nonlocalized electrons which determine the stabilization at the lower f levels. In fact the hardness falls somewhat over the series: TiC, 3000; ZrC, 2925; HfC, 2913 kg/mm^2 [22]. The electively) as a result of a reduction in the electron proportion due to a ly) as a result of a reduction in the electron proportion due to a decrease in the electron–electron interaction and to a reduction in the lattice proportion on account of greater localization of the valence electrons and a decrease in scattering at the stable configurations that have been formed.

In systems with carbon metals of groups V and VI are even weaker electron donors. It has been shown in [8] that during solution of carbon in tantalum the charge on the carbon atom becomes 2.8 and during solution in tungsten it becomes 0.6, whereas during solution of carbon in titanium, which is a strong donor, the charge on the carbon atom remains at 4. These figures, which have been established experimentally for solid solutions of carbon in the transition metals, naturally alter at the changeover to the carbide phases (in all probability they increase for carbon in tantalum and tungsten carbides), although their physical significance remains the same, indicating that tantalum and tungsten, like the weaker electron donors, are capable, though to a lesser extent than the group IV metals, of stabilizing the sp^3-configurations of the carbon atoms.

The formation of the carbide phases of the transition metals of group V is accompanied by the breakdown of the sp^3-configurations of the carbon atoms, and the delocalized electrons formed thereby can evidently be localized to a certain extent by transition-metal atoms having sufficiently large "reserves" for the formation of d^5-configurations (the SWASC is 63% for vanadium, 76% for niobium, and 81% for tantalum). This should lead to a small excess concentration of nonlocalized electrons and so to relatively high melting points, although to correspondingly lower ones than for the carbides of the group IV metals (owing to the diminishing contribution of the destroyed sp^3-configurations), but to low hardness values owing to the breakdown of the sp^3-configurations which are the determining factor in regard to hardness. In fact the melting points of the monocarbides of group V metals are: VC, 2800; NbC, 3480; and TaC, 3400°C, the microhardness values are 2094, 1960, and 1600 kg/mm^2, respectively [22]. The tendency of the carbon atoms to retain the sp^3-configurations and that of the metal atoms to show an increase in the SWASC of the d^5-configurations is reflected in the formation by the group V metals of the so-called subcarbides Me_2C which have higher hardness values than the corresponding monocarbides (Nb$_2$C, 2120; Ta$_2$C, 1700 kg/mm^2).

Returning to the monocarbides of the group V metals, we may conclude both from the SWASC values and from the significant reduction in microhardness from VC to TaC that the donor properties decrease from vanadium to tantalum in accordance with the increase in the degree of breakdown or destabilization of the sp^3-configurations of the carbon atoms. VC occupies a special place in the series of monocarbides. Vanadium is a donor, and moreover, owing to the low principal quantum number of the valence electrons, the d^5-configurations formed by the vanadium atoms have a low energy stability, and hence localization of the electrons from the destroyed sp^3-configurations of the carbon atoms at the networks of vanadium atoms takes place to only a small extent, a great part of the electrons remaining in the nonlocalized state. The latter is one of the reasons why it is impossible or difficult to obtain VC of stoichiometric composition, i.e., without a deficit of carbon atoms in the lattice.

Carbides of the group VI metals, the atoms of which have weak donor properties, cannot in any way stabilize the sp^3-configurations of the carbon atoms; at the same time they have high SWASC values, particularly in the cases of molybdenum and tungsten, which reduces the possibility of localization in the d^5-configuration, and most of the electrons from the destroyed configurations remain in the nonlocalized state.

The case of tungsten monocarbide is particularly significant. The breakdown of the sp^3-configurations in it is so great [8] and the "reserve" for the formation of d^5-configurations so small (SWASC for tungsten is 94%) that a large part of the valence electrons of carbon remains nonlocalized. This is responsible for the high energy reserve of WC, which is reflected in the low enthalpy of its formation from the elements, in its capacity for peritectic decomposition into the more stable subcarbide W_2C and carbon, in its ability to dissolve in the cubic carbides of other transition metals, and in its relatively low melting (decomposition) temperature of 2720°C. The extensive breakdown of the sp^3-configurations is responsible for the low hardness value of this carbide (1780 kg/mm^2), which would be even lower were it not for the compensating effect of the d^5-configurations of the tungsten atoms, which are very stable as regards energy and which have a higher SWASC. For the same reason the sp^3-configurations retained by the carbon atoms are insufficient for the formation of a bcc lattice, and so WC crystallizes in the less symmetrical hexagonal lattice of the nickel arsenide type [41]. With rise in temperature of WC we may expect strong localization of the valence electrons in the d^5-configurations of the tungsten atoms, with simultaneous breakdown of the sp^3-configurations of the carbon atoms. If the concentration of carbon atoms in the tungsten lattice is decreased in order to reduce the number of nonlocalized electrons formed during breakdown of the sp^3-configurations, a cubic form of the carbide WC_{1-x} can be obtained, which is based on a high SWASC of the d^5-configurations of the tungsten atoms and which is stable both at high temperatures and in the form of solid solutions, for example in the cubic carbides of group V metals. The cubic modification WC_{1-x} has been found experimentally, and the authors of [3] have shown that $\frac{2}{3} < x < 1$.

The unique concurrence between the increase in the SWASC of the d^5-configurations and the decrease in that of the sp^3-configurations with rise in temperature gives rise to a characteristic change in the mechanical strength of the carbides at high temperatures. It has been established in [23] that with rise in temperature of the carbides to $0.4-0.5T_{mp}$ their room-temperature strength, which generally is not high, is retained, while further rise in temperature leads to a considerable increase in strength with maxima at temperatures of $0.55-0.65T_{mp}$. At these temperatures appreciable signs of ductility are found, particularly in TiC. Finally, a still further rise in temperature leads to a rapid fall in strength. It may be assumed that with rise in temperature the SWASC of the d^5-configurations increases while that of the sp^3-configurations slowly diminishes (it may be recalled that the transformation of diamond into graphite is observed only at 2000-3000°C, although it begins at 1000-1500°C [35]); this causes an increase in strength, but at higher temperatures both the d^5- and sp^3-configurations break down, so leading to a reduction in strength. An increase in the SWASC of the d^5-configurations in TiC with rise in temperature occurs when there is a high concentration of nonlocalized electrons of the titanium atoms, which gives rise to ductility in this carbide in the same way as delocalization of the valence electrons of group IV elements (C —diamond, Si, Ge, Sn, Pb) leads to the appearance and increase of ductility even at room temperature.

Many of the carbide phases of the transition metals have a homogeneity range within which a change in carbon content takes place without any transformation of the crystal lattice. The width of this range is greatest in carbides of group IV metals; it narrows in carbides of group V metals, and is almost completely absent from carbides of group VI metals, except in the case of molybdenum carbide, in which the homogeneity range is about 4 at.% wide, and also that of the tungsten monocarbide deficient in carbon, which exists, as indicated above, at high temperatures. Thus the width of the homogeneity range decreases with increase in the SWASC of the d^5-configurations, particularly at the transition from the group IV metals, which are strong donors, to the metals of groups V and VI, which have weaker donor properties, and also with the concomitant reduction in the SWASC of the sp^3-configurations. It is clear that an important factor is the possibility of altering the composition of the phases without causing such a marked reduction in the SWASC of the d^5- or the sp^3-configurations as would lead to a change in structure. Such possibilities are especially great in carbides of group IV metals, since the donor capacity of the titanium atoms is sufficient to ensure stabilization of the sp^3-configurations of the carbon atoms over a wide range of carbon concentrations. In carbides of group V metals the lower donor capacity of these metals leads to a considerably more rapid breakdown of the sp^3-configurations, which is to some extent offset by the formation of d^5-configurations, which are also responsible for the existence of cubic syngony. The high SWASC of the d^5-configurations in the lattices of carbides of group VI metals cannot under ordinary conditions compensate for the low SWASC of the sp^3-configurations, and this leads either to no homogeneity range or the presence of only a very narrow one. It is characteristic that the width of the homogeneity range of carbide phases diminishes with rise in temperature, thus indicating that the principal factor governing the width of the homogeneity range is the SWASC of the sp^3-configurations.

The possibility of widely varying the SWASC of the sp^3-configurations without altering the crystal structure is due to the existence of only one carbide phase in each system of group IV metals with carbon. Metals of group V form two carbides per system — Me_2C and MeC — while metals of group VI form several carbides per system; for example, chromium forms the carbides $C_{23}C_6$, Cr_3C_2, and Cr_7C_3, though this is due not only to the SWASC of the sp^3-configurations but also to the possibility of combining the d^5- and d^{10}-stable configurations. In manganese the number of carbides increases to five — Mn_4C, $Mn_{23}C_6$, Mn_3C, and Mn_7C_3; metals of group VIII have a large number of carbide phases.

The variation in the physical properties of the phases within their homogeneity range is unambiguously represented by the model proposed in [9, 24], including the extreme values of the properties which appear as a result of the nonuniform increase of the SWASC of the d^5- and sp^3-configurations that occurs with the reduction of the carbon content in the phase.

Complex carbide systems [3-5, 14, 25] can be similarly represented, and so can the catalytic properties of carbides [2, 26], their superconductivity [27, 28], the diffusion processes associated with the formation of carbides [29, 30], the recrystallization processes in carbides [10, 11, 31], the characteristics of the fric-

tional and abrasive properties [32-34], and features of the processes that occur in the production of carbides [12].

We must also consider the oft-discussed question of the magnitude of the thermal effects of the reactions involved in the formation of carbide phase from their elements. Consideration of the
relevant values of the thermal effects [22] leads to the conclusion
that the formation of both the d^5- and sp^3-configurations contrib-
utes to these values, although the formation and stabilization of
the sp^3-configurations play the most important part. Thus, as the
SWASC of the sp^3-configuration decreases, the heat of formation
of the carbides changes very rapidly: TiC, 44; ZrC, 48; HfC, 50;
VC, 30.2; NbC, 34; TaC, 34.3; Cr_xC_y, 20-25; Mo_2C, 4.2; and WC,
9 kcal/mole, i.e., with the change from carbides of group IV metals to those of group V metals and then to those of group VI metals.
However, within the groups, the changes that can be attributed to
an increase in the energy stability of the d^5-configurations with
rise in the principal quantum number of the valence electrons,
and also on account of the localization of the electrons in the deep
f-levels, are considerably smaller. This again demonstrates the
great energy stability of the stabilized sp^3-configurations of the
carbon atoms in comparison with the most stable d^5-configurations,
for example, of such a metal as tungsten.

Conclusions

1. The application of the configuration-localization model of
the consideration of the properties of carbides of the d-transition
metals has shown that the factors which mainly determine the level
of these properties are the statistical weight of the carbon atoms
with sp^3-configurations and the conditions and degree of stabilization of these configurations.

2. With marked delocalization of the electrons which enter
into the composition of the sp^3-configurations, localization of the
electrons in the d^5-configurations begins to play a significant part,
and so does the concentration of the excess nonlocalized electrons
in the carbide lattice.

Literature Cited

1. M. P. Arbuzov and B. V. Khaenko, Poroshkovaya Met., No. 4, 74 (1966).
2. I. M. Gitis et al., in: Scientific Bases of the Choice of Catalysts [in Russian],
    Nauka, Moscow (1966), p. 230.

3. E. N. Denbnovetskaya, Author's Summary of Candidate's Thesis, IPM, Akad. Nauk UkrSSR, Kiev (1967).

4. E. N. Denbnovetskaya, Poroshkovaya Met., No. 3, 32 (1967).

5. E. N. Denbnovetskaya, Poroshkovaya Met., No. 4, 57 (1967).

6. I. R. Kozlova, V. N. Gurin, and A. P. Obukhov, Poroshkovaya Met., No. 12, 68 (1966).

7. A. E. Koval'skii and T. G. Makarenko, Zh. Tekhn. Fiz., 23:205 (1953).

8. I. I. Kovenskii, Fiz. Tverd. Tela, 5:1423 (1963).

9. A. Ya. Kuchma, Author's Summary of Candidate's Thesis, IPM, Akad. Nauk UkrSSR, Kiev (1968).

10. I. P. Kushtalova, Poroshkovaya Met., No. 8, 7 (1967).

11. I. P. Kushtalova and A. N. Ivanov, Poroshkovaya Met., No. 9, 81 (1966).

12. G. N. Makarenko and O. F. Kvas, Poroshkovaya Met., No. 8, 34 (1967).

13. I. F. Pryadko, Poroshkovaya Met., No. 12, 61 (1966).

14. V. N. Paderno, Izv. Akad. Nauk SSSR, Neorg. Mat., 3:1177 (1967).

15. G. V. Samsonov, Ukrainsk. Khim. Zh., No. 10, 1005 (1965).

16. G. V. Samsonov, Izv. Akad. Nauk SSSR, Neorg. Mat., 1:1803 (1965).

17. G. V. Samsonov, Poroshkovaya Met., No. 12, 49 (1966).

18. G. V. Samsonov, Ukrainsk. Khim. Zh., No. 12, 1233 (1965).

19. G. V. Samsonov, Yu. B. Paderno, and B. M. Rud, Izv. VUZ. Fiz., No. 9, 129 (1967).

20. G. V. Samsonov et al., in: Rare-Earth Elements [in Russian], Izd. Akad. Nauk SSSR, Moscow (1963), p. 8.

21. G. V. Samsonov, Ukrainsk. Khim. Zh., No. 8, 763 (1967).

22. G. V. Samsonov, Refractory Compounds [in Russian], Metallurgizdat, Moscow (1963).

23. G. V. Samsonov, V. K. Kharchenko, and L. I. Struk, Poroshkovaya Met., No. 3, 59 (1968).

24. G. V. Samsonov, and A. Ya. Kuchma, Izv. Akad. Nauk SSSR, Neorg. Mat., 3:1970 (1967).

25. G. V. Samsonov and V. N. Paderno, Izv. Akad. Nauk SSSR, Metally, No. 1, 180 (1965).

26. G. V. Samsonov, in: Scientific Bases of the Choice of Catalysts [in Russian], Nauka, Moscow (1966), p. 237.

27. G. V. Samsonov, in: The Metal Science and Metal Physics of Superconductors [in Russian], Nauka, Moscow (1965), p. 65.

28. G. V. Samsonov and E. N. Denbnovetskaya, in: The Metal Science, Physical Chemistry, and Metal Physics of Superconductors [in Russian], Nauka, Moscow (1967), p. 137.

29. G. V. Samsonov and A. L. Burykina, Avtomat. Svarka, No. 10 (1966).

30. G. V. Samsonov, in: High-Temperature Coatings [in Russian], Nauka, Moscow and Leningrad (1967).

31. G. V. Samsonov and I. P. Kushtalova, Dopovidi Akad. Nauk UkrSSR, Ser. A, No. 3 (1968).

32. G. V. Samsonov and V. V. Stasovskaya, Poroshkovaya Met., No. 12, 95 (1966).

33. G. V. Samsonov, A. Ya. Artamonov, and M. F. Idzov, Poroshkovaya Met., No. 11, 74 (1967).

34.   G. V. Samsonov and Yu. G. Tkachenko, this volume, p. 409.

35.   I. I. Shafranovskii, Diamonds [in Russian], Nauka, Moscow and Leningrad (1964), p. 50.

36.   H. Biltz, Z. Physik, 153:338 (1958).

37.   P. Costa and R. Conte, Compounds of Interest in Nuclear Reactor Technology (edited by I. Waber and P. Chiotti), The Met. Soc. AIME, Michigan (1964), pp. 3-27.

38.   E. Dempsey, Phil. Mag., 2:285 (1963).

39.   V. Erne and A. C. Switendick, Phys. Rev., 137:927 (1965).

40.   J. Piper, J. Appl. Phys., 33:2394 (1962).

41.   G. Pfan and W. Rix, Z. Metallk., 45:16 (1954).

42.   P. Weiss and J. J. de Marco, Rev. Mod. Phys., 30:59 (1958).

43.   J. H. Wood, Phys. Rev., 177:714 (1960).

I. METHODS OF PRODUCING CARBIDES AND CARBIDE-BASE ALLOYS

Methods of Producing Pure Carbides

T. Ya. Kosolapova and G. N. Makarenko

Methods of producing refractory carbides are classified. The most
widely used methods of carbide production are by synthesis from the
elements and by carbothermic reduction of the oxides. The formation
of carbides is discussed from the point of view of the electron structure of the metals and carbon composing them, and also from that of
a configuration-localization model which is presented.

Carbides of the transition metals of groups IV to VI possess
a high melting point and great hardness, a low rate of evaporation
and vapor pressure at high temperatures, corrosion resistance,
and other properties that put them in the class of very promising
materials for the new technology [2, 4, 6].

The known methods of producing carbides can be classified
in accordance with the diagram shown in Fig. 1. The electrolysis
of molten salts and chemical precipitation are low-output methods
which necessitate further purification of the carbides, and hence
are inefficient methods that do not find wide commercial application. The production of carbides by synthesis from their elements
can be accomplished by melting at temperatures below the melting
points of the constituents in vacuum or in a reducing atmosphere.
These methods, which have been described in detail in the literature [2, 4, 6], enable us to obtain carbides containing 0.05-0.08%
O, the amount of metallic impurities being governed by the purity
of the raw materials used.

The reaction of metals with carbon in molten low-melting-
point metals is a variant of the method of synthesis from the ele-

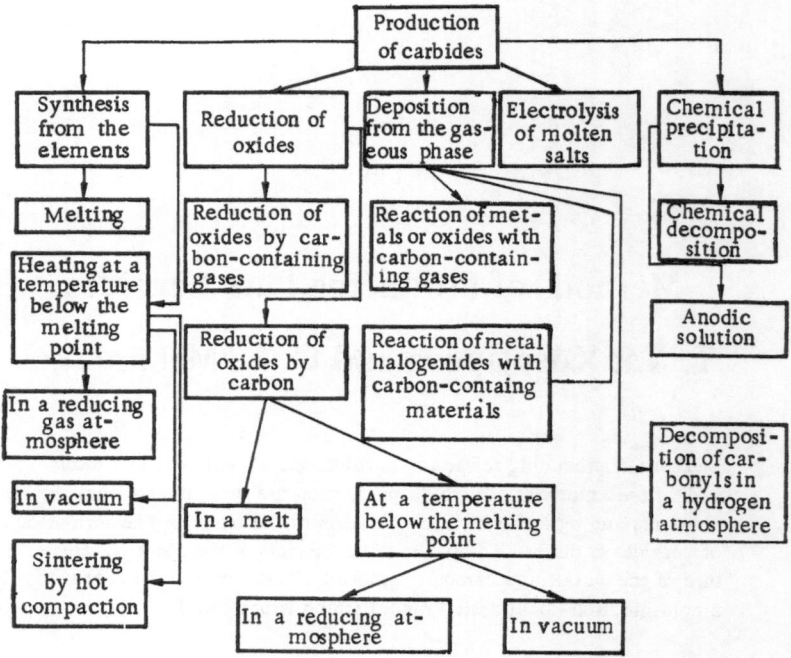

Fig. 1. Classification of methods of producing carbides.

ments. The production of carbides of titanium, zirconium, hafnium, vanadium, niobium, and chromium by a method based on the fact that these metals, when dissolved in zinc, react with carbon to form carbides has been reported in [1]. The temperature at which the process is carried out does not exceed the boiling point of zinc (907°C). This process is not efficient for the commercial production of carbides, since it is carried out in sealed ampoules, and the excess zinc has to be removed by sublimation in vacuum or by treatment with acids. Besides that, some metals form with zinc intermetallic compounds which are insoluble in acids. However, the method is undoubtedly of interest for producing pure materials and also single crystals and whiskers.

The most widely used method of producing carbides is by reduction of the metal oxides with carbon. It is based on the fact that at a certain temperature the affinity of the metals for oxygen is less than that of carbon for oxygen and so reduction of the oxides proceeds, while at a fairly high affinity of the metal for carbon the carbide is formed. The carbon dioxide gas liberated during the reduction process at temperature above 800°C reacts with excess carbon to form carbon monoxide, which is the principal reducing agent for the metal oxides.

In investigating the reduction of the oxides of the transition metals by carbon, we have studied in the present research the effects of temperature, time of heating, pressure in the system, and composition of the charge on the reduction products. A manometric analysis has also been made. The results obtained enable us to correlate features of the processes with the position in the periodic system of the metal whose oxide is being reduced.

Monocarbides of scandium and yttrium obtained by reduction of the respective oxides are formed via intermediate stages of separation of oxycarbides of the Me_2C_2O type, from which it is very difficult to obtain the carbide free from oxygen. Oxygen evidently stabilizes the structure of ScC; the monocarbide of yttrium is unstable and it is difficult to produce.

The processes of carbide formation have been considered from the point of view of the electron structure of the metals forming them and the tendency of the metal and carbon to organize stable electron configurations [3, 5]. In the isolated state atoms of scandium and yttrium have a single d-electron. It may be assumed that, in striving to form the more stable, in terms of energy, d^0-state, the scandium atom gives up the d-electron to stabilize the stable sp^3-state of carbon, the formation of which is possible as a result of the electron transition $s^2p^2 \rightarrow sp^3$. The existence of methane as the principal constituent of the gaseous products of the decomposition of the carbides concerned by water confirms that the carbon in them exists mainly in the form of the C^{4-} groups, to which the sp^3-states correspond. The stability of the d-states increases with rise in the principal quantum number of the d-electrons of the metal atoms, and so yttrium monocarbide is less stable than scandium monocarbide. It has not proved possible to obtain monocarbides of the remaining rare-earth metals.

Dicarbides of all the rare-earth metals are known. An investigation of the conditions for producing them by reduction of the oxides with carbon in vacuum has shown that the temperatures of formation of the dicarbides of all the lanthanide elements and yttrium are similar, being 1800-1900°C. This is due to the similarity of the electron structures of the lanthanides, in which filling of the deep f-shell takes place from the same outer electron shells. Electrons of the f-shells, which take no direct part in the formation of the chemical bonds, may nevertheless affect the nature of the bonds through possible $f \rightarrow$ d transitions.

Dicarbides are distinguished from mono- and sesquicarbides by their higher thermodynamic stability. An attempt to produce sesquicarbides by the reduction of oxides has shown that their formation is prevented by the formation of the thermodynamically more stable dicarbides, and it is practically impossible to obtain single-phase sesquicarbides by this method.

The liberation of acetylene during the decomposition of dicarbides by water points to the existence of strong covalent bonds between the carbon atoms forming the C_2 groups; at the same time the d-electrons of the metals, the presence of which is likely as a result of the $f \rightarrow$ d transitions, occur in the nonlocalized state, since their effect on the stabilization of the sp^3-states of carbon is less than in the monocarbide. This is confirmed by the electrophysical properties: yttrium monocarbide is characterized by a specific electrical resistivity of the order of $10^4 \, \mu\Omega \cdot$ cm, indicating a high degree of polarization of the molecule; whereas that of the dicarbide is 30 $\mu\Omega \cdot$ cm, which shows that a considerable proportion of the metallic bond is due to the presence of nonlocalized electrons. The latter is confirmed by the higher hydrogen content in the gaseous products of the decomposition of the dicarbides by water.

According to the results of the manometric studies, the reduction of the oxides of the transition metals of groups IV to VI of the periodic system takes place via the formation of the lower oxides. The nature of the reduction changes with the transition from elements of group IV to those of group VI. In the case of titanium oxide reduction proceeds via the lower oxides Ti_3O_5, Ti_2O_3, and TiO. The monoxide is directly reduced to the carbide. In the reduction of chromium oxide the formation of lower oxides, in particular CrO, was not revealed by the manometric studies, and the process takes place via the formation of metallic chromium.

In the transition from titanium to vanadium and chromium, there is a change in the electron structure of the respective metals: titanium atoms have a d^2s^2-electron configuration, vanadium

atoms d^2s^2, and chromium atoms d^5s^1. It is known from [3] that
the stable electron configurations correspond to the d^0-, d^5-, and
d^{10}-states, and that with the transition from titanium to vanadium
and chromium the statistical weight of the d^0-states decreases,
while that of the d^5-states increases.

Titanium, in tending toward the stable d^0-state, gives up electrons to oxygen to enable it to form the stable s^2p^6-state to which
the oxygen atom tends; in consequence the electron interaction between the titanium and oxygen atoms is large, and titanium monoxide is a thermodynamically very stable oxide. In the transition
to vanadium, which has three electrons in the d-state, the statistical weight of the d^5-states increases, the number of electrons given
up for the stabilization of the s^2p^6-states of oxygen decreases, interaction between the electrons weakens, and vanadium monoxide
is thermodynamically less stable than titanium monoxide. Finally,
in the case of chromium, which has five electrons in the d-state —
the most stable state — interaction between the electrons is even
less, and chromium monoxide belongs to the class of unstable
oxides. This is confirmed by the absence of decarbidization, which
is present in titanium carbide.

The reduction temperature and the temperature of carbide
formation decreases on passing from oxides of group IV metals
to those of metals of groups V and VI. Thus, the formation
temperatures of carbides of group IV metals lie in the range 2100-
2200°C, those of carbides of group V metals in the range 1800-
1900°C, and those of group VI metals in the range 1500-1700°C.
In addition to the relationship mentioned above between oxide stability and the position of the metal in the periodic system, oxides
of composition MeO_2 are reduced in the production of carbides of
group IV metals, whereas in the production of carbides of metals
of groups V and VI (chromium) oxides of the compositions Me_2O_3
and Me_2O_5 are reduced. In these oxides some of the electrons are
transferred to the covalent bonds Me−Me and O−O; as a result
the strength of the bond Mo−O diminishes, the stability of the
oxides is correspondingly reduced, and in consequence so are their
reduction temperatures.

The method of reducing metal oxides with carbon in vacuum
enables us to obtain carbides containing 99.0-99.5% of the principal component when using fairly pure raw materials. However,

present-day technology makes higher demands on the purity of materials.

Undoubted interest lies in single crystals, whose properties differ somewhat from those of polycrystalline substances. One of the methods of procuring high-purity carbides in the polycrystalline state, and also in the form of single crystals, is by deposition from the vapor phase.

In the present research we have studied the conditions required to produce the carbides of zirconium and niobium by the hydrogen reduction of a mixture of the metal-chloride vapor and carbon tetrachloride. The results demonstrate that the products of this method have a strictly stoichiometric composition.

Spectrographic analysis of the products obtained by the various methods* has shown that in the reduction, in particular, of niobium pentoxide by carbon (soot) the impurity content in the carbide is $6 \times 10^{-1}\%$; in synthesis from the elements it is $4 \times 10^{-1}\%$, and in deposition from a vapor mixture $2 \times 10^{-1}\%$. Thus, although the absolute impurity content in deposition from the vapor phase is lowest, its order of magnitude is in general the same for all the methods considered. It must be borne in mind that deposition from the vapor phase takes place on the surface of a substrate (graphite or metal) in the form of a coating and not as powder, and the productivity of this method is very low. It therefore appears to us that the production of refractory carbide powders by depositing them from a vapor mixture onto a heated surface is not efficient enough, and can be used only in effecting a reduction of a chloride mixture by hydrogen in a suspended layer. The gaseous-phase reduction method is very promising from the point of view of producing high-quality refractory-carbide coatings as well as for producing single crystals, and it should be developed in every way.

Literature Cited

1.   V. N. Gurin and A. P. Obukhov, Poroshkovaya Met., No. 12, 68 (1966).
2.   R. Kieffer and P. Schwartzkopf, Hard Alloys [Russian translation], Metallurgizdat, Moscow (1957).

---

*The spectrographic analyses were carried out in the Odessa Branch of the Institute of General and Inorganic Chemistry of the Academy of Sciences of the Ukrainian SSR, using methods devised under the direction of N. F. Zakharii.

3.   G. V. Samsonov, Poroshkovaya Met., No. 1, 98 (1965).
4.   G. V. Samsonov, Refractory Compounds [in Russian], Metallurgizdat, Moscow
      (1963).
5.   G. V. Samsonov, Ukrainsk. Khim. Zh., No. 12 (1965).
6.   G. V. Samsonov and Ya. S. Umanskii, Hard Compounds of Refractory Metals
      [in Russian], Metallurgizdat, Moscow (1967).

Production Technology and Some Properties of the Pyrolytic Carbides of Zirconium and Titanium

V. S. Davydov, B. G. Ermakov, and V. V. Sokolov

A study has been made of the conditions necessary for the production of dense deposits, uniform in composition, of titanium and zirconium carbides by deposition from a gaseous mixture. Optimum conditions have been arrived at for depositing these carbides from a gaseous phase containing the vapor of the higher chloride of the metal, methane, and hydrogen. The effect of various parameters of the process (temperature, composition of the gaseous phase, and rate of supply of the vapor mixture) on the rate of deposition, structure, and chemical composition of the carbides obtained is discussed.

The main principles of the method of chemical gas-phase deposition have been described in detail in [2, 5]. The possibility of producing refractory metals and compounds by this method has been established, and it is now used in the production of structural materials employed in various fields of high-temperature technology.

Gas-phase deposition consists in the reduction, dissociation, or exchange interaction of the halide, carbonyl, or other volatile compound of the metal on the heated surface of the component of

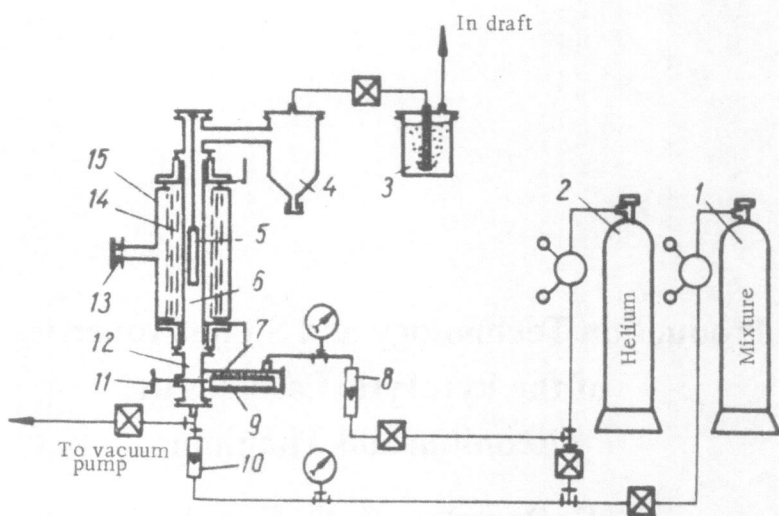

Fig. 1. Diagram of apparatus for producing carbides.

specimen. Two processes can be used for the synthesis of refractory carbides, viz.:

1) the dissociation or reduction of a volatile compound of the metal followed by carburization of the deposited metal by the carbon of the substrate;

2) an exchange reaction between the volatile metal compound and a hydrocarbon.

The first process is used only for producing coatings on graphite; the second for producing coatings on any substrate or complete carbide components.

In the present research we have used the exchange reaction between methane and the higher chlorides of zirconium and titanium in the presence of excess hydrogen, the experiments being carried out in the laboratory apparatus shown in Fig. 1. This consists of a hermetically sealed water-cooled chamber equipped with a tubular graphite heater (6) within which the components or specimens to be coated are placed (5). The gaseous mixture of hydrogen and methane is fed from the cylinder (1) through a system of valves and a flow meter (10) into the mixing chamber (12), into which is also fed the metal chloride from the evaporator (7) through the valve (11). The gaseous mixture and the chloride vapor are then passed into the reaction zone (6).

Owing to the difficulty of directly measuring the supply of solid chlorides by means of the flow meters, an inert gas was saturated with the chloride. For this purpose helium was passed through the evaporator, the temperature of which was maintained constant and in which the pressure of chloride vapor was 250–300 mm Hg. By this means the helium was saturated with the chloride, and the vapor mixture so formed was fed into the mixing chamber. The supply of chloride was controlled by the temperature of the evaporator and the rate of flow of the bearer gas — helium.

The formation of titanium and zirconium carbides can be represented by

$$MeCl_4 + CH_4 + H_2 \rightarrow MeC_{solid} + HCl + H_2. \tag{1}$$

Since the secondary reactions

$$MeCl_4 + H_2 \rightarrow Me_{solid} + HCl + H_2, \tag{2}$$

$$CH_4 + H_2 \rightarrow C_{solid} + H_2, \tag{3}$$

may occur, the reaction can be more accurately represented by

$$MeCl_4 + CH_4 + H_2 \rightarrow MeC_{\alpha\ solid} + HCl + H_2. \tag{4}$$

In this reaction α may be more or less than unity, depending on the composition of the initial gas mixture and the parameters of the process.

As a result of the experiments carried out, we established the optimum conditions for obtaining dense, uniform deposits of zirconium and titanium carbides that were close to the stoichiometric composition (Table 1).

TABLE 1. Optimum Parameters of the Process for Depositing Pyrolytic Carbides of Zirconium and Titanium

Characteristic	ZrC	TiC
Substrate temperature, °C	1250—1400	1200—1350
Composition of gas mixture, mol.%	$2\% CH_4 + 96\% H_2 + 2\% ZrCl_4$	$2\% CH_4 + 96\% H_2 + 2\% TiCl_4$
Supply of gas mixture, liter/min	10—20	8—15
Deposition rate, μ/min	5—20	3—5

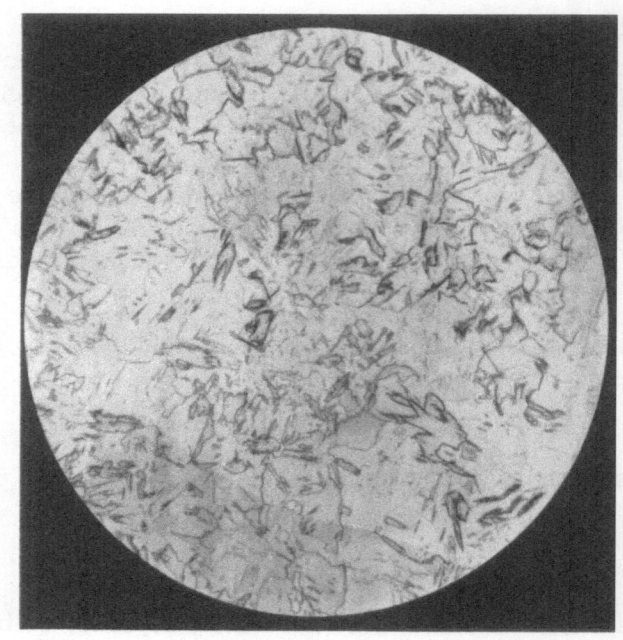

Fig. 3. Microstructure of ZrC specimens obtained at 1250°C (cut parallel to the deposition surface). x200.

Fig. 2. Microstructure of ZrC specimens obtained at 1250°C (cut perpendicular to the deposition surface). x200.

The heterogeneous nature of the process of gas-phase deposition of carbides is due to the dependence of the rate of film growth on the diffusion (rate of supply) of the components of the reaction to and from the deposition surface and on the actual kinetics of the exchange interaction at this surface, the slowest stage of the reaction being the determining factor. When the limiting stage of the process is the kinetics at the surface, the deposition rate does not depend on the configuration of the deposition surface, and this leads to the formation of a uniform carbide layer over the whole surface of the specimen. However if the limiting factors are the diffusion processes, the deposition rate is inversely proportional to the length of the diffusion path. In practice, greatest interest lies in procuring uniform coatings or components of the same thickness, and consequently it is essential either to carry out the process in the kinetic range or to secure the same length of the diffusion path by employing a rational system of presenting the components to the deposition surface. The principal factor affecting the nature and rate of the diffusion process is the substrate temperature. The kinetic range of the process is realized (other things being equal) at lower temperatures than the diffusion range is, since the rate of the chemical reaction decreases with temperature considerably more rapidly than does the rate of the diffusion processes. The temperature at which the diffusion range passes into kinetic range depends on the rate of gas flow, the concentration of the components, and the length of the diffusion path. The deposition conditions given in Table 1 were obtained during the deposition of the carbides on the inner surface of cylindrical tubes 10 mm in diameter and about 300 mm long. The lower limits of the temperature ranges correspond to the process taking place in the kinetic region and the upper limits to its occurrence in the diffusion region. At temperatures below the given range, the deposition rates fall sharply, and the degree of utilization of the chloride diminishes. At substrate temperatures above the given range, the reaction proceeds not only on the surface but also within the volume of gas, resulting in the formation of loose, porous deposits. Microsections cut perpendicular to the deposition surface reveal a characteristic columnar structure (Fig. 2). Zirconium carbide obtained at 1250°C has a crystal size considerably smaller than that of ZrC obtained at 1400°C, the difference being particularly marked in microsections cut parallel to the deposition surface (Fig. 3). The composition of the initial vapor mixture has less effect on the structure and de-

TABLE 2. Some Properties of Pyrolytic Carbides of Zirconium and Titanium

Characteristic	ZrC	TiC
Melting point, °C *	3530	3150
Density, g/cm³	6.9	4.9
Microhardness (with load of 100 g), kg/mm²	2700	3480
Chemical composition, wt.%		
metal	88.6—88.9	80.6—80.8
combined carbon	11.2—11.4	18.8—19.0
nitrogen	0.1—0.2	0.2—0.3
Conventional formula	ZrC	TiC
	0.96—0.97	0.93—0.95
	N	N
	0.01—0.02	0.01—0.02

*Published data for powder materials.

position rate, but the chemical composition of the carbides depends markedly upon it. With a $CH_4:MeCl_4$ ratio less than unity, the amount of combined carbon in the carbide decreases, whereas with a $CH_4:MeCl_4$ ratio greater than unity, carbon is deposited at the same time as the carbide to form carbide–graphite alloys. The degree of utilization of the chloride decreases with the simultaneous increase in the methane and chlorine contents, while the deposition rate falls with decrease in the methane and chloride contents.

The rate of supply of the vapor mixture has practically no effect on structure and chemical composition of the carbides, although when it is a question of producing particular components it is essential to choose the optimum rate if a deposit is to be obtained which is uniform in thickness and composition over the whole length of the specimen. The vapor mixture becomes impoverished in the original components during movement along the deposition surface, the degree of impoverishment being proportional to the rate of deposition and the time the vapor mixture is in contact with the surface. A reduction in the concentration of the original components leads to a decrease in the deposition rate, but at the same time the degree of heating of the gas mixture increases, and this in turn gives rise to an increase in the rate both of the chemical reaction and of the diffusion processes, which offsets to some extent the effect of the fall in concentration. By making a suitable choice of the rate of supply of the gaseous mixture it is possible to achieve uniform deposition over the whole length of the specimen. When altering the geometrical dimensions of the components, it is important to main-

tain constant the time of contact of the gaseous mixture with the
deposition surface, for example, by increasing the supply when the
length or diameter of the component is increased. Some properties of pyrolitic carbides of titanium and zirconium are given in
Table 2. It should be noted that in the specimens concerned the
combined carbon content is somewhat less than that corresponding to the stoichiometric composition and that a small amount of
nitrogen is present (about 1-2 at.%). This is due to the use of insufficiently pure hydrogen, containing 0.1-0.2 vol.% nitrogen.

Conclusions

As a result of the experiments we have carried out we have
determined the optimum conditions for the deposition of zirconium
and titanium carbides from a gaseous phase consisting of the vapor
of the higher chloride of the metal, methane, and hydrogen.

We have discussed the effect of the various parameters of the
process (temperature, composition of the gaseous phase, and rate
of supply of the vapor mixture) on the deposition rate, structure,
and chemical composition of the carbides produced.

Literature Cited

1.   V. S. Neshpor et al., Poroshkovaya Met., No. 2, 65-69 (1967).
2.   C. Agte and K. Moers, Z. anorg. Chem., 198:233 (1931).
3.   I. E. Campbell et al., J. Electrochem. Soc., 5:318 (1949).
4.   C. F. Powell, I. E. Campbell, and B. W. Gonser, Vapor Plating, London and
      New York (1955).
5.   A. E. Van Arkel, Reine Metalle, Julius Springer, Berlin (1939), p. 34; Edwards
      Bros. Inc., Ann Arbor, Mich. (1943).

Investigation of the Production Conditions and Some Properties of Metalloceramic Materials Based on Titanium Carbide Bonded with Alloy Steel

V. K. Narva and S. S. Kiparisov

The production conditions and some properties of titanium carbide—alloy steel materials have been investigated. It is shown that the presence of the steel bond enables the hardness after annealing to be raised to $H_{Rc} = 40$.

Metalloceramic materials based on titanium carbide are extensively used as high-temperature materials. However, the great hardness of these cermets after sintering makes their subsequent working very difficult and so limits their field of application. Work has been carried out with the object of devising and using metalloceramic materials based on titanium carbide and steel (ferro-TiC) which combine the properties of both constituents. These materials possess great hardness and wear resistance, and they can be subjected to heat treatment and mechanical working. Consequently they open up broad new fields of application.

Various steels are used for the bond, including those of the stainless austenitic class and of the alloyed martensitic class. Great hardness after quenching combined with the capacity for mechanical working after annealing has been found in these cermets.

The present work has been directed toward the development of a wear-resistant metalloceramic material based on titanium carbide bounded with alloy steel, the latter being either the chromium-molybdenum steel Kh12M or the complex alloy steel having the composition: 84.5% Fe, 8% Mo, 4% Cr, 2% Ni, and 1.5% C. Commercial TiC powder of average grain size and containing 18.5% C_{tot} was used. The Kh12M steel was produced by adding the necessary amount of molybdenum powder to Kh30 steel. The complex alloy steel was obtained by mixing the following pure powders: carbonyl iron (99.99%), reduced chromium, electrolytic nickel (99.99%), molybdenum (99.9%), and baked lamp black (ash content 0.17%). The raw materials, taken in the required proportions (50, 60, and 70 wt.% steel for the cermets bonded with Kh12M steel and 50, 60, 65, and 70 wt.% steel in the case of the cermets bonded with the complex alloy steel), were mixed with alcohol in a ball mill for 20 h and then dried.

Experiments to determine the optimum compaction pressure were carried out on samples of all the different compositions, using an 8% solution of rubber in benzene. Figure 1 gives examples of the change in relative density as a function of compaction pressure for samples containing 50 wt.% TiC + 50 wt.% Kh12M steel and 50 wt.% TiC + 50 wt.% steel containing 84.5% Fe, 8% Mo, 4% Cr, 2% Ni, and 1.5% C. In all cases the use of a P_{sp} exceeding 4 tons/cm² causes lamination of the specimens, while the relative densities of specimens compacted under specific pressures of 2 and 3 tons/cm² differ little from one another; hence for compacts of all the compositions investigated the optimum compaction pressure was taken as $P_{sp} = 2$ tons/cm².

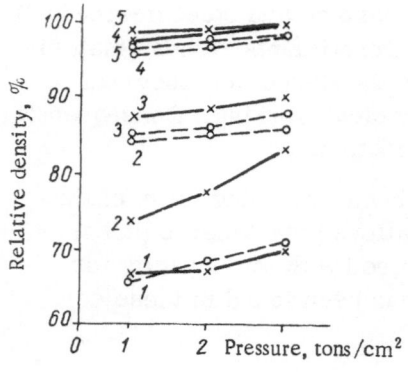

Fig. 1. The effect of compaction pressure on the relative density of the specimens: 1) after compaction; 2) sintered (1300°C); 3) sintered (1350°C); 4) sintered (1400°C, 1 h); 5) sintered (1400°C, 2 h); continuous lines) bonded with steel containing 84.5% Fe + 8% Mo + 4% Cr + 2% Ni + 1.5% C; broken lines: bonded with Kh12M steel.

TABLE 1. Sintering Conditions for Cermets of
Various Compositions and the Effect of
Cooling Rate after Sintering on Hardness
(sintering time 2 h)

Composition of cermet, %		Sintering tempera- ture, °C	Hardness H_{Rc} after sintering and cooling at a rate of (deg/min):		
TiC	Bond		100	25	10
50	50% Kh 12M	1400	67	63	63
40	60	1350	60	58	58
30	70	1300	58	50	48
50	50	1400	70	67	65
40	60	1380	69	63	60
35	65	1350	64	60	58
30	70	1300	62	55	53

The optimum temperature for sintering in the TVV-4 vacuum
furnace was chosen on the basis of relative-density checks made on
the specimens.

In studying the sintering conditions, apart from the optimum
temperature and the time of holding at that temperature, particular
attention was devoted to the rate of cooling after the specimens
had been held at the sintering temperature. This factor was important because the material had to undergo mechanical working,
and so its hardness after sintering was determined in order to discover the possibility of lowering the hardness by annealing. In
order to ascertain the dependence of the hardness on the rate of
cooling after sintering, cooling was carried out at rates of 100,
25, and 10 deg/min. Table 1 shows that a reduction in cooling
rate leads to a certain decrease in hardness, although there is
no significant fall in hardness on reducing the cooling rate from
25 to 10 deg/min; hence all the subsequent experiments were carried out at a rate of cooling after sintering of 25 deg/min.

The sintered specimens were all subjected to heat treatment
(annealing, quenching, tempering) and their hardness subsequently
measured. Annealing was carried out with the object of reducing
the hardness to a level at which the specimens could be mechani-
cally worked, of homogenizing the specimens, and of removing
internal stresses. The annealing was done under the following
conditions:

Isothermal conditions: heating to 850-870°C, holding for 1 h,
cooling to 720-740°C, holding for 7 h, followed by furnace cooling;

TABLE 2. Hardness H_{Rc} of Cermets of Various Compositions after Annealing and Quenching

Composition of cermet, %		After annealing				After quenching			
		Cooling rate = 30 deg/h	isothermal annealing	Cold treatment		Cooling rate = 30 deg/h	isothermal annealing	Cold treatment	
TiC	Bond			in N_2 liq	$t=(-10)-(-14)$			in N_2 liq	$t \sim (10)-(-14)$
50	50 12M	57	54	50	50	70	72	69	70
40	60	55	49	42	44	68	70	65	68
30	70	50	45	34	35	66	65	64	66
50	50 [84,5Fe + 8Mo + + 4Cr + 2Ni + 1.5C]	60	62	60	60	72	72	68	72
40	60	62	60	50	52	69	68	68	70
35	65	55	54	45	46	67	65	67	68
30	70	48	48	38	40	65	63	65	66

heating to 900°C, holding for 6 h, slowly cooling at a rate of 30 deg/h to 500°C, followed by cooling to room temperature in the cold chamber of the furnace;

isothermal annealing conditions, but with cooling in the cold chamber followed by cold working and then heating to 680-710°C.

Cold working was carried out in order to convert the stable residual alloyed austenite which had not broken down during prolonged heating into martensite, which is converted into sorbite during subsequent high-temperature heating below the A_{c_1} point (680-710°C); this results in a low hardness and hence makes cold working possible. The specimens were cooled in two ways: drastic cooling in liquid nitrogen (−180°C) and less drastic cooling to −10 to −14°C, the holding time in each case being 6 h; they were then reheated in a reducing furnace in a stream of hydrogen to 680-710°C and held at this temperature for 6 h. Prolonged heating is essential for complete coalescence of the carbides, which in turn produces a reduction in hardness. With drastic cooling followed by heating, the hardness of the specimens differs little from the values obtained for compacts subjected to the less drastic form of cooling. Apart from that, treating the specimens in liquid nitrogen frequently led to the appearance of cracks. For the above reasons we arrived at the conclusion that it was necessary to carry out cold treatment by the less drastic cooling procedure.

The best results in reducing the hardness after annealing were achieved by cold treatment of the annealed specimens (Table 2).

TABLE 3. Hardness of Cermets, kg/mm^2

Composition of cermet, %		After quenching from (°C):					After tempering at (°C):				
TiC	Bond	900	950	1000	1050	1100	200	300	400	500	600
50	50 Kh12M	68	70	68	65	63	70	69	67	63	58
40	60	66	68	64	60	58	68	66	65	63	55
30	70	63	66	63	60	55	63	61	58	53	43
50	50[84.5Fe + + 8Mo + 4Cr + 2Ni + + 1.5C]	63	68	70	72	70	72	70	70	68	67
40	60	62	66	68	70	69	69	68	66	65	60
35	65	60	63	66	68	67	67	66	64	63	59
30	70	53	57	60	66	64	65	64	62	58	54

In the case of specimens of certain compositions the hardness attained enabled any desired mechanical treatment of the material to be carried out.

After annealing the specimens were quenched in oil. The hardness values of specimens of various compositions after quenching from 900, 950, 1000, 1050, and 1100°C are presented in Table 3. The quenching temperature for materials bonded with Kh12M steel was chosen as 950°C, while that for materials bonded with the complex alloy steel was chosen as 1050°C. Table 3 also gives the hardness values of the specimens after tempering at 200, 300, 400, 500, and 600°C. In all cases the specimens were heated to 200°C to remove included stresses.

The results of the investigations into the effect of composition on hardness after all forms of heat treatment showed (Fig. 2) that

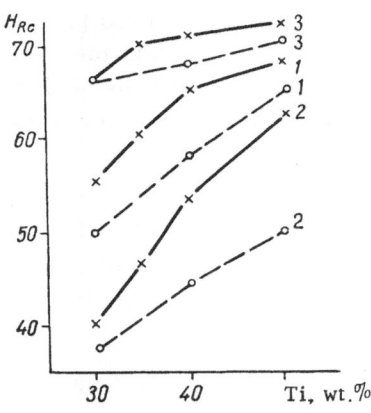

Fig. 3. Values of the coefficient of linear expansion of cermets of the compositions investigated: 1) bonded with Kh12M steel; 2) bonded with steel containing 84.5% Fe, 8% Mo, 4% Cr, 2% Ni, and 1.5% C.

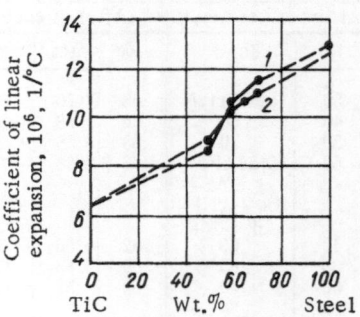

Fig. 2. Effect of composition on the hardness of cermets: 1) after sintering; 2) after
annealing; 3) after quenching; continuous
line) bonded with steel containing 84.5%
Fe, 8% Mo, 4% Cr, 2% Ni, 1.5% C; broken
line: bonded with Kh12M steel.

the greatest hardness after sintering corresponds to a higher TiC
content. Moreover, the greater the content of the refractory constituent, the less marked is the difference in hardness in the
quenched and annealed states from the values obtained after sintering, i.e., the steel bond plays the principal part in the hardness
change after heat treatment.

Measurements of the coefficients of linear expansion of the
materials in the annealed state in the temperature range 20-700°C
are given in Fig. 3. The coefficients of linear expansion of TiC
and the steels in the annealed state are included for comparison,
sintered specimens of these materials being used for the measurements. The values of the coefficients obtained for the materials
investigated lie between those obtained for TiC and the steels, although they vary in relation to the ratio between the refractory constituent and the steel bond.

Metallographic analysis of the specimens revealed the presence of two phases: TiC grains with the steel constituent uniformly distributed between them. The structures after quenching
differed from the annealed structures in ways typical of steels subjected to this form of heat treatment.

Conclusions

1. The production conditions and certain properties of materials based on titanium carbide and alloy steel have been investigated.

2. The presence of the steel bond in a particular amount
(60-70 wt.%) enables a hardness of about $H_{Rc} = 40$ to be achieved

after annealing, and this in turn makes it possible to carry out any desired mechanical working.

3. Subsequent oil quenching renders these materials very hard (H_{Rc} = 66-72). The absence of any change in the linear dimensions of the specimens after quenching is a characteristic.

Investigation of a Method of Obtaining Niobium Carbide Powder and Coatings from the Vapor Phase

T. A. Lyudvinskaya, V. S. Sinel'nikova, T. Ya. Kosolapova, and V. P. Sergeev

The possibility of obtaining niobium carbide in the form of powder and coatings by deposition from a vapor mixture in the temperature range 1500-1900°C has been investigated. The rate of deposition has been studied in relation to the concentration of the reagents, the substrate temperature, and the rate of hydrogen supply. The highest deposition rate for niobium carbide was found to occur at an optimum ratio of hydrogen to total chlorides of 10:1. The possibility of depositing coatings on graphite, tungsten, molybdenum, and niobium was also investigated.

The principal properties characteristic of niobium carbide are a high melting point and microhardness, a high resistance to the action of molten metals and metal vapors, reasonable strength at elevated temperatures, low vapor pressure, high electrical conductivity, and good emissivity. These properties determine the field of application of NbC, namely for the manufacture of heating elements for high-temperature furnaces and of evaporating plants for aluminum, for protective coatings, for coating thermionic emission elements, and for use in heat-resistant alloys [1-4].

In general, NbC can be produced by the following basic methods: synthesis from the elements, reduction of the metal oxides by carbon, electrolysis of molten salts, and deposition from the vapor phase.

However, it is impossible to obtain NbC having a high degree of purity and a strictly stoichiometric composition by using these methods. Methods of producing porosity-free specimens of refractory compounds are practically nonexistent. In consequence, the properties of NbC have been studied on porous specimens affected by the following factors: impurities, pores, complex contact phenomena, and cracks. These investigations have shown it to be essential to prepare the compounds (and in particular NbC) in a state similar to the cast state or as high-purity single crystals. Almost the only way of obtaining NbC in this state is by deposition from a vapor mixture. The method is based on the reaction between the components of a vapor mixture consisting of a halide compound of the metal, a carbon-containing gas, and hydrogen, followed by the deposition of the reaction products in the form of a finely dispersed powder on a heated surface. Hydrogen promotes the reaction and in certain cases considerably reduces the decomposition temperature of the halide.

The hydrogen reduction of niobium chlorides has been discussed in detail in [7]. The reduction of niobium pentachloride was first attempted by Roseo [8], but he succeeded in obtaining only $NbCl_3$. Agte and Moers [6], by reducing niobium pentachloride with hydrogen mixed with toluol, obtained a mixture of the metal and carbide.

The following summary reaction takes place in the deposition of NbC from a vapor mixture:

$$2NbCl_5 + 2CCl_4 + 9H_2 = 2NbC + 18HCl.$$

In investigating the reactions involved in the production of refractory metal carbides by deposition from a vapor mixture, Campbell and Blocker [7] discovered that the deposition process depends on a number of factors: temperature of the substrate, melting point of the deposited material, and concentration of the raw materials in the vapor mixture. It was found by experiment that for each compound and substrate temperature there is an optimum pressure of the vapors and gases in the system.

Carrying out dissociation for the purpose of deposition under favorable conditions or reduction by hydrogen in the presence of carbon may lead to the formation of Me, MeC, Me_2C, Me + MeC, or MeC + C, since thermodynamically it is possible for all these components to be obtained. By choosing the appropriate conditions, the required product can be obtained.

As a result of the above, we have investigated the possibility of producing NbC in the form of powder and coatings, the carbide being obtained by the reduction of niobium pentachloride and carbon tetrachloride with hydrogen. The rate of carbide deposition was studied with reference to the reagent concentrations, the substrate temperature, and the rate of supply of hydrogen. The $NbCl_5$ used gave the following spectrographic analysis:

Ta	Fe	Ti	Si	Al	W	Mo	Σ, %
0.07	<0.002	<0.01	<0.01	<0.01	<0.005	<0.005	0.112

Total impurities in the CCl_4 amounted to 3.3×10^{-3} %. The NbC was deposited in a resistance furnace of the Tammann type, into which a mixture of hydrogen and chloride vapors was fed. The hydrogen was purified by passing it over heated titanium sponge and drying with silica gel and KOH. Part of the hydrogen was introduced into a thermostat together with CCl_4, enriched in the latter, and then fed into the reaction space. The $NbCl_5$ was placed in a quartz capsule in an evaporator which was heated to the required temperature by a separate furnace. Deposition took place on a graphite substrate located in the furnace.

The rate of NbC deposition was studied in relation to temperature and the contents of hydrogen, niobium chlorides, and carbon. The results obtained are presented in Fig. 1, from which it can

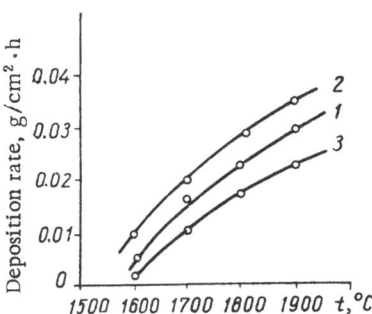

Fig. 1. Temperature dependence of the rate of NbC deposition: 1) $H_2:\Sigma(NbCl_5 + CCl_4) = 5:1$; 2) $H_2:\Sigma(NbCl_5 + CCl_4) = 10:1$; 3) $H_2:\Sigma(NbCl_5 + CCl_4) = 20:1$.

TABLE 1. Chemical Composition of NbC
as a Function of the NbCl$_5$:CCl$_4$ Ratio*

NbCl$_5$: CCl$_4$ ratio	Chemical composition, %		
	Nb	*C	Σ (Nb + C)
2 : 1	88.58	11.23	99.81
1.8 : 1	87.51	11.48	98.99
1.5 : 1	87.28	11.63	98.91

*According to x-ray analysis, the phase composition
corresponds in all cases to NbC.

be seen that the optimum ratio of the amount of hydrogen to the
amount of total chlorides, at which the deposition rate is greatest,
is 10:1. Obviously, with increase in hydrogen content some of the
chlorides are removed at the same time as the excess hydrogen.

Table 1 contains the results of investigating the chemical composition of NbC in relation to the ratio NbCl$_5$:CCl$_4$ in the gaseous
mixture at 1900°C and with a ratio H$_2$:(NbCl$_5$ + CCl$_4$) = 10:1. A
ratio of the NbCl$_5$ and CCl$_4$ contents of 2:1 ensures the production
of NbC of stoichiometric composition, which is somewhat higher
than the calculated value.

An x-ray diffraction study of the NbC obtained revealed that
in the range 1500-1900°C only NbC is formed, the deposition rate
increasing linearly with rise in temperature. The purity of the
product is higher than that of NbC obtained by other methods.
Thus, the impurity content of NbC produced by reduction of the
pentoxide is 0.6%, that of NbC produced by synthesis from the elements is 0.4%, and that of NbC produced by deposition from a vapor
mixture 0.2% (Table 2), i.e., the last method enables us to obtain
products of the highest purity, by the use of high-purity raw materials and by avoiding contamination from the material of the
components of the apparatus.

One of the principal applications of the method of deposition
from a vapor mixture is in the formation of refractory-compound
coatings on components. To provide reliable protection of the material, the coatings should possess the following properties: strength
and good adhesion to the metal surface, and the capacity to: prevent the oxidation and breakdown of the protected material, retard
the interdiffusion of the atoms of the protected material and the

TABLE 2. Spectrographic Analysis of NbC Produced by Various Methods*

Impurity	Deposition from the gaseous phase		Oxide reduction		Synthesis from the elements		
	$NbCl_5$	NbC	Nb_2O_5	NbC	Nb	C	NbC
Mg	—	$4 \cdot 10^{-3}$	$1.6 \cdot 10^{-2}$	$2 \cdot 10^{-3}$	$2.5 \cdot 10^{-3}$	$1 \cdot 10^{-2}$	$3 \cdot 10^{-2}$
Mn	—	$2 \cdot 10^{4}$	$1 \cdot 10^{-2}$	$2 \cdot 10^{-4}$	$7 \cdot 10^{-4}$	$1 \cdot 10^{-3}$	$1,4 \cdot 10^{-3}$
Pb	—	$5 \cdot 10^{-4}$	$1.5 \cdot 10^{-2}$	$5 \cdot 10^{-4}$	$1.6 \cdot 10^{-2}$	$5 \cdot 10^{-4}$	$2 \cdot 10^{2}$
Sn	—	$5 \cdot 10^{-4}$	$5 \cdot 10^{-3}$	$2.8 \cdot 10^{-3}$	$1 \cdot 10^{-3}$	$5 \cdot 10^{-4}$	$1 \cdot 10^{-3}$
Si	$1 \cdot 10^{2}$	$1 \cdot 10^{-1}$	$6 \cdot 10^{-2}$	$4 \cdot 10^{-2}$	$1 \cdot 10^{-2}$	$5 \cdot 10^{-2}$	$1 \cdot 10^{-1}$
Fe	$2 \cdot 10^{-3}$	$5 \cdot 10^{-2}$	$2.5 \cdot 10^{-2}$	$5 \cdot 10^{-2}$	$8 \cdot 10^{-2}$	$5 \cdot 10^{-2}$	$3 \cdot 10^{-2}$
Al	$1 \cdot 10^{-2}$	$2 \cdot 10^{-2}$	$1 \cdot 10^{-2}$	$2 \cdot 10^{-2}$	$1 \cdot 10^{-2}$	$2 \cdot 10^{-2}$	$2 \cdot 10^{-2}$
Ti	$1 \cdot 10^{-2}$	$2 \cdot 10^{-3}$	$2 \cdot 10^{-2}$	$3.5 \cdot 10^{-1}$	$1 \cdot 10^{-1}$	$2 \cdot 10^{-3}$	$2.8 \cdot 10^{1}$
Mo	0.10^{-3}	$2 \cdot 10^{-2}$	$5 \cdot 10^{-3}$	$1 \cdot 10^{-1}$	$5 \cdot 10^{-3}$	$2 \cdot 10^{-3}$	$5 \cdot 10^{-3}$
V	—	$5 \cdot 10^{-4}$	$4 \cdot 10^{-3}$	$1 \cdot 10^{-2}$	$2 \cdot 10^{-3}$	$2 \cdot 10^{-3}$	$3 \cdot 10^{-3}$
Ni	—	$1 \cdot 10^{-3}$	$1 \cdot 10^{-3}$	$1 \cdot 10^{-3}$	$1.5 \cdot 10^{-3}$	$2 \cdot 10^{-3}$	$1 \cdot 10^{-3}$
Cr	—	$5 \cdot 10^{-4}$	$5 \cdot 10^{-4}$	$5 \cdot 10^{-4}$	$2 \cdot 10^{-3}$	$5 \cdot 10^{-4}$	$5 \cdot 10^{-4}$
Ta	$7 \cdot 10^{-2}$	—	—	—	—	—	—
W Σ	$5 \cdot 10^{-3}$ 0.112	— 0.1993	— 0.1715	— 0.5770	— 0.3207	— 0.1405	— 0.4019.

*The spectrographic analysis was carried out in the Odessa laboratory of the Institute of General and Inorganic Chemistry of the Academy of Sciences of the Ukrainian SSR.

coating, limit volatilization of the protected material, form self-healing defects so as to prevent corrosion attack from developing, and resist cyclic thermal loading.

In studying the formation of coatings by deposition from a vapor mixture, it was found that the rate of formation of the carbide layer increases linearly with rise in temperature of the process. The rate of deposition was determined from the increase in weight of the substrate that occurred in the course of the experiment and from the area of the surface covered. The dependence of the deposition rate of the NbC coating on the supply of hydrogen has a clearly defined maximum corresponding to the ratio H_2: $\Sigma(MeCl + CCl_4) = 10:1$ at a hydrogen flow rate of 1.7 liter/min. The rate of supply of the vapor mixture and the temperature have a considerable effect on the rate of coating formation. This rate increases to a definite maximum with rise in the rate of supply of the mixture; subsequently it begins to fall, owing to the fact that at this rate of supply reaction between the gases does not proceed to completion. On increasing the ratio H_2:$\Sigma(MeCl + CCl_4)$ and con-

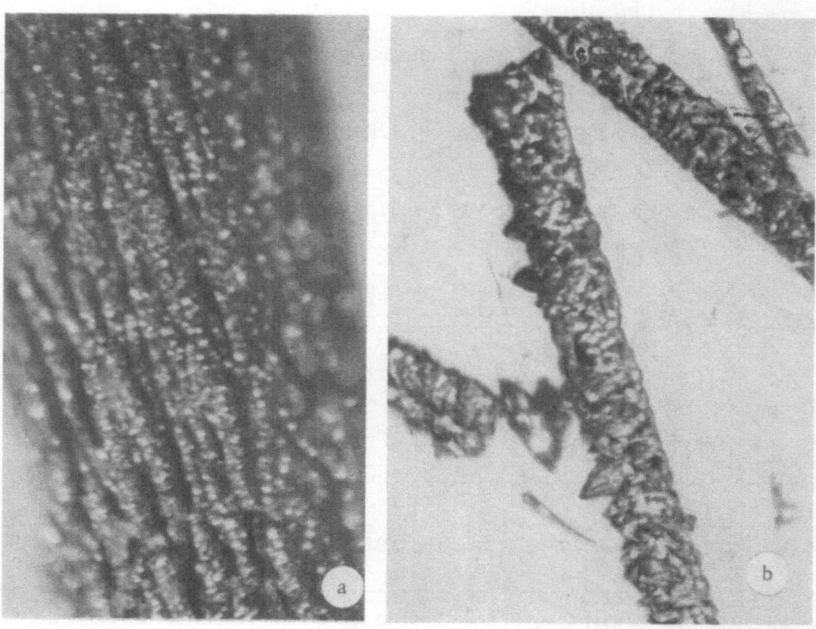

Fig. 2. External appearance of a NbC deposit on a Nb wire (a) and of well-faceted crystal (b).

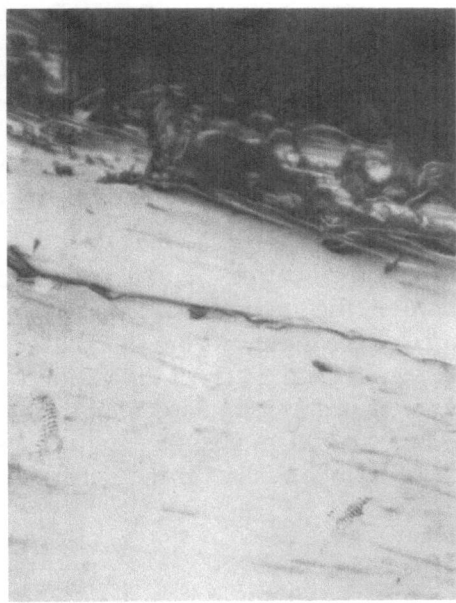

Fig. 3. Microstructure of a NbC coating on metallic niobium.

sequently the consumption of hydrogen, reduction becomes more complete. However, at a certain value of this ratio the concentration of metal chloride and chloride in the mixture decreases to such an extent that the rate of carbide deposition falls. Excess of the metal-containing vapor leads to the deposition of excess metal, and a deficiency of it leads to deposition of excess carbon. The advantage of producing the coatings by deposition from a vapor mixture lies in the fact that by this method it is possible to obtain coatings the composition of which is close to the stoichiometric; whereas it is rather difficult to obtain carbide coatings close to the stoichiometric composition by diffusion saturation of the metal surface with carbon, the problem being more complicated the wider the homogeneity range of the carbide concerned [5].

The deposition of NbC coatings on graphite, tungsten, molybdenum, and niobium was also investigated, and it was found that the structure of the coatings becomes increasingly coarse-grained with rise in temperature and the surface takes on a relief character (Fig. 2). In certain cases growth occurs only of those crystals on the surface of the deposited coatings that are favorably oriented in relation to the vapor flow. In these cases single crystals can be grown on polycrystalline conglomerates. Sometimes well-faceted crystals grow. Under certain conditions "secondary" crystals grow on the well-faceted ones.

Dense coatings that adhere well to the substrate can be obtained at lower temperatures. Metallographic examinations have revealed that in most cases there is a sharp boundary between the substrate and the coating (Fig. 3) with no appreciable intermediate layer.

The strength of the bond between the coating and the substrate depends on the conditions under which the coating was deposited, the treatment of the substrate surface, and the coating thickness. It may now be concluded that coatings deposited from the vapor phase can be formed on a substrate of practically any configuration. By varying the deposition conditions it is possible to produce thin, vacuum-tight coatings that are firmly bonded to the substrate.

Literature Cited

1.   V. N. Babich, G. V. Kurganov, and Yu. D. Strogonov, High-Temperature Coatings [in Russian], Nauka, Moscow (1967).
2.   V. P. Elyutin, B. S. Lysov, and G. I. Pepekin, Izv. VUZ. Tsvetnye Metally, No. 3 (1967).
3.   L. A. Nisel'son, Ya. M. Polyakov, and A. N. Krestovnikov, Zh. Priklad. Khim., 37:669 (1964).
4.   Ya. M. Polyakov, L. A. Nisel'son, and A. N. Krestovnikov, Zh. Priklad. Khim., 36:25 (1963).
5.   G. V. Samsonov and A. P. Epik, Refractory-Compound Coatings [in Russian], Metallurgiya, Moscow (1965).
6.   G. Agte and K. Moers, Z. Anorg. Chem., 198:233 (1931).
7.   C. F. Powell, I. E. Campbell, and B. W. Gonser, Vapor Plating, New York (1955).
8.   H. E. Roseo, Chem. News, 37(947):25 (1878).

The Production and Some Properties of Carbohydrides of Titanium, Vanadium, and Niobium

M. M. Antonova, L. N. Bazhenova, and I. I. Timofeeva

An investigation of the structure and certain properties of the carbohydrides of transition metals belonging to groups IV and V has enabled
preliminary conclusions to the drawn regarding the effect of features
of the electron structure of the d-metals on the nature of the interaction between the components.

Besides the binary hydrides of transition metals, which have
been fairly thoroughly investigated the complex systems of the
type Me−C−H and Me−N−H, about which there is extremely little
information, are also of both theoretical and practical interest.
The sparse data available in the literature relating to ternary systems of this kind are concerned primarily with determining the
positions occupied by hydrogen in the crystal lattice of the refractory compound. As a rule, the materials chosen for study have
been the carbide and nitride phases of transition metals belonging
to groups IV and V, having a stoichiometric composition or a deficiency of carbon or nitrogen; these phases have a bcc structure
which does not change on the introduction of hydrogen. From a
determination of the position of the hydrogen atoms in this type
of lattice, it is possible to make assumptions about the nature of
the interatomic bond in ternary compounds of this type, based on

47

the electron structure of the isolated atoms and the refractory compounds. It has been shown in [1-3, 11, 13-15] that, on entering titanium carbide and nitride, hydrogen occupies mainly vacant octahedral sites, whereas in zirconium carbide hydrogen is equally distributed between octahedral and tetrahedral sites.

An x-ray spectral investigation of the carbohydride phases of vanadium* reveals a similarity between the energy spectra of the carbides and carbohydrides of this metal. However, owing to small hydrogen content of the carbohydrides investigated, it proved impossible to establish the effect of hydrogen on the electron structure of the carbohydride.

In the present research we have investigated the crystal structure and certain properties of the carbohydride phases $Me-C-H$ in relation to the carbon content in the homogeneity range of the initial carbides and in the two-phase regions adjoining it.

The carbides were produced by synthesis from the elements under the conditions indicated in Table 1 in a vacuum furnace with a graphite heater at an initial pressure of 10^{-4} mm Hg. Chemical and x-ray structural analyses were carried out on the specimens obtained. According to the results of spectral analysis, the impurity contents were: in titanium, 0.28% Fe, 0.04% C, 0.06% Mg, in vanadium, 0.5% Fe, 0.2% Ca, 0.5% Al; in niobium, 0.07% Fe, 0.12% C, 0.16% Pb, 0.08% Mg, 0.07% Ti. The carbohydrides were analyzed for hydrogen content by combustion of a weighed sample in a stream of oxygen [1].

The hydrogen content in the carbohydrides of titanium and vanadium decreases with rise in carbon content, and if the indices x and y in the formula MeC_xH_y are compared, it can be seen that the hydrogen content is proportional to the number of carbon vacancies in the carbide lattice (Table 2).

Extrapolation of the curves in Fig. 1 shows that in titanium and vanadium carbohydrides the hydrogen content approaches zero at the limiting carbon content, in agreement with data published in [13].

---

*Data obtained by E. Z. Kurmaev at the Institute of Metal Physics of the Academy of Sciences of the Ukrainian SSR.

TABLE 1. Properties of Carbides Obtained

Formula of carbide phase	Production conditions		Phase composition	Lattice parameter, Å		
	Temperature, °C	Time, min		a	c	c/a
$TiC_{0.21}$	1400	60	fcc + hcf (α-Ti)	4.294	—	—
$TiC_{0.44}$	1400	20				
	1600	40	fcc	4.310	—	—
$TiC_{0.66}$	1600	40	fcc	4.318	—	—
$TiC_{0.80}$	1600	40	fcc	4.320	—	—
$TiC_{0.84}$	1600	40	fcc	4.323	—	—
$VC_{0.67}$	1500	30	fcc + hexagonal γ-VC	4.156	—	—
	1800	60				
$VC_{0.75}$	1800	60	The same	2.886	4.573	1.58
				4.156	—	—
$VC_{0.81}$	1800	60	fcc	4.159	—	—
$VC_{0.85}$	1800	60	fcc	4.161	—	—
$NbC_{0.38}$	1200	30	fcc + hexagonal Nb_2C	4.424	—	—
				3.093	4.948	1.6
$NbC_{0.56}$	1200	30	fcc + hexagonal Nb_2C	4.424	—	—
				3.100	4.961	1.6
$NbC_{0.77}$	1200	30	fcc	4.428		
	1800	60	fcc			
$NbC_{0.84}$	1800	60	fcc	4.432	—	—
$NbC_{0.87}$	1800	60	fcc	4.442	—	—
$NbC_{0.89}$	1800	60	fcc	4.459	—	—
$NbC_{0.92}$	1800	60	fcc	4.464	—	—
$NbC_{0.97}$	1800	60	fcc	4.469	—	—

The content of absorbed hydrogen in niobium carbide is considerably greater in absolute terms than the number of carbon vacancies in the initial carbide. By similarly extrapolating the curve of the dependence of hydrogen content on carbon content, it can be seen that niobium carbohydride having a limiting carbon content may have a fairly high hydrogen content.

According to the results of x-ray structural analysis, the fcc lattice is retained in titanium carbohydrides, with a small increase in parameter as compared with the carbide (Tables 2 and 3). It should be noted that as a result of hydrogenation of the two-phase specimens ($TiC_{0.21}$ + α-Ti), a two-phase carbohydride is also

TABLE 2. Properties of the Carbohydrides of Titanium, Vanadium, and Niobium

Formula of carbohydride	Sum of x + y in the formula MeC_xH_y	Phase composition	Lattice parameter, Å			Hydrogenation conditions		
			a	c	c/a	Temperature, °C	Time, h	Pressure, atm
$TiC_{0.24}H_{0.73}$	0.97	fcc + hcf (T-phase)	4.312 3.076	— 5.062	— 1.65			
$TiC_{0.46}H_{0.55}$	0.01	fcc + hcf (T-phase)	4.314 Not determined	—	—			
$TiC_{0.61}H_{0.33}$	0.94	fcc	4.320	—	—	800	3	1
$TiC_{0.78}H_{0.20}$	0.98	fcc	4.323	—	—			
$TiC_{0.82}H_{0.19}$	1.01	fcc	4.326	—	—			
$VC_{0.63}H_{0.22}$	0.82	fcc	4.155	—	—			
$VC_{0.71}H_{0.19}$	0.90	fcc	4.160	—	—	1000	3	1
$VC_{0.79}H_{0.16}$	0.95	fcc	4.162	—	—			
$VC_{0.84}H_{0.14}$	0.96	fcc	4.160	—	—			
$NbC_{0.33}H_{0.61}$	0.94	fcc + hcf	Not determined 3.067	5.009	1.66			
$NbC_{0.51}H_{0.51}$	1.02	fcc + hcf	4.43 3.112	— 4.970	— 1.60			
$NbC_{0.64}H_{0.47}$	1.11	fcc	4.430	—	—			
$NbC_{0.77}H_{0.41}$	1.18	fcc	4.436	—	—	1000	3	1
$NbC_{0.81}H_{0.40}$	1.21	fcc	4.446	—	—			
$NbC_{0.85}H_{0.39}$	1.24	fcc	4.458	—	—			
$NbC_{0.87}H_{0.38}$	1.25	fcc	4.470	—	—			
$NbC_{0.92}H_{0.36}$	1.27	fcc	4.467	—	—			

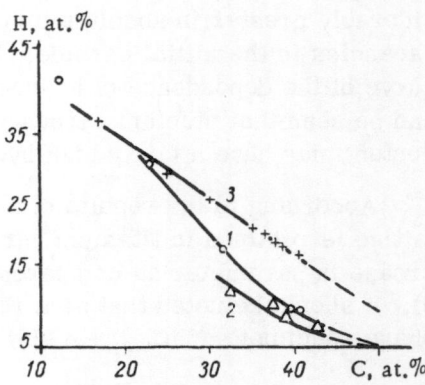

Fig. 1. Hydrogen content in carbohydrides as a function of carbon content: 1) for titanium carbohydride; 2) for vanadium carbohydride; 3) for niobium carbohydride.

TABLE 3. Interpretation of X-Ray Diagrams of HCP Titanium Carbohydride with a Lattice Parameter of $a = 3.076$ Å, $c = 5.062$ Å, $c/a = 1.65$

Indices (hkl)	d_{obs}	T-phase (according to data in [2])
010	—	2.65
002	2.48	2.48
011	2.37	2.36
012	1.84	1.83
110	1.84	1.54
013	1.43	1.42
112	1.43	1.32
021	—	1.29
022	1.18	1.18
014	1.146	1.142
023	1.039	1.046
121	1.008	0.990
114	0.970	0.978
122	0.934	0.936
030	—	0.890
123	0.862	0.866
032	0.838	0.832
0.25	0.806	0.806

formed, in which there exists, in addition to the fcc phase of titanium carbohydride, a phase having a hcp structure similar to the T-phase described in [13].

It is interesting to observe that vanadium carbohydrides produced from two-phase carbide specimens are single phase. The carbon content in them is considerably less than the lower limit of the homogeneity range of vanadium carbide, indicating a broadening of the homogeneity range of the carbohydride as compared with the carbide.

Practically no change takes place in the lattice parameter of niobium carbides whose compositions lie close to the stoichiometric $NbC_{1.00}$ when hydrogenated. Figures 2 and 3 give line diagrams of the carbides $NbC_{0.33}$ and $NbC_{0.51}$ and the carbohydrides derived from them. The observed changes in the line intensities indicate a reduction in the content of the hexagonal phase on hydrogenating the two-phase carbides concerned.

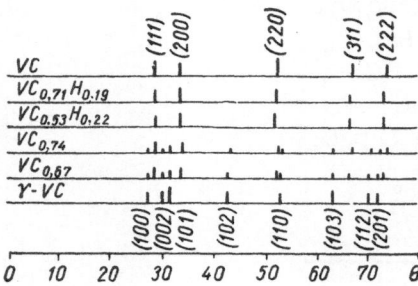

Fig. 2. Line diagrams of two-phase specimens of vanadium carbides and the carbohydrides obtained from them (for comparison, line diagrams of fcc VC and hexagonal
γ-VC are also given).

Niobium carbohydrides also display a broadening of the homogeneity range as compared with the carbides NbC_{1-x} (Fig. 3). The
minimum carbon content in the homogeneity range of niobium carbohydride can be approximately estimated. As the first two carbohydrides of niobium obtained are two-phase, the lower boundary
of the homogeneity range must fall within the range 25–30 at.%,
i.e., the lower boundary of the homogeneity range is displaced
by approximately 15% in the lower-carbon-content direction, as
compared with the carbide.

Dissociation-pressure data indicate an increase in the stability of the carbohydride phases as compared with the corresponding carbides (Table 4). The thermal stability of titanium carbohydride increases with rise in the carbon content. The carbon con-

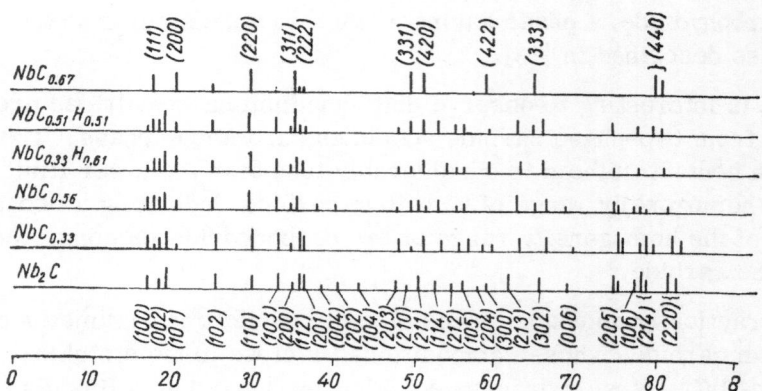

Fig. 3. Line diagrams of two-phase specimens of niobium carbides and
the carbohydrides obtained from them (for comparison, line diagrams of
fcc $NbC_{0.77}$ and hexagonal Nb_2C are also given [15]).

TABLE 4. Dissociation Pressure of Carbohydrides

Phase	Hydrogen pressure (mm Hg) at various temperatures (°C)									
	100	200	300	400	500	600	700	800	900	1000
$TiC_{0.24}H_{0.73}$	1.6	2.9	5.5	8.0	14.8	158.7	405.2	944.7	—	—
$TiC_{0.46}H_{0.55}$	0.7	1.9	4.0	6.7	11.6	110.2	241.5	687.3	—	—
$TiC_{0.61}H_{0.33}$	0.6	1.5	2.8	6.6	8.8	22.1	63.2	216.9	475.7	795.8
$TiC_{0.78}H_{0.20}$	0.5	1.2	1.7	5.6	8.4	16.3	41.8	135.7	298	468.1
$TiC_{0.82}H_{0.19}$	0.2	0.6	1.2	1.5	3.7	9.5	10.7	15.0	28.8	70.3
TiH_2[9]	1.1	3.8	10.9	15.2	230.6	1764.2	—	—	—	—
$VC_{0.63}H_{0.22}$	1.7	4	5	5.1	6.6	7.1	23.9	42.3	61.5	81.8
$VC_{0.71}H_{0.19}$	0.7	1.3	2.1	3.0	3.8	5.3	12.9	25.1	45.1	76.9
$VC_{0.74}H_{0.16}$	0.5	1.1	1.8	2.5	3.3	4.7	9.6	19.7	36.1	68.5
$VC_{0.84}H_{0.14}$	0.6	1.0	1.4	1.9	3.1	4.1	8.2	17.0	31.0	61.2
VH	4.0	5.5	7.5	8.8	84.5	—	—	—	—	—
$NbC_{0.51}H_{0.51}$	4.7	5.6	6.7	8.5	52.5	162.4	209	248	294	333.2
$NbC_{0.77}H_{0.41}$	3.8	4.8	5.7	6.6	9.9	16.5	26.4	39.5	54	71.4
$NbC_{0.81}H_{0.40}$	2.6	3.5	4.1	5.4	7.5	12.5	20.5	31	42.8	58
$NbC_{0.85}H_{0.39}$	1.7	2.9	3.4	4.5	6.6	10.2	16.4	23.8	34.5	47.6
$NbC_{0.87}H_{0.38}$	1.0	2.1	2.6	3.5	5.5	9.3	14.1	21.7	31.3	41.5
$NbC_{0.92}H_{0.36}$	0.5	16	2.2	2.8	4.6	7.6	11.6	18.3	26.5	37.2
NbH [9]	—	5.7	946.0	1528.0		—	—	—	—	—

TABLE 5. Specific Electrical Resistivity of Carbohydrides

Carbohydride	Specific electrical resistivity, $\mu\Omega \cdot cm$	Carbohydride	Specific electrical resistivity, $\mu\Omega \cdot cm$
$TiC_{0.24}H_{0.73}$	665	$NbC_{0.33}H_{0.61}$	1080
$TiC_{0.46}H_{0.55}$	645	$NbC_{0.51}H_{0.51}$	1020
$TiC_{0.61}H_{0.33}$	460	$NbC_{0.77}H_{0.41}$	990
$TiC_{0.78}H_{0.196}$	335	$NbC_{0.81}H_{0.40}$	955
$TiC_{0.82}H_{0.19}$	320	$NbC_{0.85}H_{0.39}$	795
$TiC_{0.63}H_{0.22}$	760	$NbC_{0.92}H_{0.36}$	710
$TiC_{0.71}H_{0.18}$	684	$NbC_{0.92}H_{0.36}$	510
$TiC_{0.79}H_{0.16}$	544		
$TiC_{0.84}H_{0.14}$	500		

tent in vanadium and niobium carbohydrides has no significant effect either on their hydrogen content or on their thermal stability.

The specific electrical resistivity has been measured on powder specimens by the method described in [7]. This method gives rise to considerable error in the determination of the specific electrical resistivity, and hence the figures obtained represent only rough values for the comparison of the electrical resistivities of powder specimens of carbohydrides with the original carbides [6] and for comparison among themselves. The electrical-resistivity values obtained are presented in Table 5.

The change in the specific electrical resistivity in relation to carbon content in carbohydrides is similar to that in carbides.

Discussion of Results

It is well known that in carbides the bonds $Me-Me$, $Me-C$, and $C-C$ are operative, the $Me-C$ bond being the decisive one. In considering the $Me-C$ bond, attention must be paid to the electron structure of the metal atom, which determines the nature of its interaction with carbon.

In the isolated state the titanium atom has a configuration of the valence electrons of $3d^2 4s^2$. In the formation of a metallic crystal of α-Ti, the titanium atoms tend to acquire the stable d^5-configurations of the localized part of the valence electrons, the statistical weight of which is less than 50% [8, 9].

Titanium carbide is formed at temperatures at which β-Ti is stable; in this modification the statistical weight of the stable d^5-configurations should be greater than it is in α-Ti, since the polymorphic transformation is accompanied by an increase in the lattice symmetry [10].

From the researches of I. I. Kovenskii on electrical transfer in solid solutions [5] and the x-ray structural investigations of M. P. Arbuzov and his co-workers [4] on the determination of the electron-density distribution in the titanium carbide lattice, it has been established that in these phases titanium bears a positive charge and that the electron density is shifted from titanium to carbon. Arising from this, it may be assumed that some of the valence electrons of titanium, in shifting to the carbon atoms, enter into an exchange interaction with them which leads to the stabilization of the sp^3-configurations of the carbon atoms. When hydrogen is introduced, it interacts with the metal. In titanium carbide, as a result of the shift in the electron density toward carbon [4], the sd-band in the titanium atoms is more defective and hence has a greater tendency to receive s-electrons from hydrogen.

As can be seen from the data we obtained (see Table 4), the thermal stability of the carbohydride phases of titanium is considerably higher than that of titanium hydride, indicating a stronger Me−H bond in the carbohydride of titanium than in the hydride; this is evidently also associated with the enhanced acceptor capacity of the titanium atoms in the carbide.

In the homogeneity range, the thermal stability of the carbohydride phases of titanium increases with the carbon content (see Table 4). The deficiency of the sd-band in the titanium atoms may decrease with fall in the carbon content in the titanium carbide, and a reduction in the degree of overlap with the s-electrons of the hydrogen atoms may accordingly take place.

It may be assumed that the vanadium atoms in vanadium carbide are similar in their electron properties to the titanium atoms in TiC, although acceptor properties may appear to a considerable extent in addition to the donor properties.

Niobium atoms in niobium carbide, while possessing a high statistical weight of stable d^5-configurations and a high energy stability (as a result of the increase in the principal quantum number of the valence electrons) [8-10], should exhibit predominantly acceptor properties. Hence the carbon content in niobium carbide has no great effect on the force of the Me−H interaction, which also indicates a high degree of thermal stability of all the niobium carbohydrides in the homogeneity range. The high acceptor capacity of niobium atoms in the carbide may also account for the fact that in carbohydrides close in carbon content to the stoichiometric composition NbC, there is a considerable hydrogen content, in contrast to the carbohydrides of titanium and vanadium.

According to the vapor-pressure measurements (see Table 4), the most thermally stable of the carbohydrides investigated are the niobium carbohydrides, a fact that may also be explained by the enhanced acceptor capacity of niobium atoms as compared with vanadium and titanium atoms.

On account of the greater acceptor capacity of the metal atoms in vanadium and niobium carbohydrides, as compared with titanium carbohydride, the carbon atoms in them, like the hydrogen atoms, display a donor type of interaction with the metal. This fact, together with the assumption about the preferential movement of the hydrogen atoms to the vacant octahedral sites, enables us to explain the broadening of the homogeneity range in respect of carbon in the carbohydrides of vanadium and niobium as compared with their carbides.

The specific electrical resistivity of the carbohydrides exceeds that of the respective carbides; this is evidently due to the additional scattering of the conduction electrons at the positively charged hydrogen ions.

Conclusions

1. An investigation of the carbohydrides of the transition metals of groups IV and V (titanium, vanadium, and niobium) has made it possible to draw provisional conclusions about the effect of features of the electron structure of the d-metals on the nature of the interaction between the components.

2. From the experiments that have been carried out it may be concluded that in the ternary $Me-C-H$ phases the type of chemical bond and the crystal structure remain the same as in the carbides.

Hydrogen does not affect the type of crystal structure of the carbide and has little effect on its parameters. As a consequence the NaCl type of structure is stabilized, and this is reflected in a broadening of the homogeneity range of the carbohydride phases (in comparison with the respective carbides). The latter relates to the carbohydride phases of vanadium, niobium, and above all titanium.

3. In the carbohydrides hydrogen reacts with the metal to form a stronger bond with it than that in the corresponding carbide. This points to an increase in the acceptor properties of the metal in the carbohydride, in agreement with concepts regarding the rise in carbides of the statistical weight of the metal atoms with stable d^5-configurations of the localized part of the valence electrons.

4. The most interesting of the three types of carbohydride considered as bearers of hydrogen at high temperatures are the carbohydride phases of niobium on account of their capacity to retain a considerable amount of hydrogen even in the vicinity of the stoichiometric composition in regard to carbon and also on account of their very high thermal stability, which varies little in the homogeneity range.

Literature Cited

1. R. A. Andrievskii, V. P. Kalinin, and G. I. Pepekin, Symposium on the Thermodynamics of Nuclear Materials, Vienna (1967).
2. R. A. Andrievskii and E. F. Khodosov, Poroshkovaya Met., No. 8, 65 (1967).
3. R. A. Andrievskii, E. B. Boiko, and V. P. Kalinin, Kristallografiya, 12:1068 (1967).
4. M. P. Arbuzov and B. Khaenko, Poroshkovaya Met., No. 4, 74 (1966).
5. I. I. Kovenskii, Fiz. Tverd. Tela, 5:1423 (1963).
6. A. Ya. Kuchma, Author's Summary of Candidate's Thesis, Institute of Problems in Materials Science, Academy of Sciences of the Ukrainian SSR, Kiev (1968).
7. L. I. Struk, Poroshkovaya Met., No. 3, 14 (1966).
8. G. V. Samsonov, Poroshkovaya Met., No. 12, 49 (1966).
9. G. V. Samsonov, Ukrainsk. Khim. Zh., 31:1233 (1965).
10. G. V. Samsonov, Ukrainsk. Khim. Zh., 33:763 (1967).
11. H. Bittner, Monatsh. Chem., 95:1514 (1964).
12. G. Brauer, H. Renner, and H. Wernat, Z. Anorg. Chem., 277:249 (1954).
13. H. Goretzki, H. Bittner, and H. Nowotny, Monatsh. Chem., 95:1521 (1964).
14. H. Goretzki, E. Ganglbarger, H. Nowotny, and H. Bittner, Monatsh. Chem., 96:5 (1965).
15. H. Goretzki, Phys. Stat. Solidi, 20:141 (1967).

New Methods of Comminuting Powders
of Refractory-Metal Carbides

B. D. Gurevich, L. B. Nezhevenko,
V. I. Groshev, and A. P. Gudovich

The optimum conditions for comminuting zirconium carbide by means
of ultrasound and by using a planetary centrifugal mill have been determined. The use of ultrasound leads to the production of powder
having spherical particles with a high degree of dispersivity and containing only a small amount of ground impurities. To reduce the impurities ground from the surface of the planetary centrifugal mill
during comminution, it is proposed that a zirconium carbide lining to
the mill be used.

The mechanical methods of grinding most widely used at the
present time have considerable drawbacks, such as the length of
the grinding operations and the contamination of the ground powder
by the materials of the mill lining and the balls. These methods
do not result in the production of the approximately spherical particles that are essential for the manufacture of special components.

In recent years investigators have begun to study the possibility of comminuting powders, including those of the refractory-
metal carbides, by using ultrasonic apparatus and mechanical equipment of the planetary-centrifugal-mill type [3, 4, 6].

The comminution of powders by means of ultrasonic apparatus
makes possible the production of powder having high chemical pur-

ity, a rounded shape (up to spherical), and a degree of dispersion comparable with that which can be obtained by mechanical grinding in a shorter time.

Very fine powders can be produced in only tens of minutes by comminuting powders in planetary centrifugal mills. However, as a consequence of the high rate, grinding is accompanied by contamination of the powders by materials from the surface of the mill and from mill components, and also by considerable heating of the charge.

The material chosen for investigation was zirconium carbide powders, which is extremely hard and is regarded as one of the most difficult materials to grind. The ZrC powder used was produced by the carbothermic reduction of zirconium dioxide and contained 88-88.2% Zr, 10.5-11.0% C_{total}, 0.2-0.6% C_{free}, with a specific surface of 0.3-0.6 m^2/g.

The object of the present investigation was to determine the optimum conditions for comminuting ZrC powders by means of ultrasound and by using planetary centrifugal mills. Comminution of the powders by means of ultrasound was carried out in a UZV-4D apparatus designed by B. A. Agranat (MISiS) and A. I. Kitai-

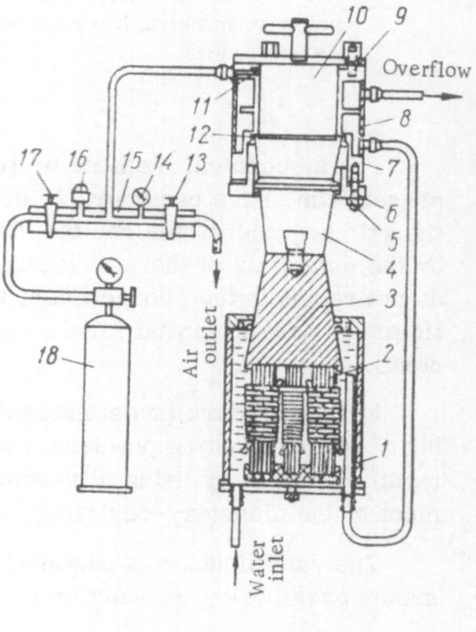

Fig. 1. Diagram of the UZV-4D apparatus: 1) concentrator housing; 2) vibratory system; 3) transformer; 4) concentrator; 5, 6) attachment of the working chamber to the concentrator; 7, 8) outer and inner walls of the working chamber, between which water circulates; 9) cover; 10) working chamber; 11, 12, 19) plastic packing; 13, 17) valves; 14) manometer; 15) casing of distributing device; 16) pressure regulator; 18) nitrogen cylinder.

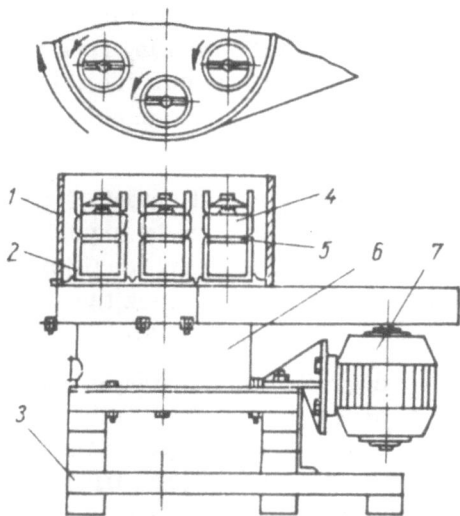

Fig. 2. General view of the M-44-L type of planetary centrifugal mill: 1) housing; 2) casing of pulleys with bands; 3) foundation; 4) barrel of the mill; 5) discharging diaphragm; 6) base; 7) electric motor.

gorodskii and V. I. Bashkirov (Scientific Research Institute of Mechanical Engineering Technology) and made at the Kutsevskii Mechanical Engineering Plant. A UZG-10M generator was used in conjunction with a PMS-15A transformer of 2.5 kW capacity. The working part of the concentrator was the bottom of the container (volume = 600 cm³), which was made of Kh18N9T stainless steel with a carefully polished inner surface. The apparatus operated at a frequency of 18-21 kHz. An outline diagram of the apparatus is shown in Fig. 1.

The effect of various factors on the ultrasonic comminution process was investigated, namely, the static pressure, comminution time, size of the charge, and ratio of solid to liquid (sol:liq). The change in shape of the powder particles and the introduction of impurities during grinding were also studied.

In investigating comminution in a planetary centrifugal mill, a mill of type M-44-L was used (Fig. 2), which had been made by the All-Union Scientific Research Institute for New Structural Materials.

The effect of the loading coefficient (K_1), the weight (volume) ratio of the load to balls (P_b), the amount of liquid (P_{liq}), and the amount of powder (P) on the rate of progress of the process was studied at a constant rate of rotation of the barrel of the mill and a constant ratio of its diameter to height. The shape of the ground particles and the amount of impurities introduced by grinding were also studied. Trichloroethylene was chosen as the working liquid for grinding ZrC, as it has no oxidizing action on the carbide [5]. Trichloroethylene is rather dangerous, as it is liable to explode on heating and has a high evaporation rate.

The optimum conditions for ensuring the greatest output of powder with maximum degree of dispersion were found by experiment for both methods. For ultrasonic dispersion these conditions were: static pressure 6 atm, volume of charge = 0.3 times the volume of the working chamber, and solid:liquid ratio = 1:4.

As can be seen from Table 1, the degree of dispersion of the powder increases from 0.3-0.6 to 8.5 m²/g with increase in the time of ultrasonic comminution, the rise being most rapid in the initial stage of the process. This may be due to the facts, first, that the particles break down along existing structural defects (cracks) in the original powder, the number of which diminishes during the process; and, second, that a change takes place in the shape of the particles produced (spheroidization), which leads to an increase in the resistance to cavitation breakdown.

TABLE 1. Degree of
Dispersion of ZrC Powders
and Amount of Impurities
Introduced as Functions
of the Comminution Time

Comminution time, h	Specific surface (Deryagin's method), m²/g	Change in impurity content, %		
		Fe	Cr	Mn
Ultrasonic comminution				
2	5.0	0.1	0.03	0.002
4	7.8	0.2	0.05	0.003
6	8.5	0.2	0.05	0.003
Planetary centrifugal milling				
4	3.1	2.0	0.026	0.005
6	4.8	2.8	0.037	0.009
14	7.8	6.8	0.09	0.017
20	8.5	8.0	0.1	0.019

The change in shape of the powder particles during comminution was studied by means of a MIM-7 microscope with a photographic attachment.

The particles of the original powder were pointed, whereas those of the comminuted product were rounded (often spheroidal) in shape. The assumption by the particles of a rounded form after ultrasonic treatment is a characteristic feature which has been noted previously [4].

The amount of impurities picked up from the surface of the working chamber was determined by chemical and spectrographic analysis, but no appreciable increase in impurity content was found. However, it was possible to detect certain increases in the impurity contents in the products, namely: up to 0.2% Fe (original content 0.1%), up to 0.05% Cr (original content 0.01%), and up to 0.003% Mn (original content 0.001%). The quantities of the other alloying elements and impurities in the steel (W, Co, Si, Ni, N, and O) remained unchanged. The impurities picked up during comminution corresponded to the composition of the steel (Kh18N9T) of which the chamber was made.

Similar ZrC powders were used for grinding in the planetary centrifugal mill, the process being carried out under the following optimum conditions: volume of charge = 0.5 volume of the barrel and weight ratio solid to liquid = 1.7 : 1. The barrel was made of steel st.3 and the balls were made of ball-bearing steel; they were 7 mm in diameter and the ratio of the weight of the balls to that of the charge was 15:1. Analysis of the data given in Table 1 reveals more reduction in the rate of grinding with time. This can be explained by the fact that during grinding of the powder and the consequent reduction in its weight per unit volume, the volume of the charge increases. The volume ratio "material:balls" chosen before grinding changes, and the coefficient of filling of the barrel becomes greater than 0.5. Under these conditions further comminution of the powder takes place more slowly, although it still proceeds very rapidly. Thus, for example, powder with a specific surface of 8.5 m^2/g was obtained after grinding for 20 min, whereas to attain the same degree of dispersion by the ultrasonic method of comminution required 6 h.

Rapid grinding is accompanied by a considerable pickup of impurities from the barrel surface and the balls. Thus, on reach-

ing a degree of dispersion of 8.5 m²/g the products of grinding in
a planetary centrifugal mill contained up to 8% Fe, 0.1% Cr, and
0.019% Mn, compared with impurities of 0.2, 0.05, and 0.003%, respectively, in the products of ultrasonic comminution (for the same
initial content). Such a large difference in impurity content is attributable to the difference in the comminution mechanism in the
two methods (as Groshev et al. [1] showed, in ultrasonic comminution the principal mechanism is cavitation breakdown, whereas in
the planetary centrifugal mill grinding takes place as a result of
the particles rubbing against one another and against the walls of
the barrel and the balls) and also to the difference in quality of
the finish of the working surfaces of the chamber and the barrel.

Pickup of impurities occurs particularly rapidly during the
initial stage of the process, when the abrasive power of the powder
is great on account of its coarseness.

Hence the possibility of using the planetary centrifugal mill,
in spite of its high productivity, is limited by the contamination
of the ground powder.

Contamination of the powder can be avoided by lining the barrel of the mill with a material similar to that which is being ground,
i.e., for grinding ZrC powders, sintered metalloceramic ZrC
should be used for the lining. In this case the grinders should also
be made of ZrC. These linings and grinders should possess high
strength and wear resistance. Containers, covers, and grinders
(cylinders) have already been produced, and barrel linings have
been made for ordinary roller mills operating at low rates of rotation (about 50 rpm) [1, 2].

Grinding powders in such mills has demonstrated the feasi-
bility of using them; the lining and the cylinders have undergone
grinding for many hours. The results of experiments in grinding
powders in a roller mill (diam. = 85 mm, h = 130 mm), lined with
ZrC and with ZrC grinders (diam. = 5 mm, h = 10 mm), presented
in Table 2, confirm the possibility of producing ZrC powders having
a given degree of dispersion [1, 2].

The strength characteristics of the metalloceramic lining and
cylinders are fairly high. Wear on them after 50 hours' operation
was 1.5 wt.%, and after 150 h it was up to 3 wt.%.

However, grinding in a planetary centrifugal mill with planetary
rotation of the barrels at a speed of 600-800 rpm makes consider-

TABLE 2. Degree of
Dispersion of ZrC as a
Function of
Comminution Time

Grinding time, h	Specific surface (Deryagin's method), m^2/g	Amount of ZrC picked up, %
15	3.7	—
30	3.9	—
50	4.5	1.5
100	7.8	2.0
150	9.8	3.0

ably greater demands on the mechanical strength and wear resistance of the grinders and the lining. The search for means of
raising the strength of the lining and balls in a planetary centrifugal mill is continuing.

Conclusions

1. The optimum conditions for comminuting ZrC by means of
ultrasound and in a planetary centrifugal mill have been determined. ZrC powders having a high degree of dispersion ($8.5 \ m^2/g$)
have been obtained in an ultrasonic field after 6 h and in the planetary centrifugal mill after 20 min (from an initial degree of dispersion of $0.3-0.6 \ m^2/g$).

2. It is shown that in comminuting ZrC powder in an ultrasonic field products are obtained that have a low percentage pickup
of impurities (up to 0.2% Fe, 0.05% Cr, and 0.003% Mn, compared
with initial contents of 0.1, 0.01, and 0.001%, respectively). In
grinding ZrC powder in a planetary centrifugal mill, materials are
obtained with a high impurity content, consisting mainly of iron
(up to 8%) picked up from the surface of the mill and the grinders.

3. The results of using roller mills lined with ZrC and equipped
with ZrC grinders have demonstrated the possibility in principle of
utilizing ZrC for lining these mills with the object of reducing the
amount of impurities picked up.

4. It is shown that the particles of powder comminuted in an
ultrasonic field have a rounded (often spherical) shape.

Literature Cited

1. V. I. Groshev, A. S. Maskaev, L. B. Nezhevenko, and N. I. Poltoratskii, Porosh-
    kovaya Met., No. 1 (1967).
2. V. I. Groshev, A. S. Maskaev, L. B. Nezhevenko, and N. I. Poltoratskii, Tsvet-
    naya Met., No. 4 (1967).
3. B. D. Gurevich, L. B. Nezhevenko, and V. S. Santi, Tsvetnaya Met., No. 4 (1967).
4. L. B. Gutnova, B. A. Agranat, and V. I. Bashkirov, in: The Development of
    the Theory and Practice of the Introduction of Progressive Ultrasonic Techniques
    in Mechanical Engineering [in Russian], Metallurgizdat, Moscow (1965).
5. B. G. Ignat'ev et al., At. Energ., 20:489 (1966).
6. S. S. Kiparisov and B. D. Gurevich, PNTPO, GOSINTI, No. 1-64-566/14.

II. METHODS OF PRODUCING COMPONENTS OF CARBIDES AND ALLOYS BASED ON THEM

Fundamental Aspects of the
Compaction of Carbides

L. I. Struk

The compaction of powders of the carbides of Ti, Zr, Nb, Cr, Mo, and W has been investigated. The following aspects of compaction were studied: the dependence of the density of the compacted samples on the pressure and the number of successive compacting operations; the effect of the time during which the powder remained under load on the compactability of refractory-compound powders; the force distribution during compaction and the ejection pressure; the effect of the charge and the plasticizer content on compactability; and the dependence of the elastic aftereffect on the compaction pressure.

Components made from transition-metal carbides and from alloys based on them are widely used in various fields of technology on account of their high melting point, heat resistance, high wear resistance, hardness, high electrical conductivity, low evaporation rate, and resistance to reducing gases, to molten metals, slags, and salts, and to the action of other chemically aggressive media [6]. Consequently the study of the compaction and sintering of transition-metal carbides is of considerable interest. Although the processes of mouthpiece compaction of refractory transition-metal carbides have been studied in detail [3, 4], information about the ordinary compaction of refractory-compound powders is meager [1, 5, 7-10].

In the present research the compaction of powders of the carbides of Ti, Zr, Nb, Cr, Mo, and W has been investigated. The following aspects of the process were studied: the effect on the density of the samples of the time during which the compacted powder was held under pressure, of successive compactions, of the moisture content of the charge, and of the plasticizer content; the distribution of forces during compaction, investigated by Unckel's method [11]; the elastic aftereffect in compacted samples; and the consolidation of the powders (studied by the electrical conductivity method).

Commercial-grade powders of the carbides of Ti, Zr, Nb, Cr, Mo, and W were used in the experiments. Tables 1 and 2 give the chemical composition of the powders and their grain size as measured by a microscopic method. The weights of the powders as poured and as shaken down were as follows:

Compound	TiC	ZrC	NbC	Cr_3C_2	Mo_2C	WC
Weight as poured, g/cm^3	1.30	2.08	2.42	3.31	2.22	7.20
Weight as shaken down, g/cm^3	2.44	3.71	3.64	4.94	3.66	8.34

An investigation of the dependence of the density of the compacted samples on the pressure and the number of successive compactions was carried out on powders of the carbides of Ti, Zr, Nb, Cr, and Mo.

The powders of the refractory compounds mentioned were compacted in a cylindrical die 8 mm in diameter in the pressure range from 1 to 10 tons/cm^2 (in stages of 1 ton), with and without the use of a plasticizer. Some of the samples were compacted under pressure and then crushed into powder, passed through a sieve, and again compacted under the same pressure.

TABLE 1. Chemical Composition of the Initial Materials

Refractory compound	Content, %			
	Me	C_{total}	C_{free}	Fe
TiC	77.6	19.5	0.8	0.3
ZrC	87.0	11.5	0.9	0.05
NbC	89.4	9.15	—	1.0
Cr_3C_2	86.0	13.2	0.2	—
Mo_2C	94.1	4.8	—	0.06
WC	93.1	6.2	0.08	0.38

TABLE 2. Grain Size of the Initial
Powders of the Refractory
Compounds

Mean parti-cle, size, μ	Content, %					
	TiC	ZrC	NbC	Cr_3C_2	Mo_2C	WC
< 2	37.4	73.7	76.6	48.0	40.0	13.6
2	28.2	14.9	12.9	18.0	20.1	17.0
4	19.1	7.4	7.3	12.4	20.5	23.4
8	9.2	2.1	2.1	13.9	15.3	9.6
13	4.4	1.3	0.7	7.7	—	13.6
13—17	—	—	—	—	—	2 6
13—21	—	—	—	—	—	5.5
13—25	—	—	—	—	4.1	3.8
17—25	1.7	—	—	—	—	—
17—29	—	0.6	0.4	—	—	—
29	—	—	—	—	—	11.9

Some of the samples were subjected to one, two, three, four,
or six successive compactions without being removed from the die.
With six successive compactions with and without intermediate
passage through the sieve, the density of the compacted sample increases from the first to the second compaction; at small compaction pressures (1 and 2 tons/cm²) the increase in density is
negligible, but at larger compaction pressures the density increases
by 2-3%, particularly with intermediate passages through the sieve.
This increase is due to the fact that brittle fracture of the particles
occurs at the higher pressures, and the fine products of this breakdown are distributed between the coarser particles; this in turn
leads to a rise in the density of the compacted sample.

Hence it is desirable to use double or triple compaction with
intermediate passage of the powder through a sieve (to secure a
more uniform distribution of the fine and coarse particles); in this
procedure the first compaction must be carried out at a pressure
of 4.0-4.5 tons/cm², which is sufficient to produce brittle
breakdown of the particles, and the second one at 1.5-2.5 tons/cm²,
since at higher pressures lamination phenomena appear.

The compactability of the powders of carbides of Ti, Zr, Nb,
Cr, and Mo can be well represented by a logarithmic relationship
between the compaction pressure P (kg/cm²) and the relative volume [2]

$$\beta = \frac{\alpha}{\gamma},$$

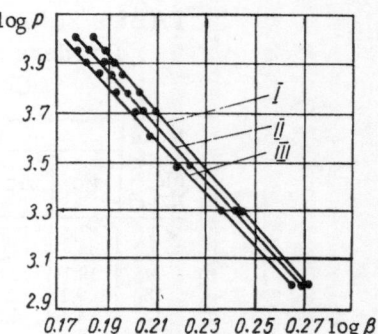

Fig. 1. Logarithmic relationship between the compaction pressure P and the relative volume β for TiC containing a plasticizer (starch): I) single compaction; II) double compaction; III) triple compaction.

where α is the density of the powder material (g/cm³) and γ is the apparent density of the sample in the pressure range from 1 to 10 tons/cm² (as an example, see Fig. 1).

It follows from the logarithmic straight lines (see Fig. 1) that the compaction of TiC plasticized with 4.5% starch can be represented by the equations

$$\log P = -11.5 \log \beta + 6.04 \quad \text{(for single compaction)},$$
$$\log P = -10.9 \log \beta + 5.91 \quad \text{(for double compaction)},$$
$$\log P = -10.8 \log \beta + 5.8 \quad \text{(for triple compaction)}.$$

Powders of TiC, Zr, and NbC were used to investigate the effect of the time of load application on compactability. Compaction pressures of 2, 4, and 6 tons/cm² were employed, and at each of these pressures the effect of the time the powder remained under load on the density of the sample was determined. The times chosen were practically instantaneous and 0.5, 1, 1.5, and 3 min. In all cases the density of the compacted sample was independent of the time under load. For example, the density of TiC samples compacted under a pressure of 2 tons/cm² was 2.89, 2.89, and 2.89 g/cm³ at compaction times of 0, 1, and 3 min, while the density of NbC samples compacted at 2 tons/cm² was 5.22, 5.22, 5.23 and 5.22 g/cm³ at compaction times of 0, 0.5, 1.5, and 3 min, respectively.

The absence of any effect on the density of the time the refractory carbide powders were held under pressure indicates their practically complete lack of plasticity and the retention of a strongly stressed state after compaction.

The effect of the moisture content of the charge on the compactability of refractory-carbide powders was investigated for the case of TiC plasticized with 1.25% of polyvinyl alcohol.

The moisture content of the charge was determined at the beginning and end of each experiment by the combustion of a weighed sample in alcohol (allowance being made for the ash content of the polyvinyl alcohol after combustion). The method of drying a sample in a desiccator until constant weight was reached was used for routine checks of the moisture content. The compactability of TiC powder was investigated at the following moisture contents in the charge: 9.2-7.4%, 4.2-3.3%, 3.6-3.3%, 1.0-0.8%, 0.6-0.5%, and 0.4-0.3% (the first figure denotes the moisture content at the beginning of the experiment and the second figure that at the end). It was not practicable to compact a charge with a moisture content exceeding 10%, since the water was squeezed out of the die and the charge stuck to the end of the punch. The results obtained on the compactability of TiC having various moisture contents are presented in Fig. 2.

The compactability of TiC is practically constant in the moisture range 3-4%. It decreases somewhat at 1% moisture, and falls sharply with reduction of the moisture content to 0.3-0.6% on account of the marked deterioration in the mobility and flow of the plasticizer itself at low moisture contents.

Thus, in compacting TiC powder the optimum moisture content may be taken as 3-4%.

NbC was used to investigate the effect of plasticizer content on compactability. The plasticizer used was polyvinyl alcohol, which was added to the charge before compaction in amounts of 0.25, 0.5, 0.75, and 1%. The experiments were carried out in the pressure range 1-6 tons/cm^2 and at a practically constant moisture content of 3.5-4%. The results obtained for the density of the samples are presented in Fig. 3. NbC powder compacted without a

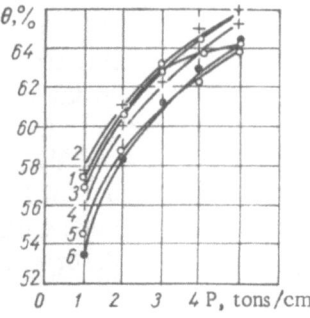

Fig. 2. Relative density θ of TiC samples as a function of the compaction pressure for various moisture contents in the charge: 1) 9.2-7.4%; 2) 4.2-3.3%; 3) 3.6-3.3%; 4) 1.0-0.8%; 5) 0.6-0.5%; 6) 0.4-0.3%.

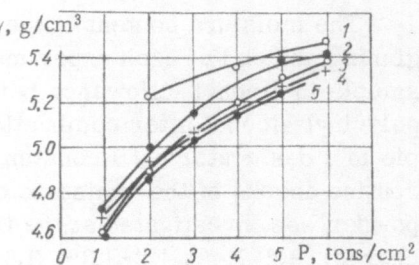

Fig. 3. Density γ of NbC samples as a function of compaction pressure for various polyvinyl alcohol contents: 1) without plasticizer; 2) 0.25%; 3) 0.50%; 4) 0.75%; 5) 1%.

plasticizer has the best compactability. The higher the plasticizer content, the worse is the compactability.

Consequently it may be concluded that polyvinyl alcohol does not improve the compactability of NbC, but only its formability. In dry compaction (without any plasticizer), there is heavy wear on the die, and the strength of the compacted sample is poor; this makes it necessary to add a minimum amount of plasticizer to the charge.

The distribution of forces during compaction was investigated by Unckel's method [11] on TiC, ZrC, and NbC powders. The previously plasticized powders were compacted in the Unckel die in the pressure range 1-8 tons/cm^2 without lubrication, with lubrication of the die walls and the end of the punch, and with a lubricant (MK machine oil) added to the powder before compaction.

As a rule, the proportion of the force conserved by friction at the die walls falls sharply with increase in pressure during the compaction of TiC, ZrC, and NbC powders. Thus, in the compaction of ZrC powder under a pressure of 1 ton/cm^2 the proportion of the force conserved by friction is 61.2%, whereas in compacting ZrC under 8 tons/cm^2 it is 49.3%. On lubricating the die with machine oil the proportion of the force conserved by friction decreases somewhat, while the compaction force and the density of the compacted sample increase correspondingly. When the lubricant is added to the charge, the frictional force falls sharply, and the density of the compacted samples increases correspondingly. Thus, in compacting TiC samples under a pressure of 4 tons /cm^2 the proportion of the force conserved by friction in the absence of lubrication is 58.6% and the density of the sample 3.08 g/cm^3; with lubricant added to the charge, the figures are 51.7% and 3.16 g/cm^3.

Such a change in the forces of friction and compaction, and likewise in the density, can be attributed to the fact that the lubricant
pressure proceeds (is squeezed out) to the die walls gradually as
the compaction increases in the case in which the lubricant is added to
the charge; whereas when the die walls are lubricated the oil is
removed by the powder and the upper punch at the start of compaction, and compaction subsequently takes place mainly without
lubrication. In consequence, in the compaction of carbides the use
of a lubricant added to the charge results in a considerable reduction in the frictional forces between the powder and the die walls,
with a corresponding rise in the density of the compact sample.
It should be noted that the distribution of forces in the compaction
of refractory-compound powder is similar to that found by Unckel
for iron powders [11].

A study of the dependence of the elastic aftereffect in compacted
powders on the compaction pressure was made by measuring the
height of the specimens under pressure and after removal from the
die. TiC, ZrC, and NbC powders were investigated in this way, and the
results are presented in Fig. 4.

The elastic aftereffect in TiC increases from 4.2 to 6.1% with
rise in compaction pressure from 1 to 5 tons/cm^2; in the pressure
ranges 2-3 and 4-5 tons/cm^2 the elastic aftereffect scarcely changes,
owing to the brittle breakdown of the interparticle bonds and of the
particles themselves and the relative strengthening of the compacted specimens that occurs as a result. The elastic aftereffect
increases sharply with rise in pressure from 5 to 10 tons/cm^2
in consequence of the lamination phenomena, which are intensified
in the pressure range 7-10 tons/cm^2. The nature of the change in

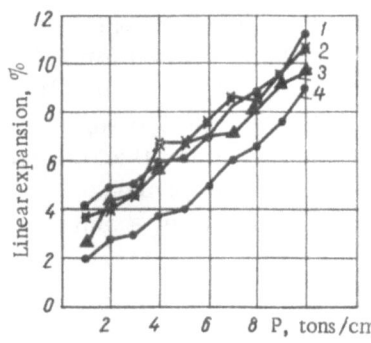

Fig. 4. Elastic aftereffect in TiC (1), NbC
(2), ZrC (3), and NbC (4) (particle size
1 μ) as a function of compaction pressure.

the elastic aftereffect in ZrC and NbC is similar to that in TiC.
It can be seen from the results reported that the higher the microhardness of the material, the greater is the elastic aftereffect in
it. However, in certain cases the particle size has a more marked
effect than the microhardness on the elastic aftereffect. Thus, in
compacting fine NbC powder ($\sim 1~\mu$) the elastic aftereffect exceeds
that of the coarser TiC powder, although the microhardness of the
latter (3000 kg/mm^2) is considerably greater than that of NbC (1960
kg/mm^2).

The relationships obtained are in agreement with data published in the literature, derived mainly from ductile and brittle
metals.

Powders of TiC, ZrC, NbC, Cr_3C_2, Mo_2C, and WC were used
to investigate compaction by the electrical-conductivity method.

The powders, with and without a plasticizer — a solution of
synthetic rubber in benzene — which was added to the charge before
compaction in the form of a 3% solution, were compacted in a hy-
draulic press having a textolite nonconducting die with an aperture
8 mm in diameter. The electrical conductivity of the powders was
measured by means of a double-bridge circuit at pressures from
0.5 to 7.5 tons/cm^2 (in 0.5-ton steps) without removing the specimen from the die, using the method proposed by Samsonov and
Neshpor [7]. The dimensions of the specimens were also measured under pressure.

The apparent specific electrical conductivity was calculated
from the formula

$$\rho = \frac{1}{\varkappa} = \frac{RS}{h},$$

where R is the resistance of the specimen (Ω), S is the cross section of the specimen (cm^2), h is the height of the specimen (cm),
and $\varkappa$ is the apparent specific electrical conductivity ($\Omega^{-1} \cdot$ cm^{-1}).
From the results obtained curves were drawn showing the rela-
tionship between the specific electrical conductivity and the relative
density for all the compounds investigated. Analysis of these
curves showed that the following relation exists between $\varkappa$ and
$\beta = 1/\theta$ (the relative volume)

$$\varkappa = ke^{-A/\theta},$$

where k and A are constants over a certain pressure range.

TABLE 3. Coefficients in the Compaction Equation of
Refractory-Carbide Powders

Carbide	Pressure range, tons/cm²	κ_1	A_1	Pressure range, tons/cm²	κ_2	A_2
TiC	0.5—4.5	3490	3.71	4.5—7.0	2150	1.59
ZrC	0.5—4.0	9030	5.21	4.0—7.0	2170	2.79
NbC	0.5—4.0	2440	4.60	4.0—7.0	2050	1.83
Cr_3C_2	0.5—4.5	3100	5.60	4.5—7.0	2090	2.75
Mo_2C	0.5—4.0	13400	4.29	4.0—7.0	2330	2.20
WC	0.5—3.5	5400	3.56	3.5—6.0	2270	2.27

Each of the compounds investigated can be represented by two
equations in the pressure range used in the experiments, each
equation having its own coefficients k_1, A_1, k_2, A_2. These four coefficients (Table 3) were determined from graphs showing the dependence of the logarithm of the specific electrical conductivity on
the relative volume (Fig. 5). As a rule, the first points, corresponding to the lowest pressures (0.5 and 1 ton/cm²) do not lie on the
semilogarithmic curves. This may be due to the fact that the total
contact surface (and hence the electrical conductivity) is so small
at low pressures that it does not conform to a definite relationship.
The points of inflection on the semilogarithmic curves show that
the rate of increase of the specific conductivity changes (slows down)
at them.

It may be concluded that as the pressure rises compaction
occurs mainly as a result of better packing of the particles and

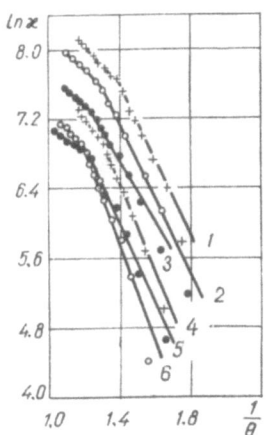

Fig. 5. Logarithm of the apparent specific
electrical conductivity $\varkappa$ as a function of
the relative volume $(1/\theta)$ for the compounds: 1) Mo_2C, 2) WC, 3) TiC, 4) ZrC,
5) NbC, 6) Cr_3C_2.

brittle breakdown of the interparticle bonds, giving rise to the relative rounding of the particle shape, with scarcely any rupture of the particles themselves. Beginning at certain pressures, the particles break down and a large number of fine grains are formed; this in turn leads to greater electrical contact resistance between the particles and to a corresponding reduction in the rate of increase of the electrical conductivity with rise in pressure. The relative rate of change of the electrical conductivity in the given range is obviously determined by the tendency of the material to brittle breakdown (in contrast to the region below the inflection, where the rate of change of the electrical conductivity is determined by the rate of compaction — the packing and the brittle breakdown of the interparticle bonds and not of the particles themselves).

By making use of the conclusion reached in [7] that A_1 characterizes the rate of brittle breakdown of the interparticle bonds and A_2 the brittle breakdown of the material itself, i.e., the brittleness, we can draw preliminary conclusions about the relative brittleness of materials, powders of which are subjected to compaction. The carbides can be arranged in the following order of increasing brittleness: $TiC-NbC-Mo_2C-WC-Cr_3C_2-ArC$. It should be noted that the addition of a plasticizer to the compacted powders has practically no effect on the character of the compaction. This can be understood if the compaction process, particularly at high pressures, is related to the brittle breakdown process, on which the plasticizer has no significant effect when it is not a surface-active substance and does not lead to breakdown of the particles as a result of physicomechanical factors (splitting forces).

Literature Cited

1. B. N. Babich, K. I. Portnoi, and G. V. Samsonov, Metal. i Term. Obrabotka Metal, No. 1, 31 (1960).
2. M. Yu. Bal'shin, Powder Metallurgy [in Russian], Metallurgizdat, Moscow (1948).
3. P. S. Kislyi and G. V. Samsonov, Poroshkovaya Met., No. 3, 31 (1962).
4. P. S. Kislyi, Poroshkovaya Met., No. 5, 68 (1962).
5. S. Ya. Plotkin and G. V. Samsonov, Vestnik Mashinostroeniya, No. 5, 53 (1959).
6. G. V. Samsonov, Refractory Compounds [in Russian], Metallurgizdat, Moscow (1963).
7. G. V. Samsonov and V. S. Neshpor, Dokl. Akad. Nauk SSSR, 104:405 (1955).
8. L. I. Struk, Poroshkovaya Met., No. 3, 14 (1966).
9. L. I. Struk, Poroshkovaya Met., No. 4, 1 (1966).
10. L. I. Struk, Izv. Akad. Nauk SSSR, Neorg. Mat., 3:624 (1967).
11. H. Unckel, Arch. Eisenhüttenwesen, 18:161 (1945).

Producing High-Porosity Materials from
Transition-Metal Carbides

G. V. Samsonov and V. K. Vitryanyuk

The conditions for producing materials with a porosity of 68-74% combined with satisfactory mechanical strength have been studied. The effect of additions of prepared carbides, zirconium powder, and particle size on the strength of the finished highly porous components has been determined.

Porous materials are widely used in various fields of technology as filters, as anodes in the manufacture of chlorine, as electrodes for electrochemical current generators, and as "sweating" materials for cooling high-temperature components [2, 9].

Transition-metal carbides are characterized by a high resistance to the action of acids, alkalies, and other aggressive media [6], and as a result it is of interest to use porous materials based on them in various chemical media in place of the porous metal or alloy materials in use at the present time.

Methods of producing high-porosity materials from metal powders have been fairly well developed, but many of these methods are unsuitable for the production of such materials from refractory compounds. It is known [1, 2] that in the production of high-porosity components additions are made to the powders of substances which decompose and escape during sintering, in this way promoting pore formation or preventing pores from being closed. Fillers and additives are used in preparing porous materials from powders of both relatively low-melting [1, 2] and re-

fractory metals or their compounds [3, 8]. However, the effec-
tiveness of their use in the latter case is small owing to the high
sintering temperature of refractory compounds.

In the present article we report the results of investigations
into the conditions for producing porous materials from carbides
of titanium, zirconium, and chromium (Cr_3C_2) simultaneously with
the formation of the carbides.

Charges consisting of oxides heated at 800°C for 2 h and carbon black heated at 400°C were made up with a view to reduction of
the oxides and formation of the carbides in accordance with the
outline reaction

$$MeO + C \rightarrow MeC + CO.$$

In investigating the effect of prepared carbides on the reduction of the oxides and sintering of the specimens, a predetermined quantity of the carbide was introduced into the charge consisting of oxide and carbon, with a view to the formation of a product in accordance with the outline reaction

$$xMeC + yMeO_2 + zC \rightarrow MeC + CO,$$

where Me is titanium or zirconium, $x = 0.2$, 0.4, 0.6, or 0.8, and
$x + y = 1$. To improve the formability of the powder, a plasticizer
was added to the charge; this consisted of a 1% solution of polyvinyl alcohol in water used in the proportion of 1.1-1.3 liter per 1
kg of charge. The charge was mixed in a ball mill for 3-4 h with
a weight ratio of balls to charge of 3:1. It was then dried in a
vacuum electric chamber at 80°C for 2-3 h till it was in a friable
state. After passing through a 60−80 mesh sieve, the mixture was
compacted into specimens 30 mm in diameter and 3-4.5 mm high
under pressures of 0.5-4 tons/cm². The compaction pressure was
found to have no significant effect on the porosity and shrinkage of
the specimens, and so in subsequent experiments the specimens
were compacted under a pressure of 0.5 ton/cm². To confer strength
and achieve complete removal of the moisture, the specimens were
dried to constant weight in a vacuum chamber at 120-140°C.

To produce components of chromium carbide the reduction-
carbidization process was carried out in an atmosphere of purified hydrogen in a furnace with a graphite heater; while in the case
of TiC and ZrC the process was carried out in a vacuum of 10^{-3}-
10^{-4} mm Hg.

The conditions of heating the specimens were chosen so as to ensure uniform evolution of both the decomposition products of the polyvinyl alcohol (which decomposes at about 400–450°C) and the CO gases formed by reduction of the oxides.

Thus, in heating specimens with the object of producing components of carbides of titanium and zirconium, reduction of whose oxides takes place in three stages [5-7], isothermal holding was carried out for 20 min at temperatures corresponding to these three stages. Then the temperature in the furnace was raised to the required temperature (1600-1900°C in 100-deg steps), and isothermal holding carried out for 60 min. By using such conditions it was intended to obtain purer carbides, since it has been shown in [4, 7] that prolonged heating of the charge has a beneficial effect on the quality of the product obtained. In order to produce components of Cr_3C_2, the rate of temperature rise in the range from 1000°C to the required temperature (1500-1600°C in 50-deg steps) was 10-15 deg/min.

During heating of the specimens the gaseous products of reduction (mainly CO) are involved, so ensuring a high porosity, and the carbide particles formed have a high activity and are sintered into the component with fairly good retention of the shape of the original compact.

The quality of the carbides formed was checked by x-ray analysis by taking photographs in an RKD camera 57.3 mm in diameter, using copper radiation, and subsequently comparing the results with those of chemical and theoretical analyses of the reaction products. In addition, the extent to which the reaction had proceeded to completion,

$$K = \frac{P_1 - P_2}{P_1 - P_3},$$

was determined, P_1 being the weight of the specimen before heating (g), P_2 the weight of the specimen after heating (g), and P_3 the theoretical weight of the specimen (g) assuming the reaction had proceeded to completion and allowance being made for the loss in weight on removal of the plasticizer.

According to the results of chemical analysis (Table 1), a rise in the carbidization-sintering temperature leads to a decrease in the free carbon content and an increase in the combined car-

TABLE 1. Results of Chemical Analysis and Degree of Completeness of the Reaction for Specimens of TiC, ZrC, and Cr_3C_2*

Charge	t, °C	Rate K, %	Chemical composition, %				
			Me	C_{total}	C_{free}	C_{comb}†	Total
$TiO_2 + 3C$	1600	96.5	77.7	20.6	1.7	19.23	98.3
	1700	99.8	79.2	20.2	1.1	19.34	99.4
	1800	100.2	79.6	20.1	0.7	19.53	99.7
	1900	100.5	79.8	20.0	0.4	19.78	99.8
$ZrO_2 + 3C$	1600	97.0	87.1	11.7	0.9	10.89	98.8
	1700	100.0	87.9	11.4	0.5	10.96	99.3
	1800	100.8	88.1	11.4	0.4	11.05	99.5
	1900	101.2	88.5	11.3	0.3	11.04	99.8
$3Cr_2O_3 + 13C$	1500	99.2	86.4	13.2	0.2	13.03	99.6
	1550	99.5	86.5	13.3	0.2	13.13	99.8
	1600	98.9	86.4	13.1	0.3	12.84	99.6
$0.2Zr + 0.8ZrO_2 + 2.6C$	1900	100.8	88.4	11.0	0.3	10.75	99.4
$0.4Zr + 0.6ZrO_2 + 2.2C$	1900	101.2	88.5	11.2	0.2	11.02	99.7
$0.6Zr + 0.4ZrO_2 + 1.8C$	1900	101.0	88.5	11.2	0.4	10.84	99.7
$0.8Zr + 0.2ZrO_2 + 1.4C$	1900	100.6	88.6	11.3	0.5	10.90	99.9

*Holding time = 1 h.

†The combined carbon content was determined for the carbide phase from the formula
$C_{comb} = (C_{total} - C_{free})/(100 - C_{free}) \cdot 100\%$.

bon content. The maximum combined carbon contents is reached at 1900°C in TiC and ZrC and at 1550°C in Cr_3C_2.

X-ray analysis has shown that the reaction products (for TiC and ZrC) obtained at all the temperatures investigated contain a single phase, TiC or ZrC, with lattice parameters of 4.321–4.322 and 4.692–4.695 Å, respectively.

In the case of chromium carbide the reaction products consist of a single phase, Cr_3C_2, but at 1600°C the x-ray diffraction diagram reveals very weak lines from the Cr_7C_3 phase.

With rise in the carbidization-sintering temperature, a fall occurs in the porosity of the specimens of all the charges investigated, while the volume shrinkage increases. In all cases the addition of prepared carbides to the original charge consisting of oxides and carbon leads to a reduction in porosity and – at small contents – to an increase in the volume shrinkage. This is due to the fact that the carbide particles introduced act as centers of formation and crystallization for the carbides produced by reduction of the oxides and lead to an increase in the rate of formation of the latter, which considerably activate the sintering process.

At the same time the porosity falls owing to a decrease in the quantity of gases formed during oxide reduction.

The addition of prepared carbide particles activates sintering more strongly in the case of ZrC components than in that of TiC components (Fig. 1). Evidently the particles of the added carbides differ in activity from the newly formed carbide particles, although their dimensions are approximately the same. In order to determine the effect of the degree of dispersion of the particles of added carbide (which controls the number of available centers in the volume of the component), there was added to the initial charge, consisting of zirconium dioxide and carbon black, ZrC with a particle size less than 1μ, produced by the wet grinding of the 44μ fraction of the carbide in a barrel lined with hard-alloy plates and using hard-alloy balls as the grinding agent; there was also added carbide of the fraction $-160 + 100 \mu$ and metallic zirconium powder with the amount of carbide required to produce in the end carbide of the stoichiometric composition.

It can be seen from the course of the curves in Fig. 2 that specimens obtained from the various charges differ in porosity and volume shrinkage. Such an effect of the particle size of the carbides can evidently be explained in the following way:

1. When coarse carbide particles are added to the charge, the distance between them is large and their number within the body of the charge is small. In consequence the rate of carbide formation

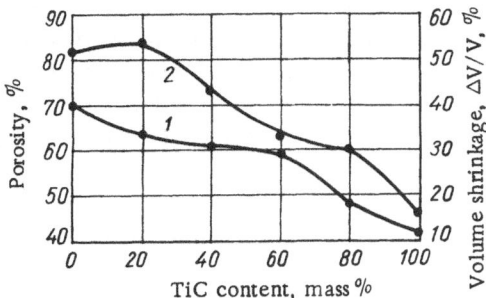

Fig. 1. Dependence of the porosity (1) and volume shrinkage (2) of TiC specimens on the content of prepared carbide, of fraction -44μ, in the charge (calculated on the final products; sintering temperature = 1900°C, holding time = 60 min).

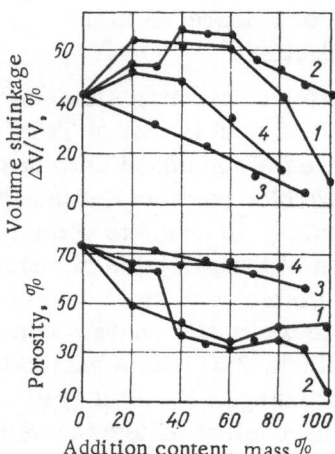

Fig. 2. Porosity and volume shrinkage of ZrC specimens as functions of the content in the charge of (calculated on the final product): 1) carbide of fraction −44 μ); 2) carbide with a particle size $\leq 1\,\mu$; 3) carbide of the fraction −160 + 100 μ; 4) zirconium of the fraction −44 μ (sintering temperature = 1800°C; holding time = 60 min).

from the charge (denoted by V_1) will obviously be somewhat higher than the rate in the absence of carbide added to the charge, since prepared centers already exist. The porosity falls a little, but owing to the large distance between the particles the "bonding" between them will be negligible, and consolidation of the material does not occur (shrinkage decreases), since it is difficult for "bridges" to form between the particles, and the particles themselves have a very low activation for sintering. On raising the carbide content above 20 mass%, the rate of carbide formation from the charge increases, but at the same time there is a rise in the number of "inactive" particles, and this leads to a further decrease in the shrinkage (sintering is deactivated), while the porosity falls as a result of the sharp decrease in the volume of gases formed during reduction of the oxide. Confirmation of the poor "connection" between the particles of the added carbide is provided by the considerable reduction in the strength of the components even when the charge contains only 40% of prepared carbide, while with 80% of prepared carbide the specimens are almost friable when a slight load is applied.

2. With the addition of carbide particles 44 μ in size (Fig. 2, curve 1), the particles are situated closer to each other and their content in the volume of the specimen is greater than in the first case; this favors an increase in the rate of carbide formation from the charge (in the given case the rate of carbide formation $V_2 > V_1$) and also promotes the formation of "bridges" between the indi-

vidual particles. This in turn favors activation of sintering, a considerable reduction in porosity, and an increase in the volume shrinkage of specimens containing 20% of prepared carbide. Compaction diminishes somewhat with rise in the prepared carbide content of the charge (shrinkage at first remains almost constant and then decreases, while the porosity also decreases very slowly), since there is an increase in the number of particles that have no great activity for sintering (although much greater than in the first case). With 80% of prepared carbide in the charge the porosity is approximately the same as that of specimens made from carbide powder of the same fraction.

3. With the addition of carbide particles having a mean size of about 1 μ (curve 2), the number of particles within the specimen is very large. As a result one might expect a considerable rise in the rate of carbide formation from the charge ($V_3 > V_2 > V_1$). However the very fine carbide particles added to the charge have a very defective crystal structure, as a result of which the formation and crystallization of the carbide probably takes place not only on the prepared centers but also on those formed by reduction, which in general activate the sintering somewhat less than in the preceding case.

With increase in the content of fine particles ($> 30\%$) having a high activity (the porosity of specimens prepared from such carbide is only 13% and the shrinkage 40%, compared with figures of 40% and 10%, respectively, for specimens prepared from carbide having a particle size less than 44 μ), sintering is activated more strongly; as a result shrinkage increases while porosity falls more steeply. Sintering of specimens at 1900°C showed that even with a prepared carbide content of 30% the porosity falls sharply while the shrinkage increases. Moreover, owing to the large number of fine particles within the specimen and their closeness together, compaction of the specimens can continue right up to the end of oxide reduction.

Confirmation of the above considerations regarding the effect of the presence of prepared centers on the formation of carbide from the charge and the simultaneous activation of the sintering process is provided by the course of curve 4 (Fig. 2), which shows the variation in porosity and volume shrinkage of specimens prepared from charges consisting of zirconium powder, zirconium dioxide, and carbon (see Table 1). With rise in the zirconium con-

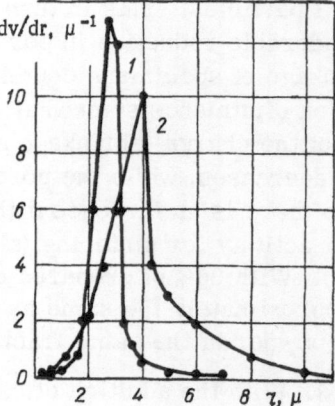

Fig. 3. Differential curves showing the
pore-radius distribution for Cr_3C_2 specimens: 1) produced directly by the reduction of Cr_2C_3 with carbon in a hydrogen
atmosphere; and 2) produced by the compaction of fine-grained Cr_3C_2 powder followed by sintering.

tent to 20 mass %, the porosity decreases somewhat, while the
shrinkage increases, since on account of the presence of already
reduced metal some acceleration of the carbide-formation process
is possible throughout the volume of the specimens. However, with
further rise in the zirconium content the porosity becomes practically constant and remains fairly high, while the shrinkage begins gradually to fall, since the activity of the carbide particles
formed by reaction between zirconium and carbon is obviously
lower than that of particles formed only by oxide reduction. X-ray
analysis revealed that in this case also the reaction products consist solely of a single phase, zirconium carbide with a lattice parameter 4.695–4.696 Å.

A study of the effect of the length of the isothermal holding
time during sintering showed that the porosity and volume shrinkage hardly change. Evidently sintering of the carbide particles
produced in a porous body takes place immediately they are formed
and ends at the same time as the oxide-reduction process.

The structure of the porous materials was also investigated
in this research (the pore sizes being determined by mercury
porometry at the Institute of Electrochemistry of the Academy of
Sciences of the USSR). It was found (Fig. 3) that components produced directly by carbide formation (curve 1) possess a very uniform pore size, in contrast to components produced by compaction
of chromium carbide powders followed by sintering (curve 2).

Conclusions

1. The conditions for producing porous materials consisting of the carbides of titanium, zirconium, and chromium have been investigated, and it is shown that components having satisfactory strength and a porosity of 68-74% can be made in this way.

2. The effect of additions of prepared carbides to a change comprising oxide and carbon black on the sintering of the components has also been studied. These added carbide particles act as crystallization centers for the carbides formed by oxide reduction, and they considerably activate the sintering process in the specimens and permit the porosity to be controlled over a wide range (33-74%).

3. The effect of the number of prepared centers added in the form of zirconium carbide particles of various sizes to a charge consisting of zirconium dioxide and carbon has been investigated. It is shown that coarse carbide particles practically deactivate the sintering process in the specimens, and the latter have poor strength; whereas fine-grained and very fine carbides activate the process and increase the shrinkage of the specimens. The strength of the specimens is fairly high.

4. The effect of the addition of zirconium powder to a charge consisting of its dioxide and carbon has been studied as regards carbide formation. The porosity of the specimens was found to remain fairly high at all zirconium contents; this is attributed to the absence of any prepared centers, since the centers are formed during the reduction of only a single oxide in the process of heating the charge, as a result of which sintering of the specimens is not activated. In view of the small shrinkage and satisfactory strength of the specimens, this method may be regarded as suitable for the production of porous components.

5. The uniformity of pore size of these materials is demonstrated in the case of Cr_3C_2 components having a porosity of about 70% obtained directly by the reduction of the oxide with carbon.

Literature Cited

1. Otsetek K. Agtek, Metalloceramic Filters [in Russian], Sudpromgiz, Leningrad (1959).

2. R. A. Andrievskii, Porous Metalloceramic Materials [in Russian], Metallurgiz-
    dat, Moscow (1964).
3. V. K. Vitryanyuk and V. B. Ordenko, Poroshkovaya Met., No. 4, 34 (1967).
4. R. Kieffer and P. Schwartzkopf, Hard Alloys [Russian translation], Metallurgiz-
    dat, Moscow (1957).
5. G. A. Meerson and G. V. Samsonov, Zh. Priklad. Khim., 25:744 (1953).
6. G. V. Samsonov and Ya. S. Umanskii, Hard Compounds of Refractory Metals [in
    Russian], Metallurgizdat, Moscow (1957).
7. G. V. Samsonov, Ukrainsk. Khim. Zh., 23:287 (1957).
8. V. M. Sleptsov, E. M. Prshedromirskaya, and Yu. P. Kukota, Poroshkovaya Met.,
    No. 10, 84 (1965).
9. E. Justi and A. Winsel, Fuel Elements [Russian translation], Mir, Moscow (1964).

The Effect of the Conditions of Producing Carbide Powder on the Properties of Sintered Specimens

L. B. Nezhevenko, V. I. Groshev, B. D. Gurevich, and O. V. Bokov

The effect of the content of incompletely reduced oxides on the sinterability of components made of zirconium carbide powders has been investigated. It was found that increasing the oxygen content in ZrC powders from 0.2 to 7 wt.% reduces the rate of grain growth in the finished components and promotes stabilization of the grain size. A method for producing high-density ZrC specimens has been developed.

The development of new fields of technology has made very heavy demands on materials, namely a combination of high temperatures and pressures with vibration and rapid loading. Of the materials that can satisfy these requirements, the most promising are refractory compounds, and in particular the transition-metal carbides. However, the application of these is restricted by their low strength characteristics and by the considerable variation in properties of specimens prepared from these materials.

The physicomechanical and high-temperature-strength properties of components depend on their structure; grain size and shape and the ratio of coarse to fine grains [2, 3, 5-7]. Preliminary investigations we have carried out into the effect of grain size on the strength of ZrC specimens (for example, on the diametric compression) have shown that with an increase in grain size from 20 to 50 μ the strength of the specimens falls from 15 to 11 kg · mm^{-2}. The formation of the structure in the specimens is affected by the degree of dispersion and shape of the original powders, the compaction pressure, the sintering temperature, the holding time, and the impurity inclusions.

The object of the present research was to determine the possibility of controlling grain growth in ZrC components:

a) by arranging for the reduction reaction to take place during sintering, which can be brought about by the presence in the powders of particles of zirconium oxides and an equivalent amount of free carbon; and

b) by the sintering conditions (temperature and time).

In order to obtain powders containing various quantities of incompletely reduced oxide and free carbon, experiments were carried out on the carbothermic reduction of a ZrO_2 + C mixture at 1400-2200°C in a vacuum furnace having a graphite heater. Four batches of powder were chosen as most typical as regards chemical composition; these were produced by reduction at 1600, 1800, 2000, and 2200°C. From a consideration of the physicochemical properties of these batches of powder (Table 1), it can be seen that by altering the temperature of reduction it is possible to vary the oxygen content in the powders (as a criterion of the degree of incomplete reduction of the product) and the free carbon content from 0.2 to 7% wt.% and from 0.1 to 6.7 wt.%, respectively.

In order to obtain powders with the same mean particle size (to eliminate the effect of their factor on sintering), suitable grinding conditions were chosen for all four batches of carbide. After

TABLE 1. Physicochemical Properties of Powders
Obtained at Various Carbidization Temperatures

Batch No.	Reduction temperature, °C	Grinding time, h	Chemical composition, wt.%					Pycnometric density, g/cm³	Weight as poured, g/cm³
			Zr	C_{total}	C_{free}	O	Total impurities		
1	1600	15	78.5	14.4	6.71	7.0	0.08	4.83—5.1	0.79
2	1800	30	87.1	11.3	1.0	2.0	0.08	5.77—5.8	0.963
3	2000	50	88.7	10.6	0.1	0.4	0 09	5.93—6.2	0.999
4	2200	50	88.8	10.8	0.1	0.2	0.08	5.79—6.2	1.05

the powders had been ground, their specific surface, as determined by the gas-adsorption method, was identical in all cases, being 12 m²/g. To obviate the pickup of impurities from the surface of the mill and the grinding agents, the grinding of the ZrC powder (and also the mixing of the original materials) was carried out in a mill having the lining and grinding agents made of ZrC [1]. Specimens 6 mm high and 10 mm in diameter were compacted from the powders so produced. Paraffin to the extent of 2 wt.% was used as a binding agent in compaction. The relative density of the compacted specimens was 60%.

The effect of reduction during sintering on the amount of shrinkage and on the structure formation was investigated in the temperature range 2200-2500°C (at 100-deg intervals) for holding times of 5, 15, 30, 60, and 120 min. The process was carried out in a vacuum furnace equipped with a graphite heater at a rate of temperature rise of 50 deg/min with subsequent furnace cooling. Up to 2000°C the specimens were sintered in vacuo and above that in an argon atmosphere, the argon being introduced into the furnace in order to maintain the efficiency of the heating elements.

The sinterability of the zirconium carbide powders of various compositions was assessed from the change in the chemical and phase compositions, shrinkage, density, and microstructure of the specimens. In spite of the difference in oxygen content of the original powders, the sintered specimens had the same chemical composition and structure of ZrC with a lattice parameter of 4.697 Å. The chemical composition of the specimens after sintering at

TABLE 2. Chemical Composition (wt.%) of ZrC Specimens after Sintering*

Reduction temperature, °C	Zr	C_{total}	O
1600	88.8	10.5	0.25
1800	88.9	10.5	—
2000	88.9	10.4	—
2200	88.9	10.5	0.2

*C_{free} = 0.1 wt.%.

2500°C for 50 min is shown in Table 2. This fact is important because it confirms the reaction between the incompletely reduced
zirconium oxides and the remains of the free carbon, which in turn
leads to activation of the sintering process [4].

A study of the nature of the shrinkage of the specimens showed
that shrinkage proceeds more rapidly as the oxygen content of the
initial powders increases; maximum shrinkage, amounting to 58%
(sintering temperature 2500°C and holding time 30 min), occurs in
specimens of batch No. 1, and minimum shrinkage (45%) in those of
batch No. 4.

The chosen sintering conditions enabled fairly high and approximately uniform (97-94 rel.%) densities to be obtained in all
the specimens.

Analysis of the microstructure of the specimens (Fig. 1) sintered in the range 2200-2500°C for 30 min revealed that, irrespective of the composition of the original powders, grain growth takes
place in the specimens with rise in the sintering temperature.
Raising the oxygen content in the powders leads to a fall in the rate
of grain growth. For batch No. 4 (containing 0.2 wt.% O) the mean
grain size increases from 5 to 50 μ with rise in sintering temperature from 2200 to 2500°C, while for batch No. 1 (containing 7
wt.% O) the grain size increases from 5 to 20 μ, other conditions
being the same. In specimens prepared from powders having a
high oxygen content, the structure is more uniform as regards
grain composition (Fig. 2). The majority of grains in specimens

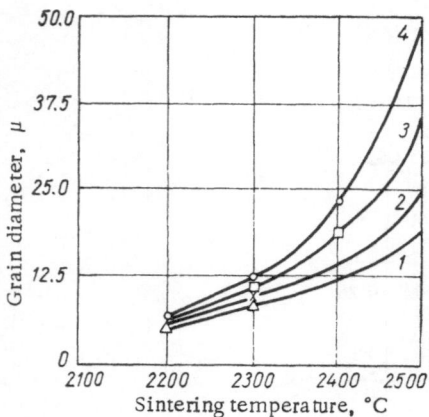

Fig. 1. Variation in grain size in ZrC specimens
as a function of the sintering temperature and
the amount of oxygen in the original powders:
1) 7 wt.% O; 2) 2 wt.% O; 3) 0.4 wt.% O; 4) 0.2
wt.% O.

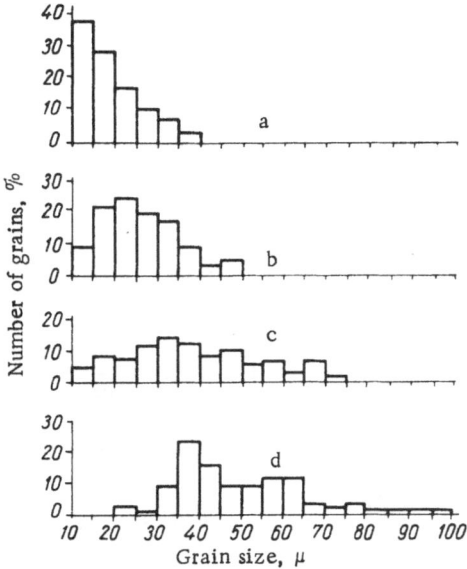

Fig. 2. Histograms of the grain-size distribution in ZrC specimens compacted from powders produced at the following reduction temperatures: a) 1600°C; b) 1800°C; c) 2000°C; d) 2200°C (sintering temperature = 2500°C; holding time = 30 min).

of batch No. 1 lie in the range 10-25 μ, whereas in specimens of batch No. 4 the grain size varies from 20 to 100 μ.

Investigation of the effect of the sintering time, in the range 5-120 min at the chosen temperatures, showed that increasing the oxygen content of the original powder is accompanied by a reduction in the rate of grain growth (Fig. 3). For example, grains with a mean size of approximately 45 μ were obtained for specimens made from powder of batch No. 4 by sintering for 30 min at 2500°C; whereas for specimens made from powder of batch No. 1 the mean grain size reaches only 30 μ even after sintering for 60 min at the same temperature. In both cases the increase in grain size takes place as a result of a rise in the number of grains of the coarse fraction (40-100 μ) and to the disappearance of the grains belonging to the fine fraction (5-15 μ). The latter occurs for the most part in specimens compacted from powders having a small oxygen content.

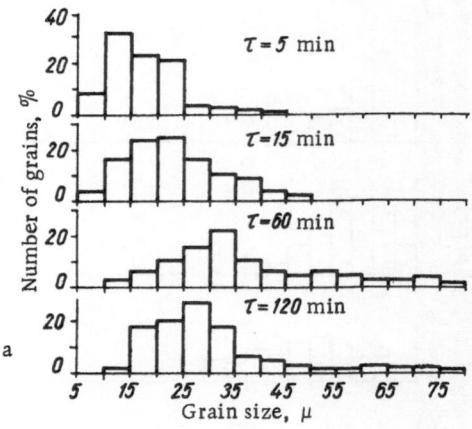

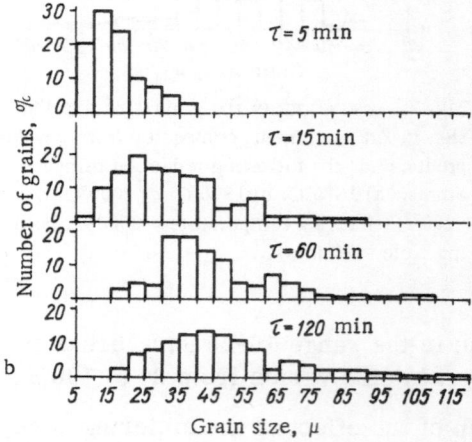

Fig. 3. Histograms of the grain-size distribution in ZrC specimens sintered at 2500°C for various times after being compacted from powders of batch No. 1 (a) and batch No. 4 (b).

The variation in the rate of grain growth with rise in temperature from 2200 to 2500°C is due to the appearance in the specimens of additional porosity resulting from the reduction reaction. These pores, which are formed and uniformly distributed throughout the volume of the specimens, retard grain growth and promote the stabilization of the grain size.

Conclusions

1. A study has been made of the effect of the content of incompletely reduced oxides in ZrC powders on the sinterability of components made from them. It is shown that with increase in the oxygen content of the ZrC powders from 0.2 to 7 wt.% the rate of grain growth in the finished components falls and stabilization of the grain size is promoted.

2. ZrC specimens having a high density and uniform chemical composition are obtained, irrespective of the content of incompletely reduced oxides in the original powders.

3. The experimental results obtained enable us to predict the possibility in the future of controlling the physicomechanical properties of the specimens.

Literature Cited

1.   V. I. Groshev et al., Poroshkovaya Met., No. 1, 49 (1967).
2.   W. D. Jones, The Fundamentals of Powder Metallurgy [Russian translation], IL, Moscow (1967), pp. 218-226.
3.   D. McLean, The Mechanical Properties of Metals [Russian translation], Metallurgiya, Moscow (1965), pp. 52-65.
4.   I. M. Fedorchenko, Poroshkovaya Met., No. 2 (1962).
5.   Ya. F. Fridman, The Mechanical Properties of Metals [in Russian], Oborongiz, Moscow (1952).
6.   E. N. Shilova, O. G. Nikitaeva, and E. N. Vasil'eva, Izv. Akad. Nauk SSSR, Metally, No. 6 (1966).
7.   E. M. Passmore, R. M. Spriggs, and T. Vasilos, J. Am. Ceram. Soc., 48:1 (1952).

Porosity-Free Polycrystalline Silicon Carbide and Its Application in High-Temperature Technology

G. G. Gnesin

The procedure for producing porosity-free polycrystalline silicon carbide and components made from it is described. A mechanism by which the structure of the material is formed is suggested. The results of determining the physical properties of porosity-free polycrystalline silicon carbide are reported, and some of the properties are compared with those of other carborundum materials and refractory compounds. Recommendations are made in regard to the industrial application of components made of porosity-free polycrystalline silicon carbide.

In modern power installations the chemical apparatus and metallurgical plant have joints and components which operate under conditions of aggressive media, high temperatures, thermal stresses, and erosion wear. Often these components must also be capable of making rapid heat exchange possible. Consequently, the materials from which thermally stressed joints and components are made should possesses high strength at elevated temperatures, high resistance, resistance to thermal shock, and also good thermal conductivity.

The creation of new high-temperature materials to satisfy the above requirements is based as a rule on the methods of powder metallurgy and ceramic production and also on the use of refractory compounds that do not contain oxygen (carbides, borides,

and silicides). To this class of compounds belongs silicon carbide, which possesses high chemical stability, strength at elevated
temperatures, hardness, and thermal conductivity. However, the
carborundum materials that are in wide use at the present time do
not possess the high qualities that distinguish pure SiC, since they
consist of porous multiphase mixtures [1]. In these materials the
individual SiC grains are cemented together by a ceramic or carbon bonding agent which differs considerably in its properties from
SiC, and this leads to a reduction in the mechanical and other
properties of the material.

It is natural to expect that a porosity-free material containing
almost 100% SiC will best meet the demands made on structural
materials operating in aggressive atmospheres at high temperatures.

Porosity-free polycrystalline SiC can be produced either by
hot compaction or by reaction sintering. The first method results
in the production of practically single-phase material with a density approaching the theoretical value. However, hot compaction
has serious limitations owing to the low efficiency of the method,
the difficulty of producing large components, and the high cost
arising from the heavy consumption of graphite dies.

The method of reaction sintering, which is free from the above
drawbacks, makes it possible to produce components having fairly
large dimensions (up to 550-600 mm), with a uniform structure
and density throughout the whole volume, and with a SiC content
up to 95-98% [2].

The basis of the reaction-sintering method is that a porous
sample compacted from a mixture of SiC and C powders is impregnated with molten silicon in an inert gas atmosphere at 1900-
1950°C.

Under these conditions there takes place chiefly solution of
carbon in the molten silicon [9] and its isothermal passage through
the melt under the influence of the concentration gradient of a
certain excess interphase energy, and of capillary pressure in the
pores. As a result of this process the surface of the SiC particles
becomes supersaturated with carbon, and secondary SiC is precipitated. The increase in the amount of the carbide phase without
change in the volume of the sintered specimen leads to the formation of a continuous skeleton of SiC with a large area of contact

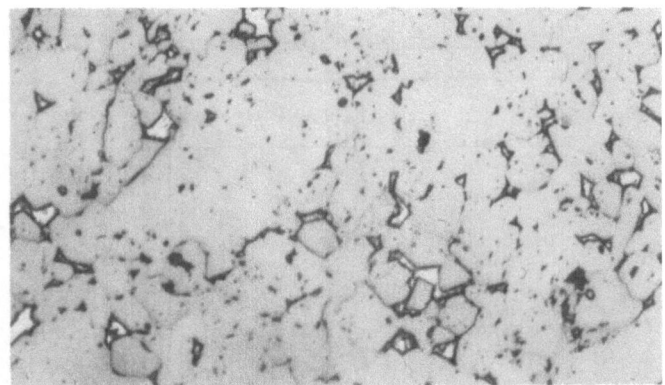

Fig. 1. Microstructure of porosity-free polycrystalline SiC
(×400): gray phase) SiC; white phase) Si; black inclusions)
free carbon.

regions in which are dispersed inclusions of the residual silicon
phase consisting of a solid solution of carbon in silicon [3] (Fig. 1).

The completeness of the reaction of secondary-SiC formation
depends on a large number of different factors, the most important
of which are the temperature and time of heat treatment, the extent of the reaction surface, and the reactivity of the carbon.

The compacts are formed by various methods depending on
the size and shape of the components being made. Long rods and
round and rectangular tubes are made by the mouthpiece compaction of a mixture of SiC and coke powders, plasticized with starch.
Components of simple shape and small height (lining plates and
bricks) are formed in steel dies under a pressure of 300–500 kg ·
cm^{-2}. Components of complex shape (crucibles and muffles) can
be formed by hydrostatic compaction.

The most important operation in the whole technological process is sintering, as it is during this operation that the final structure of the porosity-free polycrystalline material is formed. Sintering is carried out under conditions of oxygenless heating so as
to ensure impregnation of the porous compacts with molten silicon
and complete formation of secondary SiC.

To realize these conditions we used induction melting furnaces of types MGP-252 and OKB-936, commercially made in the
USSR; these were equipped with a graphite crucible into which

TABLE 1. Physical Properties of Monolithic
Polycrystalline Silicon Carbide

Property	Numerical value	Notes
Density, g/cm³	3,05—3.10	—
Porosity, %	Less than 1.0	—
Hardness, kg/mm²	3100—3200	—
Ultimate strength in bending, kg/mm²	25—35	At 20° C
	25—35	At 1200° C
Ultimate strength in compression, kg/mm²	110—130	At 20° C
	110—130	At 1200° C
Specific heat, kcal/kg·deg	0.20—0.22	Mean over the range 20—1000° C
Thermal conductivity, kcal/m·h·deg	71	At 300° C
	29	At 1400° C
Coefficient of linear expansion, 1/deg	$(4.5—4.7) \cdot 10^{-6}$	Mean over the range range 20—1200° C
Maximum working temperature, °C	1650	In air
	1900	In vacuum
	2200	In inert atmosphere

the protective atmosphere was passed. Components with a maximum linear dimension of 550-600 mm can be made in furnaces
of these types.

In order to raise the operating temperature of the furnace
to 2000°C, a refractory inductor packing consisting of zirconium
dioxide or boron nitride was used.

TABLE 2. Thermal Conductivity of Oxygen-Containing
and Oxygen-Free Refractory Compounds

Material		Melting point, °C	Thermal conductivity, kcal/m·h·deg		Porosity, %
			20° C	1000 °C	
Al_2O_3		2030	24.8	5.1	3—7
BeO		2570	180.0	16.6	3—7
MgO	Oxygen-contain-	2800	29.5	5.8	3—7
ZrO_2	ing materials	2550	1.8	1.8	3—10
Fireclay		1750	1.44	1.8	10—20
Magnesite					
Graphite		2100	14.4	3.2	10—20
$MoSi_2$	Oxygen-	3600	108.0	36.0	20—30
TiC	free	2030	27.0	10.8	2—10
B_4C	materials	3140	28.8	7.2	3—10
SiC		2450	25.2	18.0	2—5
		2800 (dissociation)	71.0 (300° C)	35.3	1.5—2.5

Determination of the phase composition of the material by x-ray structural and chemical analyses showed that the principal phase is SiC of the α- and β-modifications, amounting to 95-98%. It should be noted that the raw material used consisted of SiC produced at an abrasive plant; it had a hexagonal crystal lattice (α-SiC). The secondary SiC formed in the course of sintering is of the cubic β-modification, which has a diamond-type lattice.

The second constituent in the material is a silicon phase in amounts not exceeding 4.5-5%, which consists, as we showed in previous work [3], of a solid solution of carbon in silicon. Porosity-free SiC possesses very high mechanical strength, which is retained unchanged up to 1200°C (Table 1). The great hardness of the material makes it suitable for use under conditions of highly erosive action of the external medium.

Monolithic SiC is characterized by a high thermal conductivity, approaching that of metals. It should be noted that this value is considerably higher than that of almost any other ceramic material.

Monolithic polycrystalline SiC, with BeO and graphite, is one of the best thermal conductors at elevated temperatures (Table 2). The high thermal conductivity of porosity-free SiC determines the porosity of its use as a structural and lining material in the parts of apparatus where there is intense heat transmission (the lining of reactors and furnaces, heat exchangers, checkers, and muffles).

It is of interest to compare the thermal conductivity and mechanical strength of monolithic SiC and of carborundum materials bonded with various agents. These properties are given in Table 3,

TABLE 3. Thermal Conductivity and Mechanical
Strength of SiC and Carborundum Materials
Bonded with Various Materials

Material	Thermal conductivity, kcal/m·h·deg		Porosity, %	σ_B, kg/cm²	
	20° C	1000° C		20° C	1 000° C
SiC with siliceous bond	7.3	6.2	18.0	195	175
SiC with siliceous bond	11.7	6.9	17.2	235	135
SiC with a silicon nitride bond	14.2 (300° C)	7.3	21.0	310	235
Monolithic SiC	71.0 (300° C)	35.3	1.5—2.5	1680	1680

based on data in [1, 4, 8]. The table shows that monolithic SiC has
the highest thermal conductivity and mechanical strength. The
presence of a bonding agent and porosity considerably reduce these
properties. It must be borne in mind that porosity greatly increases
the reaction surface of the material and that as a result, during
prolonged operation in an aggressive medium, the material will
change its properties considerably.

In its chemical resistance, monolithic polycrystalline SiC approaches single crystals, confirming our previous investigations
into the oxidizability of SiC at elevated temperatures [5].

Figure 2 shows the change in weight of single-crystal and polycrystalline specimens of SiC with time during their oxidation in air
at 1770 and 1920°K (curves 4, 5, 10). The fairly close agreement
between these cufves provides evidence that the polycrystalline
specimen is practically free from pores and that the oxidation reaction takes place only on the surface. For the same reason there
is also good agreement between the oxidation isotherms for the sin-

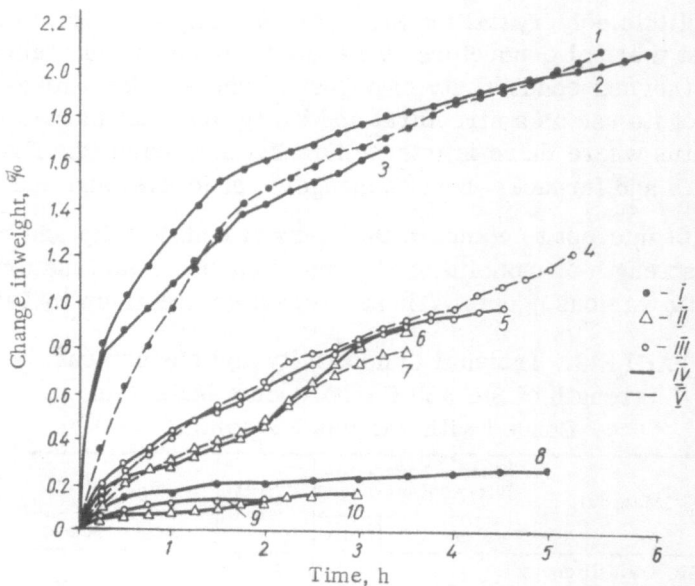

Fig. 2. Isotherms of the oxidation of single crystals (V) and polycrystalline specimens of SiC (IV) in atmospheres of carbon dioxide
(I), water vapor (2), and air (III): 1-7) at 1920°K; 8-10) at 1770°K.

gle-crystal and polycrystalline specimens in the case of the reaction of SiC with carbon dioxide (curves 1, 2, 3, 8).

Monolithic SiC has, in addition, a high chemical resistance to acids. On boiling in the following acids, specimens of monolithic SiC showed no change in their original weight after treatment for 2 h: HCl (conc.), HCl (1:1), HNO_3 (conc.), HNO_3 (1:1), H_2SO_4 (conc.), H_2SO_4 (1:4), H_3PO_4 (conc.), and H_3PO_4 (1:3). Only in a 1:1:1 mixture of HF, HNO_3, and H_2SO_4 is slight solution of the material observed on boiling.

As the polycrystalline material consists of a mixture of the α- and β-modifications of SiC, it may be assumed that selective solution of the β-SiC occurs. This assumption is confirmed by the fact that pure β-SiC, which we synthesized at 1900°C from silicon and carbon, dissolves appreciably on boiling in a mixture of HF, HNO_3, and H_2SO_4 as follows:

Time of treatment in acid, min	30	60	120
Amount of SiC dissolved	5.06	5.9	6.3

Pure α-SiC is practically insoluble under the conditions described.

Consequently, to raise the acid resistance of polycrystalline SiC, the material must undergo additional heat treatment at 2100–2150°C in order to convert the β-modification into α-SiC.

According to [6], SiC has an extremely low resistance to alkaline media, dissolving rapidly even in a boiling 50% solution of NaOH. Intense corrosion attack of the material takes place in its reaction with molten NaOH and Na_2CO_3.

Laboratory tests and industrial trials that we have carried out on SiC as a lining for aluminum electrolyzers [7] indicate its high resistance to molten fluoride salts at temperatures up to 1000°C. The industrial production of bricks and large plates of SiC for lining the furnaces of a fluidized bed, aluminum electrolyzers, and other metallurgical plants is now routine at the Brovarskii powder-metallurgy plant. A technique and operating conditions have been developed which enable SiC components to be produced in the form of bricks, plates, tubes, and crucibles having a maximum dimension up to 550–600 mm.

Literature Cited

1. I. S. Kainarskii and E. V. Degtyareva, Carborundum Refractories [in Russian], Metallurgizdat, Moscow (1963).
2. I. N. Frantsevich, Avt. Svid., No. 176,070 (February 26, 1964); Byull. Izobr., No. 21 (1965).
3. G. G. Gnesin and A. V. Kurdyumov, in: Silicon Carbide [in Russian], Naukova Dumka, Kiev (1966), pp. 83-91.
4. W. D. Kingery, High-Temperature Measurements [Russian translation], IL, Moscow (1962).
5. I. A. Yavorskii et al., Poroshkovaya Met., No. 1 (1968).
6. G. E. Mangsen and R. E. Diel, Corrosion, 17:1 (1961).
7. I. N. Frantsevich et al., Tsvetnye Metally, No. 6 (1965).
8. V. A. Zeigarnik and G. G. Gnesin, Poroshkovaya Met., No. 9, 40-45 (1969).
9. R. I. Scace and G. A. Slack, J. Chem. Phys., 30:1551 (1959).

III. REFRACTORY-CARBIDE COATINGS

The Vacuum Carbidization of Transition Metals of Groups IV and V

G. L. Zhunkovskii

The carbidization in vacuum of transition metals of groups IV and V
has been investigated. A relationship has been established between the
thickness of the coatings formed and the pressure, temperature, and
time of impregnation; the optimum temperatures for producing dense,
pore-free layers that are well bonded to the substrate have also been
determined. The two-phase nature of the structure of coatings on
group V metals has been demonstrated, and a proportionality has been
found between the thickness of the phases and the width of the phase
regions in the constitutional diagrams.

The study of the reaction diffusion of carbon into titanium, zirconium, hafnium, vanadium, niobium, and tantalum is interesting in
connection with the development of a technique for producing coatings whose physical and chemical properties are considerably better than those of the underlying coated materials.

The fullest investigations of the carbidization of some of the
transition metals have been carried out in [3, 4, 7], in which a
study was made of the kinetics of impregnation, the phase composition, thickness, hardness, and a number of other properties
of the diffusion zones over a wide range of temperatures and times,
and the diffusion coefficients and activation energies of the process were also determined. The researches reported in [1, 5, 8]
are concerned with the search for effective methods of carbidizing
titanium. Various authors have investigated the effect of different

carbon-containing media [5, 8] and heating conditions [1] on the
kinetics of growth, properties, phase compositions, and the strength
of the bond between the substrate and the carbide layer formed.

It has been found that the cementation of titanium can be carried out from a solid-phase bath only in vacuum; it can be done
from the gaseous phase on condition that the cementation gas is
mixed with argon in accurately controlled proportions and that only
very thin TiC layers are produced. In all the articles cited above
except [5], hardly any attention was paid to the important factor of
the capacity of the metals studied to absorb large quantities of hydrogen, oxygen, and nitrogen at elevated temperatures [6]. By
dissolving in the diffusion layers and the substrate, these elements
as a rule cause a considerable deterioration in the mechanical
and other properties of the coatings obtained. To a certain extent
it is clearly possible to avoid such saturation, and thereby improve
the properties of the coatings, by carbidization in vacuo. Such experiments have already been made on some metals [3, 6], but no
systematic investigation of the kinetics of vacuum carbidization
over a wide range of temperatures and times has yet been undertaken. Hence we have studied the carbidization of the above-
mentioned metals in vacuum.

The raw materials for preparing the specimens consisted of
titanium of grade VT-1, iodide zirconium, iodide hafnium, vanadium, and niobium with an impurity content not exceeding 0.2%,

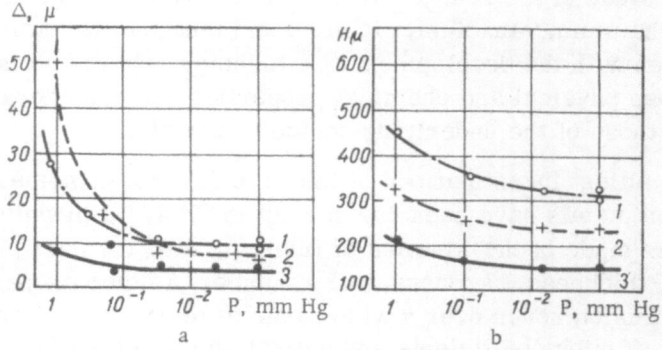

Fig. 1. Thickness of the layer (a) and hardness of the substrate
(b) as functions of the degree of vacuum in the furnace (impregnation temperature = 1300°C; time = 2 h); 1) Ti, 2) Zr, 3) Nb.

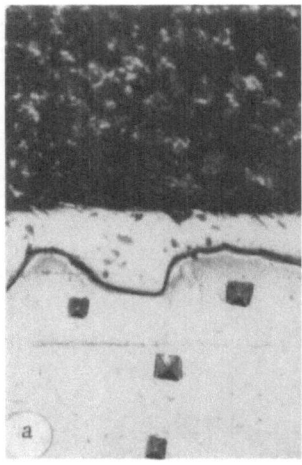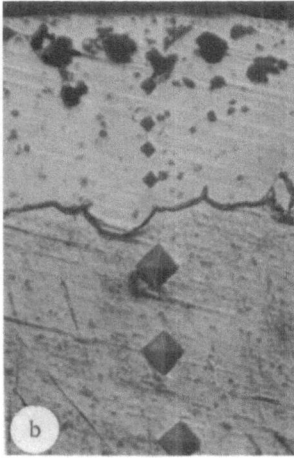

Fig. 2. Microstructures of coatings on titanium (temperature = 1300°C, time = 2 h) under pressures of 1 mm Hg (a) and 1×10^{-2} mm Hg (b). ×300.

and also tantalum rolled into sheets 0.5 mm thick and containing not more than 0.1% of impurities.

In order to determine the effect of evacuation on the nature of the change in the substrate and on the formation and growth of the carbide layers, a series of experiments was carried out on specimens of titanium, zirconium, and niobium for a period of 2 h at 1300°C in the pressure range from 1 to 1×10^{-3} mm Hg. The dependence of the thickness of the phases formed and the hardness of the substrate on the degree of vacuum is shown in Fig. 1. The considerable growth of the layers at pressures of 1×10^{-1} mm Hg and over is evidently due to the presence of the gaseous phase, which is responsible for the transfer of the active carbon atoms to the surface of the specimen. Coatings produced in this way are porous and easily break away from the substrate (Fig. 2a). The latter is to a considerable extent saturated with gases, and this leads to a rise in its hardness. At pressures below 1×10^{-1} mm Hg the thickness of the carbide layers diminishes, but the density and the strength of adhesion to the substrate increase, though the hardness of the substrate hardly increases. On the basis of these results, a vacuum of 1×10^{-2} mm Hg was chosen for investigating the kinetics of the carbidization process.

TABLE 1. Dependence of the Thickness of the Carbide Phases on the Temperature and Time of Impregnation

Metal	Temperature, °C	Time, h	Total thickness of layer, μ	Metal	Temperature, °C	Time, h	Total thickness of layer, μ	Metal	Temperature, °C	Time, h	Total thickness of layer, μ
Ti	1200	1	3	Hf	1100	1	40	Nb	1300	1	0.5
		2	6			2	60			2	21
		3	10			3	80			3	3.0
		5	16			5	103			5	8.0
	1300	1	10		1200	1	55		1400	1	1.5
		2	14			2	80			2	4
		3	24			3	100			3	7
		5	37			5	125			5	13
	1400	1	14		1300	1	75		1500	1	4
		2	30			2	110			2	9
		3	44			3	135			3	19
		5	70			5	158			5	28
Zr	1200	1	1	V	1200	1	1	Ta	1400	1	2
		2	30			2	3			2	3
		3	3			3	7			3	4
		5	5			5	15			5	7.0
	1300	1	4		1300	1	3		1500	1	6
		2	6			2	12			2	8
		3	9			3	26			3	10
		5	13			5	40			5	15
	1400	1	9		1400	1	6		1600	1	11
		2	16			2	21			2	16
		3	22			3	42			3	18
		5	35			5	68			5	21

The method of preparing the specimens and the atmospheres before carbidization has been described in detail in [2]. The temperature and holding time are given in Table 1. In the carbidization of titanium under the conditions described above, significant growth of the new phase begins at about 1200°C.

The microhardness of the layer varies from 1834 ± 123 kg · mm^{-2} at the boundary with the substrate to 2636 ± 148 kg/mm^2 at the outer edge of the specimen; the latter figure is too low, owing to the fact that the increased brittleness of the layer leads to the hardness measurements being accompanied by crack formation. These results are in accord with the existence in the Ti−C constitutional diagram of a region of homogeneous TiC in the range 18-50 at.% C. The carbon content across the thickness of the layer clearly varies from 18 at.% at the boundary with the substrate to almost 50 at.% at the surface of the specimen, and this leads to a considerable increase in the hardness and brittleness of the coating.

The growth of the layer proceeds for the most part frontally, and so the boundary of separation of the phases is relatively smooth (Fig. 2b). The hardness of the substrate is 290 ± 10 kg/mm^2; it increases slightly with rise in the impregnation temperature and is practically constant across the section of the specimen.

Carbidization of zirconium gives a completely different picture. The layer is discontinuous and has a strongly developed boundary with the substrate, particularly at short impregnation times. The dependence of the thickness of the layer on the temperature and time is given in Table 1.

Cracks form as a rule around the indentations when the microhardness of the layer is being measured, and this reduces the true results. Hence the hardness of the layer obtained, of the order of 2000 kg/mm^2 is too low. The microhardness of the substrate is 230 ± 12 kg/mm^2. There is no significant variation in hardness across the section of the specimen.

In the carbidization of hafnium, certain special features are observed which do not appear in the first two metals. After impregnation, even at 1100°C, the external appearance of the specimen is markedly changed. In spite of careful preliminary polishing, the surface takes the form of a gray "scale" with projections to which carbon has been "baked"; the sharp facets become somewhat elongated and are overgrown by this "scale," which, according to the results of x-ray phase analysis, consists of hafnium carbide. No substantial change is found in the microhardness across the depth of the carbide phase, in spite of the fairly wide homogeneity range (36-50 at.% C). The hardness depends only on the time and, particularly, the temperature of impregnation. Thus, for 1100°C and 1 h, the hardness of the layer is 1540 ± 83 kg/mm^2; at 1100°C and 3 h, it is 1600 ± 91 kg/mm^2, while at 1300°C, and 1 h it is 2229 ± 129 kg/mm^2, remaining at this figure throughout the depth of the phase being formed. In consequence, the carbon content is likewise approximately the same throughout the thickness of the coating, increasing only with rise in temperature of the process. The thickness of the carbide phase formed on hafnium is several times greater than that formed on titanium and zirconi-

um under similar conditions. This can clearly be explained by the
fact that during carbidization hafnium, unlike titanium and zirconium, exists in the low-temperature modification which has a hexagonal crystal lattice. Titanium and zirconium have bcc lattices in
the temperature range concerned, and the interstitial phases formed
have fcc lattices of the NaCl type. The transformation of the hexagonal (as compared with the bcc) lattice into the fcc lattice is more favorable in terms of energy and promotes the diffusion of carbon into the
metal during the transformation. The hardness of the substrate
increases with rise in the carbidization temperature; after being
held for 5 h at 1300°C the substrate varies from 326 ± 12 kg/mm^2
at the center to 470 ± 20 kg/mm^2 immediately below the carbide
layer.

The relationship between the thickness of the layer and the temperature and time of impregnation is presented in Table 1. The
carbide layers formed are dense and adhere well to the substrate,
but with increase in thickness beyond 50-60 μ partial flaking off
occurs (Fig. 3a) owing to the difference in the coefficients of thermal expansion of the substrate and coating.

Appreciable formation of carbide phases begins on vanadium
specimens at 1200°C, the layer consisting of individual "needles"
and a film of the order of 1-2 μ thick on the surface. The number,

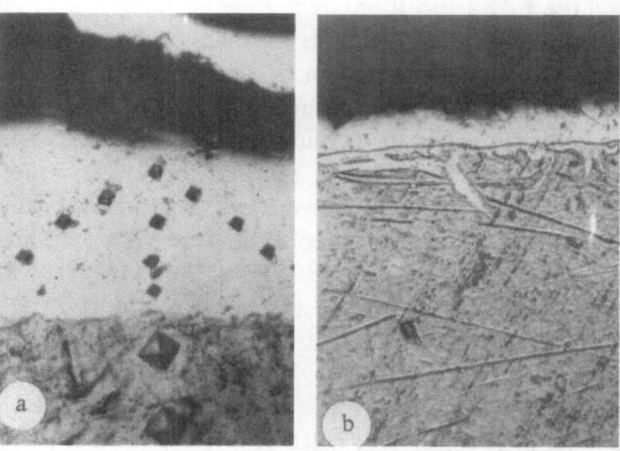

Fig. 3. Microstructures of coatings on hafnium impregnated at 1200°C (a) and on tantalum impregnated at 1600°C
(b) (pressure = 1×10^{-2} mm Hg). ×300.

thickness, and length of these "needles" increase with the duration of the process to form a single continuous carbide layer. It is characteristic that the acicular growth of the phase being formed occurs not only along the boundaries, but also throughout the body of the grain. Such a structure of the layers indicates that during their formation the carbide phases grow rapidly in the direction perpendicular to the surface, while growing only slowly along the surface itself. The carbide V_2C has a hexagonal lattice deficient in carbon. Obviously diffusion of this carbide along the c axis will differ to a certain extent from diffusion in the remaining directions owing to the anisotropy of the diffusion coefficients of carbon in the various crystallographic directions.

The results of x-ray phase analysis showed that impregnation at 1300-1400°C is accompanied by the formation on the surface of a carbide layer consisting of the two phases VC and V_2C which have microhardness values of 1970 ± 150 and 1524 ± 116 kg/mm^2, respectively. Both phases are very brittle and crumble to a considerable extent during preparation of the microsection, thus making it difficult to measure the thickness and hardness of the coating. Moreover, at an impregnation temperature of 1400°C part of the carbide layer formed easily flakes off and separates itself from the specimens. Appreciable layers begin to appear on niobium specimens at 1200°C; they are very thin (1.5-2 μ), and there is no significant dependence on the impregnation time. Greater carbidization occurs at 1300°C and above. The carbide layer formed consists of the two phases NbC and Nb_2C; both are very brittle, which makes measurement of the microhardness very difficult. When the process is carried out in the region of 1500°C and higher, particularly with impregnation times of more than 1 h, the surface of the specimens becomes covered with a network of cracks, the depth of which depends on the temperature and time of holding. The microhardness of both phases increases somewhat with rise of the carbidization temperature; for 1500°C it is 2313 ± 333 kg/mm^2 for NbC and 2636 ± 422 kg/mm^2 for Nb_2C. The hardness of the substrate is 264 ± 9 kg/mm^2, becoming somewhat less with rise of the impregnation temperature.

Appreciable carbide layers begin to be formed on tantalum at 1300°C, the specimens taking on the yellowish tinge characteristic of TaC. X-ray phase and metallographic analyses revealed that the layer consists of the two phases TaC and Ta_2C. The boundary between them is relatively smooth; the boundary between Ta_2C and the substrate has a more developed surface owing to the difference in shape of the projections. Of the two phases formed on tantalum, TaC predominates; Ta_2C takes the form of a thin film which is closely attached to the substrate (Fig. 3b).

It is interesting to note that the coatings formed on all the metals of group V consist of two phases whose thickness is approximately proportional to the width of the corresponding homogeneity ranges in the constitutional diagrams. This is possible only if the coefficients of diffusion of carbon through the two phases are about the same.

The microhardness of the tantalum substrate varies from 273 ± 2 kg/mm^2 at 1300°C to 339 ± 3 kg/mm^2 at 1600°C. No relationship was found between the hardness of the substrate and the impregnation time.

Conclusions

1. The vacuum carbidization of transition metals of groups IV and V of the periodic system has been investigated, and a relationship established between the thickness of the carbide coatings formed and the temperature and time of impregnation. The optimum temperatures for producing dense and porosity-free layers that adhere firmly to the substrate have been determined; these are 1500°C for titanium, 1400°C for zirconium, and 1100°C for hafnium. At higher temperatures the quality of the layers deteriorates, cracks appear, scaling occurs, and brittleness increases considerably.

2. The carbide layers formed on metals of group V consist mainly of two phases. It is shown that such a structure of the coatings formed correspond to the Me−C constitutional diagrams, and that the thickness of the individual phases is proportional to the width of their homogeneity ranges on the constitutional diagrams, thus indicating that carbon diffuses at about the same rate through the two phases.

3. The characteristic acicular structure of the layers formed on vanadium is attributed to the crystallographic directionality of carbon diffusion through the carbide V_2C. It was found that the optimum temperatures for producing satisfactory layers are: 1300°C for vanadium, 1400°C for niobium, and 1600°C for tantalum.

Literature Cited

1. Yu. V. Grdina, A. T. Gordeeva, and L. T. Timonina, Izv. VUZ. Chernaya Met.,
    No. 4, 129 (1963).
2. G. L. Zhunkovskii, in: Diffusion Coatings on Metals [in Russian], Vol. 2, Naukova
    Dumka, Kiev (1968).
3. G. V. Samsonov and V. P. Latysheva, Fiz. Metal. i Metalloved., 2:309 (1956).
4. G. V. Samsonov and A. P. Epik, in: Reports on the Theory of the High-Temper-
    ature Strength of Metals and Alloys [in Russian], Inst. Met. im. A. A. Baikova,
    Akad. Nauk SSSR, Moscow (1963), p. 129.
5. A. V. Smirnov and A. D. Nachinkov, Metal. i Term. Obrabotka Metal., No. 3,
    22 (1960).
6. M. A. Filyand and E. I. Semenova, The Properties of Rare Elements [in Rus-
    sian], Metallurgiya, Moscow (1964).
7. A. V. Shcherbedinskaya and A. N. Minkevich, Izv. VUZ. Tsvet. Met., No. 4,
    123 (1965).
8. K. Bungardt and K. Rüdinger, Z. Metallk., 47:577 (1956).

Some Problems in the Thermodynamics
and Kinetics of the Formation of
Carbide Coatings on Graphite
from the Vapor Phase

A. V. Emyashev, I. I. Chernenkov,
and L. N. Panshin

Thermodynamic calculations for the systems $Si-Cl-C-H$ and $Zr-Cl-C-H$ in the course of carbide formation have been made on an electronic computer. The kinetics of formation of the carbides SiC and ZrC has been studied experimentally in the temperature ranges 1400-1700°K and 1500-2400°K. The data calculated on a thermodynamic basis and the experimental results on the kinetics of the process have been compared, and it has been found that with a supply of the necessary components from the vapor phase the carbide coatings grows at a constant rate up to large thicknesses.

Vapor-phase metallurgy is now being successfully utilized in the fields of electronics (production of semiconductors), atomic energy (production of fuel elements), and rocketry. However, the thermodynamic and kinetic aspects of the process have scarcely been studied, mainly owing to the complexity of multicomponent and multiphase processes, though there are some individual articles [4] dealing with the thermodynamic analysis of the system $Si-Cl-H$.

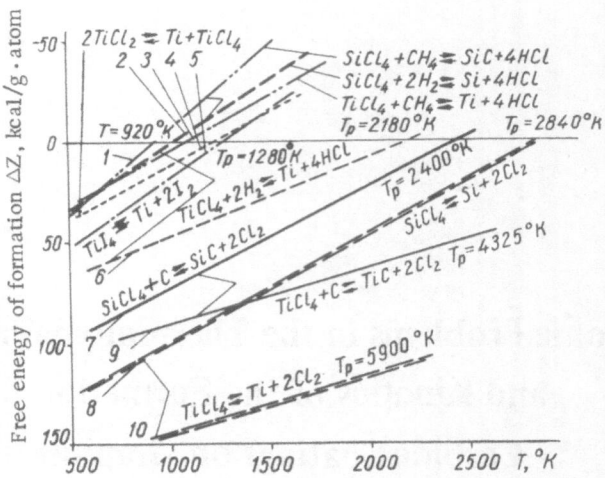

Fig. 1. Temperature dependence of the thermodynamic potential of the formation of silicon, titanium, and their carbides from the halogenides for the standard conditions:

1 — $SiCl_4 + CH_4 \rightleftarrows SiC + 4HCl$, $T_p = 920°$ K; 2 — $SiCl_4 + 2H_2 \rightleftarrows Si + 4HCl$; 3 — $TiCl_4 + CH_4 \rightleftarrows Ti + 4HCl$; 4 — $2TiCl_2 \rightleftarrows Ti + TiCl_4$; 5 — $TiI_4 \rightleftarrows Ti + 2I_2$, $T_p = 1280°$ K; 6 — $TiCl_4 + 2H_2 \rightleftarrows Ti + 4HCl$, $T_p = 2180°$ K; 7 — $SiCl_4 + C \rightleftarrows SiC + 2Cl_2$, $T_p = 2400°$ K; 8 — $SiCl_4 \rightleftarrows Si + 2Cl_2$, $T_p = 2840°$ K; 9 — $TiCl_4 + C \rightleftarrows TiC + 2Cl_2$, $T_p = 4325°$ K; 10 — $TiCl_4 \rightleftarrows Ti + 2Cl_2$, $T_p = 5900°$ K.

We have not found in the literature any researches concerned with the study of the thermodynamic and kinetic aspects of producing carbides from the vapor phase, so we have thought it desirable to investigate some aspects of the formation on graphite of carbides of silicon, titanium, and zirconium from their chlorides.

An elementary calculation of the temperature dependence of the thermodynamic potential of the formation of silicon and titanium and their carbides and halogenides shows (Fig. 1) that addition of HCl or CH_4 to the vapors sharply reduces the temperature at which production of the element and its carbide can take place. Thus, the production of silicon and titanium is possible thermodynamically at temperatures of 1000 and 2180°K, respectively, and that of their carbides at 1100 and 920°K.

In order to obtain significant reaction rates, the actual temperatures for carrying out the processes should be higher. Naturally, this elementary thermodynamic analysis cannot solve the more complex problem of the quantitative determination of the output of the respective reactions in relation to the ratio of components. For this purpose it is necessary to solve a system of many equations with many unknowns. A full thermodynamic analysis of the production of silicon single crystals from a mixture of chloride and hydrogen was first made by Lever [4], who determined the partial pressures of the nine components in equilibrium with solid silicon in relation to the total pressure and the Cl/H ratio and temperature.

Swedwick [5] attempted to interpret the experimental data in terms of a quasi-equilibrium model, using Lever s calculations.

Deposition from the vapor phase is a high-temperature process, and we may expect that the composition of the reaction products under actual conditions will be close to the equilibrium composition, and a thermodynamic calculation can indicate the choice of the optimum conditions for the process.

Calculations of the equilibrium composition of the reaction products are complicated and laborious, while reactions involved in deposition from the gaseous phase are even more complicated by reason of the need to take into account the phase transformations.

In order to find n_i components of the reaction products at a given temperature and pressure, n_i independent equations must be constructed and solved. In choosing the n_i components we attempted to take into account all possible components under the conditions considered.

In determining n_i components in equilibrium, we can always write:

1) $n_i - m$ equilibrium-constant equations, where m is the total number of chemical elements in the system;

2) $m - 1$ material-balance equations;

3) the modified Dalton equation

$$\sum_{i=1}^{n} n_i (g) = 1.$$

Even for three chemical elements the system of independent equilibrium-constant and material-balance equations is a power

one, and the number of roots is equal to the power of the equation.
In solving physical problems the appropriate root of the equation
is chosen in accordance with the physical significance of the expected result. As regards the problem we are concerned with,
all the roots of the true solution should be positive and real, since
they represent the components of the mixture of reaction products. It has been demonstrated in [2] that the system of equilibrium-
constant and material-balance equations has only one solution corresponding to physical reality. Hence the solution of the system
of equations concerned is reduced to a search for this unique system of roots. Moreover, the system of nonlinear equations has no
analytical solution and can be solved only by one of the approximate methods.

For three or more chemical elements, we solved the system
of nonlinear equations by Newton's method [1]:

$$\Phi(x_i) + \Phi'(x_0)\, \Delta x = 0,$$

by limiting ourselves to the first derivative. All the equations
were differentiated with respect to all the variables, using the
relationship

$$\Phi'(x) = K\Phi \int_x (x),$$

where K is a factor with which the variable x enters the given
function.

A first approximation is needed in the solution by Newton's
method, i.e., it is necessary to assign "m" components. The calculations were carried out on a BESM-2M electronic computer.
The method of choosing the first approximation is a special feature of the method. We assigned the components whose content is
a maximum at equilibrium in the system. It is desirable to have
among the given variables those which can also exist in the condensed phase, since this provides the formulation of the arithmetic
part of the computer program. The equilibrium constants were
taken from [3] and were fed into the computer in logarithmic form.
The arithmetic operations were carried out by taking logarithms,
as a result of which it was possible to avoid overloading the dis-
charge network of the computer by varying the digital material
over a wide range.

The equilibrium composition of the gases can be expressed in terms of partial pressures or volume or molar proportions, and that of the condensed phase in terms of molar or relative weight proportions. The solution of the system of equations when the gaseous components are expressed in terms of partial pressures is similar to the equilibrium constant given in the handbooks, while that of the condensed phase expressed as molar or weight proportions is complicated, particularly when numerical methods and the computer are used. Consequently we expressed the whole composition of the reaction products as molar proportions, by introducing appropriate corrections into the values of the equilibrium constants.

We have made a thermodynamic analysis of the more complex $Si-Cl-C-H$ system with the solution of 23 equations and the determination of the partial pressures of 21 components under various conditions.

Maximum production of SiC occurs in the temperature range 1400-1700°K, i.e., several hundred degrees above the "equilibrium" temperatures. The kinetics of the formation of SiC has been studied in this temperature range. A rise in temperature from 1400 to 1700°K is accompanied by a sharp (thirtyfold) increase in the rate of growth of the carbide coating on graphite (a change in the logarithm by 1.5 orders). An increase in the total flow rate of the components by twice does not affect the kinetic relationship. The process takes place with a large apparent activation energy —72.8 kcal/mole. Raising the temperature from 1400 to 1700°K leads to scarcely any increase in output, although the rate of carbide growth increases thirtyfold. It should be noted that these results require further checking, because under the experimental conditions heating of the reagents to the equilibrium temperatures proceeds slowly and this distorts the actual conditions somewhat.

A similar thermodynamic calculation of the production of ZrC was carried out on the BESM-2M computer for the quaternary system $Zr-Cl-C-H$ at pressures of 760, 100, and 10 mm Hg in the temperature range 1500-2400°K.

In this case the positive effect of lowering the pressure in the system on the output of condensed carbide is clearly revealed. There is no doubt about the usefulness of this type of calculation, taking account of the kinetic data, for determining the optimum conditions for conducting the process. However, under actual conditions additional correction of the optimum parameters is necessary in order to obtain coatings of the required quality, i.e., dense and fine-grained, with good adherence to the substrate.

It is interesting to note certain distinctive features of growth from the vapor phase of carbide films formed 1) as a result of diffusion of carbon from the substrate into the titanium deposited from the vapor phase, and 2) as a result of the supply of both carbon and metal from the vapor phase. The rate of growth of the coating in the first case is as a rule several times greater than in the second case. The difference in the time dependence of the growth rate in the two cases is also interesting.

When all the components needed for the formation of the carbide — carbon and titanium — come from the vapor phase, the coating grows at a constant rate up to large thicknesses (of the order of 2-3 mm). Growth occurs quite differently when one of the components of the coating — carbon — is supplied by the graphite substrate. In this case the rate of growth of the coating gradually begins to decrease from a certain temperature until finally there is complete cessation of growth. Thus, after a carbide layer of a particular thickness has been attained, the process is to a large extent limited by the diffusion of the coating components. In the low-temperature region, where the rate of deposition is comparable with the rates of diffusion of the components, the coating does not cease to grow.

Carbide of almost stoichiometric composition (hardness = 3150 kg/mm^2 and lattice parameter = 4.32 Å) forms over the whole depth of penetration by the reaction of titanium deposited above the melting point of the metal, the titanium penetrating to a depth of 1.5-2 mm into the porous graphite. In the deposition of the carbide directly from the vapor phase surface deposition of the coating takes place and the open pores in the surface are filled. Undesirable separation of excess carbon in the form of layers can be avoided by heat treatment of the carbide.

It should be noted that the coating has a definite texture in relation to the crystallographic planes and that this affects the microhardness of pyrolytic SiC, the values being 2800 kg/mm^2 in the plane of deposition and 3400 kg/mm^2 in the plane perpendicular to it.

Conclusions

1. A thermodynamic analysis has been made of the formation of carbides from the vapor phase in the systems $Si-Cl-C-H$ and $Zr-Cl-C-H$.

2. The basic kinetic relationships of the process have been derived.

3. The data calculated on a thermodynamic basis have been compared with experimental results obtained from the kinetics of the process with a view to discovering the limiting parameter of the process.

4. The results of a metallographic examination of the coatings are reported, and the textured character of the carbide coatings has been established.

Literature Cited

1. I. S. Berezin and N. P. Zhidkov, Methods of Computation [in Russian], Fizmatgiz, Moscow (1962).
2. Ya. B. Zel'dovich, FKhT, 11(5):685 (1938).
3. Thermodynamic Properties of Individual Substances [in Russian], Izd. Akad. Nauk SSSR, Moscow (1962).
4. R. F. Lever, J. Res. Dev., 8:460 (1964).
5. T. O. Sedwick, J. Electrochem. Soc., 111(12):138 (1964).

Investigation of the Effect of Boron Additions
on the Carburization of Silicon

A. N. Shurshakov, V. S. Dergunova,
G. A. Meerson, and B. A. Sizov

The effect of boron additions on the rate of carburization of molten silicon and on the growth of the carbide layer at the graphite–melt interface has been investigated. It is shown that on the introduction of 14 wt.% B the thickness of the carbide layer at the interface increases and so does the carbon content in the melt.

The degree of saturation of molten silicon with carbon has a considerable influence on the kinetics of saturation of porous graphite bases by silicon and also on the extent of the conversion of the silicon filling the pores into carbide.

The literature contains only data on the diffusion of carbon and boron into pure silicon at temperatures below the melting point of the latter [8] and there are no data on the coefficient of diffusion of carbon into molten silicon and silicon carbide.

In the present research we have studied the effect of boron additions on the rate of carburization of molten silicon and on the growth of the carbide layer formed at the graphite–melt interface. Charges of six different compositions were melted under an argon atmosphere in a TVV-5 furnace, using crucibles made of dense graphite (specific gravity = 1.94 g/cm^3), 25 mm in diameter and 70 mm high. Dense graphite was chosen for the crucible because of the need to secure a sharp boundary of separation between the graphite and the melt, so as to avoid penetration of the former by the latter.

The raw materials used consisted of semiconductor silicon of grade KS-1 and amorphous boron of grade D-6 containing 95.03 wt.% B. Boron in powder form contained 4 wt.% chlorine adsorbed on its surface.

The specific surface of the boron powder was 130 m^2/g. Spectrographic analysis revealed the presence of traces of magnesium and 0.001% silicon. No other impurities were found. The silicon ingots were crushed in a steel mortar to a particle size of 0.1-0.3 mm. Spectrographic analysis of the crushed silicon revealed the presence of traces of iron.

In the first series of experiments the effect of temperature was studied, temperatures of 1600, 1800, and 2000°C being used in conjunction with a constant holding time of 30 min. In the second series of experiments the effect of holding times of 15, 60, and 90 min at a constant temperature of 1800°C was investigated. Charges of all six compositions were melted simultaneously at each temperature and time. Heating was carried out in two stages: up to 1000°C in 35 min and then up to desired temperature in 15 min. The specimens were cooled over a period of 60 min, the temperature being measured on the crucible lid by means of an OPPIR-017 optical pyrometer. The heating and cooling conditions were maintained constant in all the experiments. The height and weight of the ingots obtained were measured, and the ingots were subjected to chemical (see Table 1) and metallographic analysis.

The phase composition of the ingot and the thickness of the carbide layer were determined metallographically. The microhardness of the phases was measured on a PMT-3 apparatus under a load of 50 g.

The structure of pure silicon ingots containing up to 0.38 wt.% C was single-phase and consisted of a solid solution of carbon in silicon which had a microhardness of about 1300 kg/mm^2. Data given in the literature for the solubility of carbon in solid silicon are conflicting. Scace and Slack [9] found that less than 3×10^{-3} at.% C dissolves in silicon below 1400°C. Brokhin and Funke [2] showed that at room temperature Si−C alloys are single-phase

TABLE 1. Results of Chemical Analysis and Measurement of the Thickness of the Carbide Film on the Specimens Investigated

Temperature, °C	Holding time, min	No. of charge	B content in charge, wt.%	B content in alloy, wt.%	C content in alloy, wt.%	Thickness of carbide layer, μ
1600	30	1	0.0	—	0.20	17
		2	1.9	6.0	0.37	—
		3	4.1	7.4	0.37	—
		4	8.8	10.2	0.57	22
		5	14.2	12.5	0.47	28
		6	27.8	25.0	0.77	39
1800	15	1	0.0	None	0.15	28
		2	1.9	3.2	0.18	—
		3	4.1	4.6	0.36	—
		4	8.8	9.5	0.47	56
		5	14.2	15.6	0.53	67
		6	27.8	26.4	1.80	78
	30	1	0.0	None	0.26	56
		2	1.9	4.2	0.47	—
		3	4.1	5.4	0.47	—
		4	8.8	9.05	0.48	70
		5	14.2	13.9	0.93	92
		6	27.8	22.2	1.72	112
	60	1	0.0	None	0.23	72
		2	1.9	4.5	0.09	—
		3	4.1	6.0	0.42	—
		4	8.8	8.6	0.62	98
		5	14.2	14.6	0.67	112
		6	27.8	20.2	1.90	123
	90	1	0.0	None	0.38	78
		2	1.9	5.4	0.47	—
		3	4.1	5.1	0.43	—
		4	8.8	10.9	0.52	112
		5	14.2	15.2	1.10	129
		6	27.8	22.5	2.20	140
2000	30	1	0.0	None	0.28	67
		2	1.9	5.4	0.35	—
		3	4.1	6.0	0.48	—
		4	8.8	10.8	6.74	84
		5	14.2	16.7	1.52	106
		6	27.8	24.5	3.24	118

up to a carbon content of 0.7 wt.%. Gnesin and Kurdyumov [4] established that 1.3 at.% C (0.56 wt.%) remains in solid solution after molten silicon has been held for 5 h in a graphite crucible. As in the case of [2, 4], we did not attain the solubility of carbon in pure silicon.

The structure of boron–containing ingots in which the carbon content exceeded 0.4 wt.% consisted of a solid solution of carbon

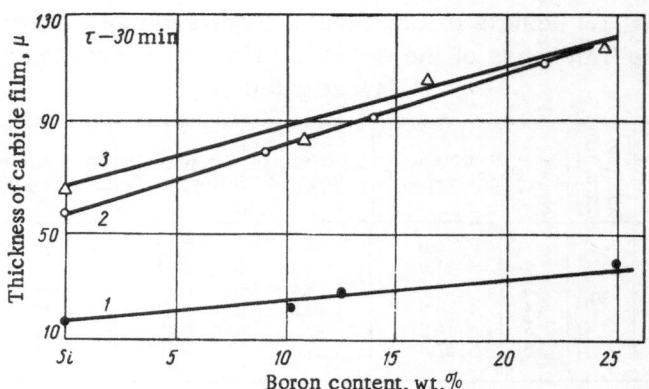

Fig. 1. Thickness of the carbide film as a function of boron content: 1) 1600°C; 2) 1800°C; 3) 2000°C.

and boron in silicon, which had a microhardness of about 1400 kg/mm^2; boron silicides (probably BSi_3 and B_3Si), with microhardness values of about 1850 and 2250 kg/mm^2, respectively (in agreement with data given in the literature for the microhardness of these silicides); and a solid solution of boron in SiC with a hardness of about 5000 kg/mm^2. A hardness of 5000 kg/mm^2 was found by Meerson et al. [6] for a solution of boron in SiC. As the boron content rises, the amount of the silicide phases and of the phase based on SiC increases and so does the size of the crystals of these phases.

Analysis of the results obtained (see Table 1) shows that on adding up to 10 wt.% B to the melt the carbon content of it increases only very slightly and that it is independent of the holding time and temperature. Nor is there any change in the rate of carburization. With further addition of boron (15 wt.% and more), the carbon content of the melt begins to rise significantly, and the rate of carburization increases sharply with rise in temperature. Although the rate of carburization of pure silicon scarcely changes in the range 1600-2000°C, on addition of about 20 wt.% B it increases almost fourfold in the same temperature range.

Addition of boron to silicon produces an increase in thickness of the carbide layer in accordance with a straight-line relationship (Fig. 1) with a holding time of 30 min at all three temperatures investigated. Thickening of the carbide layer with time of holding evidently takes place in accordance with a parabolic relationship (Fig. 2). The microhardness of the carbide layer increases from 3400 to 5000 kg/mm^2 with the addition of boron. This considerable increase in hardness of the carbide layer is probably due to the formation of a substitutional solid solution of boron in SiC.

The results of our investigation have shown that on melting pure silicon a very thin carbide layer is formed in spite of the fact that a minimum amount of carbon enters the melt. The thickness of the carbide layer increases on addition of boron, and so does the carbon content of the melt. Consequently, a boron-containing layer prevents the passage of carbon into the melt to a lesser degree than does a layer of pure SiC.

Silicon carbide is a compound of strictly determined chemical composition, it having no homogeneity range. The concentration of carbon is constant throughout the thickness of the layer of SiC; consequently, there is no concentration gradient of diffusing carbon in the SiC layer, and hence it cannot be regarded as a factor affecting diffusion.

It has been found that for metallic systems the assumption that a concentration gradient provides the driving force for diffusion is an incorrect simplification. In such cases [7] the driving force of diffusion should be expressed as the gradient of the chemical potential of the components of the system at the two boundaries of the layer. However, there are no equations for determining in practice the diffusion coefficient which take account of the chemical potential gradient.

Borisov et al. [1] have suggested determining not the diffusion coefficient but the so-called diffusion parameter $\delta D = D(C_1 - C_2)$ of compounds that have no homogeneity range. Using the method

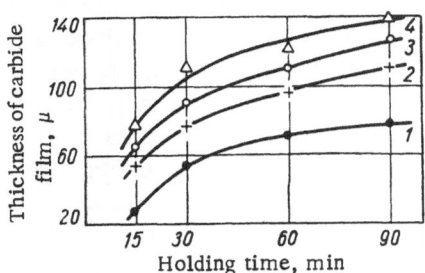

Fig. 2. Thickness of the carbide film as a function of the holding time at 1800°C for melts containing various amounts of boron: 1) pure Si; 2) Si + 10% B; 3) Si + 14% B; 4) Si + 20% B.

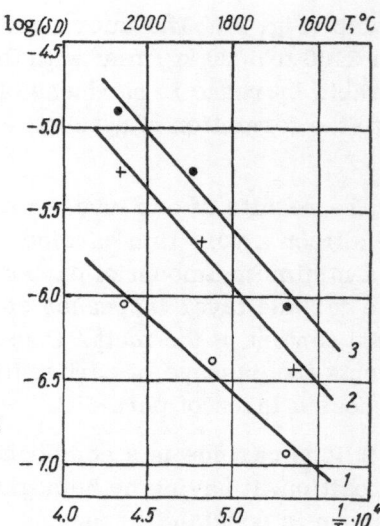

Fig. 3. Diffusion parameters for carbon passing through pure SiC and SiC alloyed with boron: 1) SiC (43); 2) Si + 14% B (53); and 3) Si + 24% B (53 kcal/mole).

they have proposed, we have calculated the diffusion parameters for carbon passing through pure SiC and SiC containing boron additions (Fig. 3). On adding 14-24 wt.% B to a silicon melt, the diffusion parameter increases by an order of magnitude; the activation energy of the diffusion of pure SiC is about 43 kcal/mole and that of boron-containing SiC about 53 kcal/mole.

Dekhtyar [5] has shown that in substitutional solutions the activation energy is a definite part of the binding energy in the crystal lattice. The greater the binding energy, the higher is the activation energy. The addition of boron evidently produces an increase in the binding energy of the lattice based on SiC, and this leads to a significant rise in the activation energy of the diffusion process.

The effect of a third element on the diffusion coefficient has been studied in [3], where it was shown to be more accurate to consider the effect of the third element not directly on the diffusion coefficient, but on such parameters affecting the diffusion process as the activation energy and activation entropy. Here we understand by the activation entropy the probability of a diffusing atom passing into the activated state. The diffusion coefficient is regarded as being determined by the joint influence of the energy and entropy of activation, which in turn depend on the size, valence, and amount of the third element. In the opinion of the authors of [3], there are two possible ways in which the diffusion coefficient increases: by an increase in the activation entropy with constant activation energy and by a decrease in the activation energy with constant activation entropy.

In our case the passage of carbon through the boron-containing carbide layer is evidently facilitated in two ways:

1. The boron-containing carbide layer consists of a substantial solid solution of boron in SiC. With the formation of the solid solution the number of lattice defects (vacancies and dislocations) considerably increases in the silicon sublattice within the grains, and this greatly facilitates diffusion through the crystal lattice.

2. Grains of the carbide phase crystallize and grow while the melts are being held at constant temperature. The boron present in the melt evidently has a catalytic action on the crystallization and growth of the carbide phase within the melt, since metallographic examination reveals large crystals of the carbide phase, having great hardness, within all boron-containing ingots in which the carbon content exceeds 0.4 wt.%. The hardness of the phase based on silicon is practically independent of the boron and carbon content in the melts. In consequence, the composition of the phase based on silicon varies little with increase in the boron content of the melt. During crystallization of the carbide grains a directional flow of carbon is probably set up through the liquid melt from the carbide layer to the carbide phase precipitated within the melt. This, in turn, leads to a rise in the diffusion flow of carbon through the carbide layer.

The factors indicated above are responsible for the considerable increase in the rate of carburization of silicon on the addition of boron to it.

Literature Cited

1. V. T. Borisov, V. M. Golikov, and G. N. Dubinin, Diffusion Coatings on Metals [in Russian], Naukova Dumka, Kiev (1965), p. 26.
2. I. S. Brokhin and V. F. Funke, Zh. Neorg. Khim., 3:847 (1958).
3. S. D. Gertsriken and I. Ya. Dekhtyar, Zh. Tekh. Fiz., 20:38 (1950).
4. G. G. Gnesin and A. V. Kurdyumov, Silicon Carbide [in Russian], Naukova Dumka, Kiev (1966), p. 83.
5. I. Ya. Dekhtyar, Zh. Tekh. Fiz., 20:1015 (1950).
6. G. A. Meerson et al., Izv. Akad. Nauk SSSR, Otd. Tekh. Nauk, Metallurgiya i Toplivo, No. 4, 90 (1961).

7.   B. Chalmers, Physical Metallurgy [Russian translation], IL, Moscow (1963).
8.   Metallurgy of Semiconductive Materials (1962), p. 201.
9.   R. I. Scace and G. A. Slack, Silicon Carbide, a High-Temperature Semiconductor, Pergamon Press, London and New York (1960), p. 24.

A Study of the Soldering of Graphite Materials with the Formation of a Carbide Layer

L. T. Anikin and G. A. Kravetskii

Aspects of the carbidization that occurs during the soldering of graphite by means of zirconium and molybdenum foil have been studied. Optimum soldering conditions — a temperature of 2200°C and a holding time of not less than 3 min — have been determined. Under these optimum conditions the strength of the soldered joint is not less than that of the base material.

Because of the valuable properties of graphite (high melting point and high specific strength), this material is widely used in industry particularly in high-temperature technology [3].

In many cases it is necessary to join graphite materials together in such a way that the joints can operate at elevated temperatures. Existing methods of joining graphite components (mechanical jointing, cementing, and soldering by means of various metals) are not without drawbacks that restrict their use (either the size and weight of the component are increased, or its airtightness is destroyed and the maximum operating temperature of the soldered component is reduced) [1, 5, 6]. Methods of producing purely carbon joints by welding [7] are of great interest, although the production of joints between components by the methods that have been suggested is fairly complicated, and the strength of the welded joints is not great at elevated temperatures.

The most interesting of the methods is that in which the graphite components are joined by the formation of a carbide layer, patented in the USA [4]. This method is based on the fact that such metals as titanium, zirconium, niobium, tantalum, and hafnium, on reacting with carbon, form carbides which have melting points considerably above those of the metals themselves. To form a soldered joint between the surfaces of two graphite components, a piece of foil or a layer of powder of the carbide-forming metal is introduced between the surfaces. The components are then pressed together and heated above the melting point of the metal, so that the layer melts and the molten metal flows into the pores of the graphite and becomes carburized.

Scarcely any investigation has been made of the properties of the soldered joints at high temperatures or of aspects of the carbide-formation process during soldering.

In the present research we have studied aspects of the carbidization of the soldered joint, when using mainly zirconium and molybdenum foil as the solder material. The investigation was carried out on a laboratory soldering apparatus. The components to be joined were heated by a contact method, the source of current being a PSM-1000 dc generator. The apparatus was suitable for heating components 10 mm in diameter up to a temperature of 3000°C. The temperature was measured with an accuracy of ±50°C by means of an OPPIR-17 optical pyrometer. Specimens were heated to the desired temperature in 20-60 sec. The soldering chamber was first evacuated to a pressure of 1×10^{-1} mm Hg and then filled with argon to an excess pressure of 0.2-0.5 atm. The compression force was constant in all cases (0.2-0.3 kg/mm^2).

Optimum soldering conditions should ensure fairly deep penetration of the solder metal into the pores of the graphite and the greatest possible degree of carbide formation. The latter process, particularly in the case of carbides having a large homogeneity range (of the ZrC type), limits the high-temperature strength of the soldered joint (the melting point of ZrC varies from 2500 to 3000°C, depending on its carbon content). The completeness of carburization of the molten metal in contact with the graphite depends on the temperature and duration of heating and on the porosity and chemical activity of the graphite.

The kinetics of carburization was studied mainly on the widely used graphites of grades MG and GMZ, which differ in porosity and in the nature of the pore distribution. The kinetics of carburization was determined by metallographic (hardness measurements), electron-diffraction, and chemical analyses. The soldering temperature was varied between 1400 and 1800°C and the time between 0.5 min and 5 h.

At soldering temperatures above 2500°C and a time of 3 min, the zirconium joint is completely carburized to the carbide of stoichiometric composition. Such a joint has maximum strength. However, in the case of large components it is technically difficult to heat the surfaces to be soldered to 2500°C. Hence an attempt was made to increase the degree of carbide formation by heating the area to be soldered to lower temperatures and by increasing the holding time. It was then found to be practically impossible to achieve a high elevated-temperature strength with a soldering temperature of 1400°C (increasing the time within reasonable limits). At a soldering temperature of 2000°C carbidization ceases after 2 h. The effect of the degree of carburization on the high-temperature strength of the soldered joint is well illustrated in Fig. 1, which shows the change in ultimate strength at a test temperature of 2500°C as a function of the soldering time (at 2000°C). The strength of the soldered joint remains almost constant beyond a soldering time of 2h.

Examination of the microstructure of a joint soldered with zirconium in relation to the soldering temperature (holding time = 4 min) showed that at 1800°C (solder not yet melted) the structure of the joint consists of a solid solution of carbon in zirconium. Such a joint has an extremely low strength. Zirconium carbide of nonstoichiometric composition and containing inclusions of zirconium is formed on soldering at 2000°C. Soldering at 2500°C results in the production of stoichiometric ZrC, and the joint possesses a high strength at elevated temperatures. Finally, at a soldering temperature of 2900°C the solder melts on account of the formation of the $ZrC-C$ eutectic, which has a typical fan-shaped structure [1].

The coefficient of diffusion of carbon in liquid zirconium was determined from the formula [11]:

$$D = \frac{\bar{x}^2}{2t} \, ,$$

where x is the thickness of the carbide layer (determined metal-

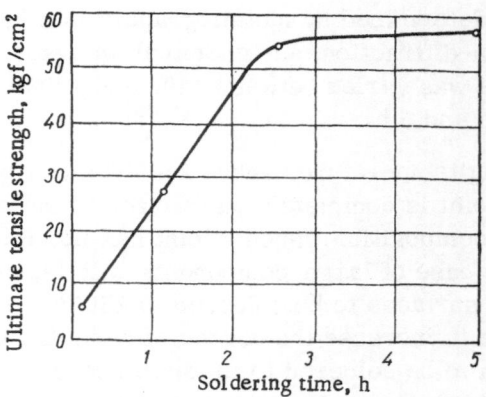

Fig. 1. Strength of a soldered joint in graphite
of grade GMZ soldered with zirconium as a
function of the time of holding at 2000°C.
Test temperature = 2500°C.

lographically). As can be seen from Table 1, the results we obtained are in good agreement with those reported in [9].

The carburization of molybdenum was investigated in the range
of soldering temperatures from 1600 to 2600°C and times of 0.5-3
min. Carburization takes place considerably more rapidly in
molybdenum than in zirconium. Thus, even at 1800°C and a time
of 3 min, carbidization is complete in a molybdenum foil 0.3 mm
thick, the carbide Mo_2C being formed. The strength of a joint made
in this way is low owing to the lack of penetration of the solder
into the pores of the graphite. The molybdenum solder melted at
2200°C (the melting point of the $Mo-Mo_2C$ eutectic). At soldering
temperatures above 2400°C the structure of the joint consists of
single-phase MoC. The strength is highest in these joints.

The authors consider that the soldering of graphite components by means of a carbide layer is not more widely used owing
to the thermal resistance of the soldered joint. To confirm or re-
fute this belief, tests were carried out on the thermal resistance of
soldered joints in graphite of grades MG and GMZ made with molybdenum, zirconium, and niobium, by using a method developed in the
Mechanical Testing Laboratory of the Scientific Research Institute
for Graphite. The essential features of the method are as follows.
A temperature gradient is set up between the internal and surface
layers of a cylindrical specimen by heating it with an electric

current and cooling it outside by a stream of a nitrogen-water mixture, and this gradient causes thermal stresses to be set up. Because graphite materials have good thermal resistance and a steep
temperature gradient across the section of the specimen is required to bring about failure, there is applied to the specimen an
additional axial tensile load which is selected in such a way that its
action in conjunction with the thermal stresses leads to failure
of the specimen. The center of the specimen was heated to 2200°C,
while the temperature of the outer surface was maintained constant
(at about 700-800°C).

It was found that the thermal resistance of soldered joints
made in MG graphite by means of molybdenum, zirconium, and
niobium was no worse than that of the graphite itself. In all cases
failure occurred at a distance from the joint. The thermal resistance of soldered joints made in GMZ graphite by means of niobium and molybdenum was the same as that of the graphite or somewhat lower, failure occurring both in the graphite and in the vicinity of the joint. On soldering GMZ graphite with zirconium, failure occurred mainly in the region of the joint, no doubt as a result of the low strength of the zirconium joints.

The completeness of carburization, and consequently the high-
temperature strength of the soldered joint, was assessed on the
basis of short-term tensile tests carried out at temperatures from
20 to 2800°C. For each material the specimens were prepared
under the optimal conditions for securing the most complete carburization.

TABLE 1. Values of the Coefficient
of Diffusion of Carbon in Zirconium,
D (cm^2/sec) at Various
Temperatures

Tempera-ture, °C	Experimental results	From $D = 3.44 \cdot 10^{-2} \exp \times \left(-\frac{20,500}{T}\right)$ [9]
1880	$5 \cdot 10^{-7}$	$6.2 \cdot 10^{-7}$
2000	$2 \cdot 10^{-6}$	$1.16 \cdot 10^{-6}$
2200	$4 \cdot 10^{-6}$	$3.1 \cdot 10^{-6}$
2400	$2.2 \cdot 10^{-5}$	$6.88 \cdot 10^{-6}$

TABLE 2. Ultimate Tensile Strength of Soldered Joints in
Specimens of Various Grades of Graphite

Grade of graphite	Solder	Ultimate tensile strength (kg/cm^2) at test temperatures (°C)						
		20	800	1000	1500	2000	2500	2800
GMZ P≈25%	Zr	54+		65+		80+	55+	
	Mo	59				134	83+	9+
	Hf	41				140	184	108+
	W	55				95	89+	51+
Fine porosity, Mo	Nb					75+	106+	
		116						
P≈20%	Zr	77			151	155	127+	
	Nb	74				129		
MG P≈30%	Zr					120	125+	38+
	Mo	79		81		86	71+	
Dense P≈15%	Zr	90		135		156	167+	134+
	Mo	103		162		166	160+	131+
	W						132+	79+
	Nb					19+	60+	86+
3 OPG P≈18%	Zr						54+	
EEG P≈30%	Zr					166		
	Mo					198		

The tests were made on an MR-05 machine, the specimens
being heated in a stream of argon in a resistance furnace. The
rate of heating specimens 10 mm in diameter was 400-600 deg/min,
and the rate of movement of the moving grip was 10 mm/min. The
results of high-temperature-strength tests on soldered joints in
specimens of various grades of graphite are given in Table 2. In
addition to joints made with zirconium and molybdenum solders,
specimens soldered with hafnium, niobium, and tungsten were also
tested. On the left-hand side of the heavy line in Table 2 are given
data on the strength of specimens which failed through the graphite (at a distance from the joint), while on the right-hand side are
data for cases in which failure occurred at the joint. Table 2
gives the mean values of the ultimate tensile strength of at least
five specimens. The following conclusions can be drawn from an
analysis of the table:

On soldering under optimum conditions and heating the specimens to 2000°C, the strength of the joint is not less than that of

the base material (excepting the pair: grade-GMAZ graphite — zirconium solder). In tests at 2500°C the strength of soldered butt joints was 70-80% of that of graphite, while in tests at 2800°C it was 60-65% of that of graphite.

When specimens fail at the joint, the failure load increases with rise in temperature and reaches a definite maximum at a temperature of the order of 2000-2500°C, i.e., the strength of the carbides (in particular those of zirconium and niobium) increases with temperature. One of the possible reasons for this increase may be a reduction in the stress concentration at microdefects. According to [2, 8], the maximum tensile strength of brittle metallic joints occurs at a temperature of $0.5-0.8T_{mp}$. In our investigations the maximum strength of joints made with NbC corresponds to a temperature of $0.7T_{mp}$ and that of joints made with ZrC to about $0.6T_{mp}$.

Conclusions

1. The formation of zirconium carbide having a composition close to the stoichiometric — and consequently also a maximum high-temperature strength of the soldered joint — occurs at a soldering temperature of 2500°C and a time of not less than 4 min. Similar results are obtained by reducing the soldering temperature to 2000°C and increasing the time at this temperature to at least 2 h.

2. The optimum conditions for soldering with molybdenum are temperature above 2200°C and a time of not less than 3 min.

3. In soldering under optimum conditions, the strength of the joint when the specimen is heated to 2000°C is as a rule not less than that of the graphite.

Literature Cited

1. M. M. Angelovich, Carbon Graphitized Components [in Russian], Catalog, GNTIL of Ferrous and Nonferrous Metallurgy, Metallurgizdat, Moscow (1961), pp. 17-19.
2. Problems of High-Temperature Strength in Mechanical Engineering [in Russian], Izd. Akad. Nauk UkrSSR, Kiev (1963).
3. Graphite-Base Structural Materials [in Russian], Vol. 2, Metallurgiya, Moscow (1966).
4. U. S. Patent No. 3,101,403 (1963).

5. French Patent No. 1247.673 (1960).
6. Federal Republic of Germany Patent No. 1,030,482 (1958).
7. Federal Republic of Germany Patent No. 1,072,178 (1960).
8. E. M. Savitskii, The Effect of Temperature on the Mechanical Properties of Metals and Alloys [in Russian], Izd. Akad. Nauk SSSR (1957).
9. G. V. Samsonov, Refractory Compounds [in Russian], Metallurgizdat (1963).
10. A. P. Epik, in: Surface Phenomena in Melts and Powder-Metallurgy Processes [in Russian], Naukova Dumka, Kiev (1963).

IV. PHASE DIAGRAMS OF SYSTEMS
CONTAINING CARBON

The Phase Diagram of the System Titanium–Molybdenum–Carbon

V. N. Eremenko, T. Ya. Velikanova, and S. V. Shabanova

The phase diagram of the boundary system Mo–C has been redetermined at high carbon contents. A higher carbide of molybdenum (cubic, NaCl type of lattice, homogeneity range 37-40 at.% C) has been found to form by a peritectic reaction at 2560°C; at 2000-2200°C this undergoes a polymorphic transformation into the low-temperature hexagonal form, which subsequently, at 1630°C, decomposes into the lower carbide and carbon. A phase transformation occurs in the lower carbide of molybdenum at 1170°C. The phase diagram of the ternary Ti–Mo–C system has been determined. Its principal features are: the system is divided into two triangles by the Mo–TiC section; the cubic carbides of molybdenum and titanium form a continuous series of solid solutions; at low temperatures titanium stabilizes the phase based on the higher hexagonal carbide of molybdenum.

This article presents the results of an investigation into the phase equilibria in the ternary system Ti–Mo–C over the whole composition range from 1400°C to the solidus temperature. Some properties of the components are given in Table 1.

The boundary binary systems Ti–Mo and Ti–C have already been studied in detail. The former contains a continuous series of solid solutions at high temperatures. In the latter a congruently melting refractory compound TiC, with a wide homogeneity range, is formed [1, 5, 10, 13, 16, 19-23].

TABLE 1. Structure and Properties of the Components of the
Ternary System Ti−Mo−C

Property	β-Ti	Mo	C(graphite)
Crystal structure	Body-centered cubic	Body-centered cubic	Hexagonal and rhombic forms
Space group	O_h^9	O_h^9	—
Lattice parameters, Å	3.282	3.146	—
Melting point, °C	1668	2620	—
Boiling point, °C	3170	4820	—
Volatilization temperature, °C	—	—	4200

Nowotny et al. [24] and Sykes [29] suggest that a congruently melting higher carbide of molybdenum with equiatomic proportions is present in the Mo−C system. However, according to Rudy et al. [26] this higher carbide of molybdenum corresponds to a carbon content of about 40 at.%. The carbide exists in two modifications: α (NaCl type) and η (hexagonal lattice, characteristic type) with a transformation temperature in the range 2000-2200°C. The higher carbide of molybdenum undergoes a eutectoid decomposition $\eta \rightarrow \gamma$ + C in the range 1400-1800°C. The properties of the phases are given in Table 2.

TABLE 2. Properties of Intermediate Phases in the Boundary
Binary Systems, Based on Data Available in the Literature

Property	TiC	Mo_2C	MoC_{1-x}	Mo_3C_2
Designation of phase	δ	γ	α	η
Crystal structure	Face-centered cubic, B1-type	Hexagonal, L_3'-type	Face-centered cubic, B1-type	Hexagonal, characteristic type
Homogeneity range, at.% C	30—50 at the temperature of the peritectic reaction L + $\delta \rightleftharpoons \beta$	30—33.75	40	40
Lattice parameters, Å	4.287—4.327	$a = 3.005$—3.008 $c/a = 1.58$	4.28	$a = 3.01$ $c/a = 4.86$
Mode of formation	Congruent melting	Incongruent melting	Incongruent melting	Transformation in the solid state
Melting point, °C	3150	2420 ± 20	—	—

The System Mo − C

Alloys of the Mo−C system were investigated in order to determine more accurately the phase diagram of this system in the high-carbon region.

The raw materials used consisted of molybdenum powder (99.7% Mo); high-purity compact molybdenum having a melting point of 2620°C and a lattice parameter $a = 3.147$ Å; and reactor graphite.

The alloys were prepared by melting in an arc furnace under a protective atmosphere of argon or by sintering (in vacuo or under argon).

The Mo−C system was investigated by metallographic and x-ray phase analysis, microhardness measurements, chemical analysis, differential thermal analysis up to 1800°C, measurements of the temperatures of the beginning of melting and remelting (the solid−liquid state region) using Alterthum and Pirani's method, and electrical-resistivity measurements.

The results of the investigation are presented in Fig. 1. The eutectic formed by the β- and γ-phases lies close to 19 at.% C at 2180 ± 20°C. The lower carbide of molybdenum, Mo_2C, is formed

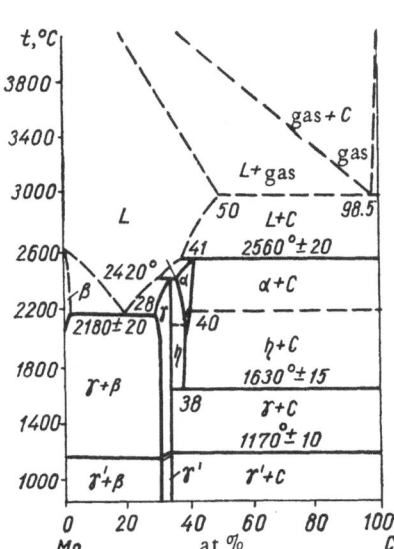

Fig. 1. Phase diagram of the binary Mo−C system, based on the results of the present investigation.

by the peritectic reaction

$$\alpha + L \rightleftarrows \gamma,$$

at $2420 \pm 20°C$; it has a homogeneity range of 28.5-33.75 at.% C at the eutectic temperature and of 31-33.75 at.% C at temperatures below 2000°C.

It was first shown by thermal analysis and x-ray phase analysis that the lower carbide of molybdenum undergoes a phase transformation at $1170 \pm 10°C$, accompanied by a thermal effect and a reduction in the symmetry of the crystal lattice. If the strongest reflections from the low-temperature form are indexed in hexagonal syngony, the values of a and c for the crystal lattice coincide accurately up to the second decimal place with the parameters of the original high-temperature form Mo_2C.

It was shown that the higher carbide of molybdenum MoC_{1-x} is formed at $2560 \pm 20°C$ by the peritectic reaction

$$L + C \rightleftarrows \alpha\text{-}MoC_{1-x}.$$

No dystectic points were found in the region of 40 or 50 at.% C. The specimens can be heated considerably above the solidus temperature. Needles of primary graphite were observed in the structure of cast alloys containing 40 at.% C and more. No eutectic structures were found.

From the shape of the solidus curve in the region of the α-phase, the lower boundary of the homogeneity region of the α-phase appears to lie at 37 at.% C, in agreement with Wallace's results [30]. This conclusion was confirmed metallographically.

In alloys containing 40 at.% C and more (up to 50% C_{total}), cast and annealed at 2200, 2100, 2000, and 1900°C, the combined carbon content was shown by chemical analysis to be 40 ± 2 at.%, which corresponds with the upper boundary of the homogeneity range of the higher molybdenum carbide.

The $\alpha \rightarrow \eta$ transformation takes place extremely rapidly, and we did not succeed in retaining the high-temperature modification MoC_{1-x} either by quenching into molten tin or by pouring into a water-cooled copper mold. In all cases the x-ray diffraction diagrams revealed a system of lines of the η-phase and possibly lines of the η- and α-phases. It is difficult to determine the latter point, as the strongest reflections from the η-phase merge into the lines of the cubic phase.

The parameters of the cubic sublattice of the η-phase in the alloys as quenched from the molten state and after being annealed at 2000°C are given in Table 3.

The values of the lattice parameters of the cubic higher carbide of molybdenum were obtained by extrapolation of the concentration dependence of the lattice parameter of $(Ti, Mo)C_{1-x}$ solid solutions saturated with carbon at various temperatures and zero Ti content (the data were obtained in the course of an investigation of alloys of the ternary system $Mo-Ti-C$). The combined carbon content in this carbide was also obtained by extrapolation of the solubility isotherms of carbon in the solid solution $(Ti,Mo)C_{1-x}$ at zero Ti content (in the $Mo-Ti-C$ concentration triangle). It emerged that the parameter of the cubic lattice for alloys in the cast state is 4.267 ± 0.003 kX at a combined carbon content of 40 at.% and that this hardly changes when the temperature is reduced to 2200°C. In the temperature range 2000-2200°C the parameter decreases to 4.243 ± 0.003 kX with fall in the combined carbon content to 38 at.%.

The results obtained are in agreement with the data available in the literature. Thus Rudy and his coworkers [27] obtained a combined carbon content of 40.0-40.8 at.% for $Mo-C$ alloys rapidly quenched from the liquid state (these alloys were identified as α-

TABLE 3. Lattice Parameters of the Hexagonal Form of $MoC_{1-x}(\eta)$ [*]

State of specimen	Lattice parameters, kX			
	η-phase			cubic subcell
	a	c	c/a	a
Suddenly cooled from the molten state	3.01 ± 0.01	14.67 ± 0.01	4.86	4.267 ± 0.001
Annealed at 2000°C and below	2.99 ± 0.01	14.52 ± 0.01	4.86	4.243 ± 0.001

[*] Total carbon content = 40.0 at.%.

phase by x-rays), and a content of 38-40 at.% (more often 38 at.%) for alloys produced by arc melting (η-phase). Wallace and his coworkers [30] found 40 at.% of combined carbon in cast alloys produced by arc melting and 38 at.% in alloys annealed at 2100°C. In both cases the alloys were identified at η-phase by x-rays.

Thus, the lattice parameter of the cubic higher carbide of molybdenum at the upper boundary of the homogeneity range (in equilibrium with graphite) is identical at high and low temperatures with that of the cubic sublattice of the hexagonal form η-MoC_{1-x}. If the crystallographic similarity between the α- and η-phase lattices is taken into account, this identity points to the polymorphic nature of the $\alpha \rightleftharpoons \eta$ transformation. According to these data, the temperature range of the transformation is 2000-2200°C.

It was found by differential thermal analysis of arc-melted alloys and by metallographic analysis of alloys sintered at 1550, 1650, 1700, 1820, 2100, and 2300°C, and also by chemical analysis, that the η-phase undergoes a eutectoid decomposition η-$MoC_{1-x} \rightleftharpoons \gamma + C$ at 1630 ± 15°C.

In attempts to melt in an arc furnace alloys containing more than 50 at.% C it was found that, irrespective of the amount of carbon added to the charge (up to 95%), the total carbon content in the ingots did not exceed 50 at.%. Marked loss of carbon was observed during melting. In consequence the equilibrium

$$\text{gas} \rightleftharpoons \text{liquid phase + graphite}$$

may be considered to exist at high temperatures, the composition of the liquid phase lying close to 50 at.% C. Carbon to the extent of 98.5-99 at.% was found in the condensate from the gaseous phase. Similar equilibria in the systems $Ti-C$ and $Nb-C$ have been described in [18].

The phase diagram we have constructed for the system $Mo-C$ differs substantially at high carbon contents from the diagram given by Hansen and Anderko in their handbook [13].

The System Ti — Mo — C

The solubility of molybdenum carbide in titanium carbide has been studied in [6, 7, 11]. Albert and Norton [14] have published the phase equilibria in the $Ti-Mo-C$ system at 1710°C, based on the assumed existence of an equiatomic higher carbide of molybdenum, an assumption which we have shown above to be untrue.

We have investigated the phase diagram of the ternary system Ti−Mo−C over the whole composition range from 1400°C to the solidus temperature (more than 200 alloys were studied).

The raw materials consisted of molybdenum and carbon (as used in the investigation of the Mo−C binary system) and titanium in the form of powder containing 99.7% Ti and iodide titanium. The methods of preparing and investigating the alloys were the same as those employed for the binary alloys.

From the thermodynamic analysis of the likely direction of flow of the reactions

$$\langle Mo_2C \rangle + \langle Ti \rangle \rightleftarrows \langle TiC \rangle + 2 \langle Mo \rangle, \tag{1}$$

$$\langle MoC_{0.67} \rangle + 0.67 \langle Ti \rangle \rightleftarrows 0.67 \langle TiC \rangle + \langle Mo \rangle, \tag{2}$$

$$2 \langle MoC_{0.67} \rangle + 0.34 \langle Ti \rangle \rightleftarrows 0.34 \langle TiC \rangle + \langle Mo_2C \rangle \tag{3}$$

in alloys situated at the points of intersection of the Mo−TiC and Mo_2C−TiC sections with the Ti−Mo_2C and Ti−$MoC_{0.67}$ sections, it was concluded that the above reactions must proceed from left to right almost to completion, since the difference in the free energies of formation of the titanium carbide and the molybdenum carbide is negative and its absolute value is considerable [17, 25] over the whole temperature range investigated. Hence it is to be expected that, as TiC is a congruently melting compound, the Mo−TiC polythermal section is a quasibinary section of the ternary Ti−Mo−C system, and that the phase diagram of this system is divided by this section into two partial diagrams, Mo−TiC−C and Mo−TiC−Ti. An experimental study made by Guertler's method of intersecting sections confirmed this conclusion [2].

A detailed investigation of the polythermal section Mo−TiC [3] has shown that the TiC−Mo system is of the eutectic type with a wide homogeneity range of the phase based on titanium carbide and small solubility of TiC in molybdenum.

Preliminary results of an investigation into the polythermal sections Mo_2C−TiC and $MoC_{0.67}$−TiC have been reported in [4]. From the results of a study of the structure of alloys covering the whole composition range, the probable projection of the liquidus

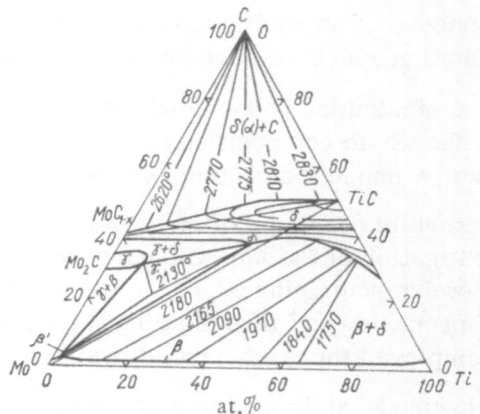

Fig. 2. Projection of the solidus surface on the
concentration triangle of the system Mo−Ti−C.

surface on the concentration triangle has been constructed, together with the projections of the solidus surface and isothermal
sections at 2180, 2000, and 1400°C.

The projection of the solidus surface (Fig. 2) reveals the
formation of a continuous series of solid solutions of the isostructural carbides α-MoC_{1-x} and TiC. A wide field of primary crystallization of this phase is connected with the carbide solid-solution
field at the liquidus surface. On the boundary curve of the joint
crystallization of graphite and the solid solutions $(Mo,Ti)C_{1-x}$ there
is a transition from the congruent method of joint crystallization
of the δ-phase and graphite, $L \rightleftharpoons C + \delta$ to the incongruent method,
$L + C \rightleftharpoons \delta(\alpha)$ at 2620°C, 4 at.% Ti and 41.5 at.% C. In accordance
with the Van Rein−Van Alkamad law, the eutectic point on the
straight line connecting Mo and TiC is a temperature maximum
on the boundary curve of joint crystallization of the δ- and β-phases.
From this point the temperature falls to the ternary eutectic point
E (2130°C, 17 at.% Ti, 21 at.% C) and to the temperature of the
peritectic crystallization of β-titanium in the binary Ti−C system. A transition occurs in the vicinity of 2165°C from the con-
gruent method of crystallization $L \rightleftharpoons \delta + \beta$ to the incongruent
method $L + \delta \rightleftharpoons \beta$. On the solidus surface of the NaCl-type phase
there is a valley in the maximum temperatures at which the alloys
begin to melt, which coincides with the section $MoC_{0.67}$−TiC. The
upper boundary of the homogeneity region of the NaCl-type phase

protrudes toward the carbon corner, while the lower boundary proceeds uniformly at 38 at.% C. The solubility of titanium in the γ-phase is about 9 at.%. The lattice parameters of the phases in the three-phase field are given in Table 4.

A break occurs in the NaCl-type solid solution field in the ternary Mo−Ti−C system as a result of the polymorphic transformation $\alpha \rightarrow \eta$ in the binary Mo−C system, and below 2000°C these solutions are in equilibrium with the η-phase.

A feature of the phase equilibria at 1400°C (Fig. 3) is the disappearance of the η-phase in the binary Mo−C system as a result of the eutectoid decomposition at 1630°C. However, small additions of titanium or TiC stabilize the η-phase in the ternary system down to 1400°C and evidently down to room temperature. It has been shown by an experiment specially carried out for the purpose that the η-phase is thermodynamically stable at 1400 and 1200°C. The NaCl-type solid-solution field contracts both from the carbon side and along the equilibrium boundary with the metallic phase (at 40-42 at.% C). Thus, the form of the phase diagram of the Ti−Mo−C system is determined by the presence of the strong compound TiC which exists over a wide homogeneity range. All the phases present in the system at a particular temperature

TABLE 4. Lattice Parameters of the Phases Coexisting in the Three-Phase Field at the Solidus Surface and at Temperatures of 2000 and 1400°C

Tempe-rature,°C	Phase field	Lattice parameters, kX			
		β-phase	δ-phase	γ-phase	η-phase
Solidus surface	$\beta + \gamma + \delta$	3.140 ± 0.004	4.265 ± 0.003	$a = 2.998 \pm 0.002$ $c/a = 1.58$	—
2000	$\beta + \gamma + \delta$	3.138 ± 0.002	4.273 ± 0.003	$a = 2.998 \pm 0.002$ $c/a = 1.58$	—
	$\eta + \delta + C$	—	4.255 ± 0.002	—	$a = 3.001 \pm 0.002$ $c/a = 4.85$
1400	$\beta + \gamma + \delta$ $\gamma + \delta + \eta$	3.138 ± 0.002 —	4.292 ± 0.002 4.257 ± 0.004	3.002 ± 0.002 3.002 ± 0.002 $c/a = 1,58$	— 3.005 ± 0.002 $c/a = 4.86$
	$\delta + \eta + C$	—	4.263 ± 0.002	—	3.005 ± 0.005 $c/a = 4.86$
	$\gamma + \eta + C$	—	—	$3.002 \pm 0,002$ $c/a = 1.58$	$a = 3.005$ $c/a = 4.86$

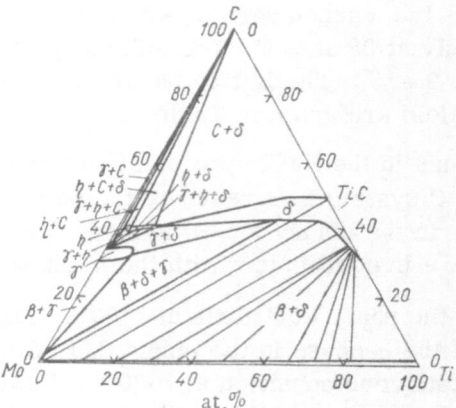

Fig. 3. Isothermal section of the phase diagram
of the ternary Mo—Ti—C system at 1400°C.

are in equilibrium with the phase based on this compound. As is
to be expected, the isomorphous carbides TiC and MoC_{1-x} form a
continuous series of solid solutions at high temperatures. Titanium carbide broadens the temperature range over which the α-
and η-modifications of the higher molybdenum carbide exist right
down to room temperature. In this research we have made the
first detailed study of the homogeneity range of the solid solutions of the higher molybdenum carbide with another cubic carbide (solidus surface, concentration dependence of the lattice parameter of the NaCl-type phase at various temperatures at the
upper boundary of the homogeneity range, and limits of the homogeneity range).

A thermodynamic appraisal of the Mo—Zr—C and Mo—Hf—C
systems made on the same lines as that for the Mo—Ti—C system leads to the conclusion that there is a possible similarity in
behavior between the three systems. Actually, ZrC and HfC have
even more negative free energies of formation than TiC, and the
homogeneity ranges of ZrC and HfC are narrower than that of TiC.
Consequently, it is to be expected that the solubility of Mo in ZrC
and HfC will also be less than it is in TiC, and clearly the reduction in the free energy of formation of the respective solid solutions (Zr,Mo)C and (Hf,Mo)C as compared with the stoichiometric
carbides ZrC and HfC, for which we have data, will be less than
in the case of TiC. From this it follows that the systems Mo—

Zr − C and Mo − Hf − C must be divided into triangles by the quasibinary sections Mo − ZrC and Mo − HfC. This has been confirmed
experimentally by the investigations of Ordan'yan, Avgustinik, and
their coworkers [8, 9], who found that the sections Mo − ZrC and
Mo − HfC of the ternary systems Mo − Zr − C and Mo − Hf − C, respectively, are quasibinary sections of the eutectic type.

Size relationships allow the formation of continuous series
of solid solutions between the isomorphous carbides MoC_{1-x} and
ZrC or HfC. In fact the metallographic examination of several
cast Mo − Zr − C alloys in the vicinity of the composition of the
higher molybdenum carbide has led Wallace and his coworkers
[30] to postulate the existence of continuous series of solid solutions of the higher molybdenum carbide and zirconium carbide.

The ternary system Mo − Hf − C has not been investigated.
Fedorov et al. [12] have found the η-phase to be stabilized by additions of ZrC at 1400°C.

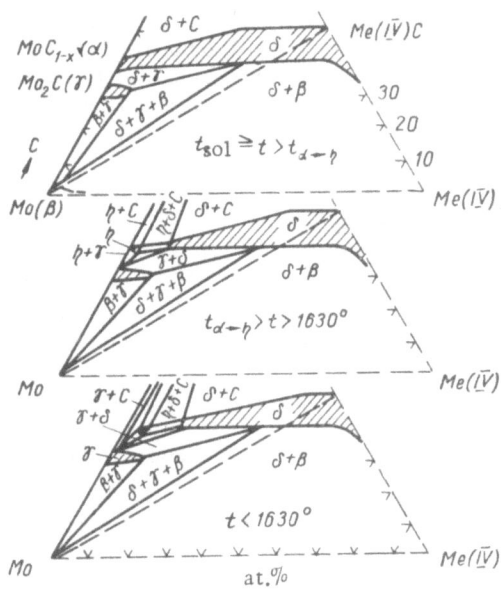

Fig. 4. Diagrams showing the phase equilibria at
various temperatures in the systems Mo − Me(IV) − C,
where Me(IV) represents Ti, Zr, or Hf.

Thus diagrams of the phase equilibria in the ternary systems Mo — Me(IV) — C as a function of temperature can be depicted in the way shown in Fig. 4. Similarities in the shape of the liquidus surfaces are to be expected in a view of the similarity between the solidus surfaces.

In the partial diagrams Mo — Me(IV)C — Me(IV) special features appear that are due to the difference in structure of the phase diagrams of the boundary binary systems Mo — Me(IV) and to the form of the Me(IV) — C phase diagrams in the vicinity of Me(IV).

Literature Cited

1.   V. N. Eremenko, Titanium and Its Alloys [in Russian], Izd. Akad. Nauk UkrSSR, Kiev (1960).
2.   V. N. Eremenko, Z. I. Tolmacheva, and T. Ya. Velikanova, in: Investigations of High-Temperature Alloys [in Russian], Vol. 8, Izd. Akad. Nauk SSSR, Moscow (1962), p. 95.
3.   V. N. Eremenko and T. Ya. Velikanova, Poroshkovaya Met., No. 5, 3 (1963).
4.   V. N. Eremenko, T. Ya. Velikanova, and S. V. Shabanova, in: New Studies of Titanium Alloys [in Russian], Nauka, Moscow (1965).
5.   R. Kieffer and P. Schwartzkopf, Hard Alloys [Russian translation], Metallurgizdat, Moscow (1957).
6.   A. E. Koval'skii and Ya. S. Umanskii, Zh. Fiz. Khim., 20:769, 773, 929 (1946).
7.   L. P. Mol'kov and I. V. Vikker, Vestn. Metalloprom., 16:75 (1936).
8.   S. S. Ordan'yan, A. M. Avgustinik, and V. S. Vigdergauz, in: Researches in the Field of the Chemistry of Silicates and Oxides [in Russian], Nauka, Moscow (1965), p. 220.
9.   S. S. Ordan'yan, A. A. Kraskovskaya, and A. I. Avgustinik, Izv. Akad. Nauk SSSR, Neorg. Mat., 2:299 (1966).
10.   G. V. Samsonov and Ya. S. Umanskii, Hard Compounds of Refractory Metals [in Russian], Metallurgizdat, Moscow (1957).
11.   Ya. S. Umanskii, Izv. Akad. Nauk SSSR, Fiz.-Khim. Anal., 16(1):127 (1943).
12.   T. F. Fedorov, Yu. B. Kuz'ma, and L. V. Gorshkova, Poroshkovaya Met., No. 3, 69 (1965).
13.   M. Hansen and K. Anderko, The Structure of Binary Alloys (Russian translation), Metallurgizdat, Moscow (1962).
14.   H. Albert and J. Norton, Planseeber, Pulvermet., 4:2 (1956).
15.   I. Cadoff and J. P. Nielsen, Trans. Am. Inst. Min. Metal. Eng., 197:248 (1953).
16.   M. Hansen, E. Kamen, H. Kessler, and D. McPherson, Trans. Am. Inst. Min. Metal. Eng., 191:881 (1951).
17.   M. Gleiser and J. Chipman, J. Phys. Chem., 66:1532 (1963).
18.   H. Kimura and J. Sasaki, Trans. Japan Inst. Metals, 2(1):98 (1961).
19.   Y. Muracami, H. Kimura, and Y. Nishimura, J. Japan Inst. Metals, 21:665 (1957).

20.   Y. Muracami, H. Kimura, and Y. Nishimura, J. Japan Inst. Metals, 21:712
        (1957).
21.   Y. Muracami, H. Kimura, and Y. Nishimura, Mem. Fac. Engng. Kyoto Univ.,
        19:302 (1957).
22.   Y. Nishimura and H. Kimura, J. Japan Inst. Metals, 20:528 (1956).
23.   Y. Nishimura and H. Kimura, J. Japan Inst. Metals, 20:589 (1956).
24.   H. Nowotny, E. Parthe, R. Kieffer, and F. Benesovsky, Monatsh. Chem., 85:255
        (1954).
25.   Physical Chemistry of Metals (Metallurgy and Metallurgical Engineering
        Series), McGraw-Hill, New York (1953).
26.   E. Rudy, F. Benesovsky, and L. Toth, Z. Metallk., 54:345 (1963).
27.   E. Rudy, El. Rudy, and F. Benesovsky, Planseeber. Pulvermet., 10(1/2):42
        (1962).
28.   E. Rudy, F. Benesovsky, and Sedlathek, Monatsh. Chem., 92:841 (1961).
29.   W. P. Sykes, K. R. Van Horn, and C. M. Tucker, Trans. Am. Inst. Min. Metal.
        Eng., 117:173 (1935).
30.   T. C. Wallace, C. P. Guttierez, and P. L. Stone, J. Phys. Chem., 67:796
        (1963).

Study of the Phase Equilibria in the System Tantalum–Vanadium–Carbon in the Range 0-33 at.% C

L. A. Tret'yachenko, S. A. Komarova, and V. N. Eremenko

The phase equilibria in the system Ta–V–C have been investigated at 1600°C in the range from 0 to 33 at.% C, and a projection of the solidus surface in this range has been constructed. It was found that in the region investigated the ternary system Ta–V–C is characterized by the existence of equilibrium between the carbide solid solutions $(Ta,V)_2C_{1-x}$ and the metallic phase based on the Ta–V solid solutions. The carbide phase is richer in tantalum than the metallic phase coexisting with it. The change in the lattice parameters in the continuous series of solid solutions $(Ta,V)_2C_{1-x}$ has been determined.

The phase diagram of the ternary system Ta–V–C has never been studied, although the binary systems that form the boundaries of the ternary system have been investigated a number of times.

In the V–C system [8] two phases exist in the range 0-33 at.% C: a solid solution of carbon in vanadium and the lower carbide of vanadium V_2C_{1-x}. This carbide has a hexagonal crystal structure, with a homogeneity range of 30–33.3 at.% C at 1650°C and 31.5-33.3 at.% C at 1450°C. In this concentration range the lattice parameters of the carbide phase vary at 1450°C from $a = 2.886 \pm 0.003$ Å, $c = 4.576 \pm 0.003$ Å to $a = 2.904 \pm 0.003$ Å, $c = 4.576 \pm 0.003$ Å. V_2C is formed at 2165°C by the peritectic reaction

$$Liq + VC_{0.63} \rightarrow V_2C.$$

The maximum solubility of carbon in vanadium is 4 at.% C at 1650°C [8], decreasing with fall in temperature. This value is close to the data given by Gebhardt et al. [11], who found in a special investigation of the solubility of carbon in the transition metals of Group V, that the maximum solubility of carbon in vanadium is 5 at.% C, in contrast to the value of 9 at.% C reported in [21]. The vanadium-base solid solution and V_2C_{1-x} form a eutectic at 1650°C and 14 at.% C.

The phase diagram of the system Ta−C is similar to that of the V−C system [2, 9, 18]. Ta_2C is formed by a peritectic reaction. Ta_2C_{1-x} and the solid solution of carbon in tantalum form a eutectic at a temperature which has been variously reported as 2750-2902°C [9, 11, 14, 18]. The homogeneity range of Ta_2C_{1-x} is 26-33.3 at.% C at the eutectic temperature; below 2000°C it is 30-33.3 at.% C [2, 11]. The maximum solubility of carbon in tantalum is about 7 at.% C [10, 11, 17], falling steeply with temperature.

In the Ta−V system a continuous series of solid solutions is formed at high temperatures [1, 4, 5]. A minimum occurs on the solidus curve. Below 1300°C a phase based on the compound TaV_2 is formed. A neutron-diffraction study of the structure has shown [6] that this compound belongs to the class of Laves phases having a cubic structure of the $MgCu_2$ type, with $a = 7.16$ Å. The existence of a high-temperature modification of this compound with a hexagonal structure of the $MgZn_2$ type has been suggested. Teslyuk [7] has reported lattice parameters for two modifications of TaV_2: a high-temperature form of the $MgZn_2$ type and a low-temperature form of the $MgCu_2$ type.

It is known from x-ray diffraction studies [15, 16] that the isomorphous monocarbides of tantalum and vanadium, which have a cubic structure of the NaCl type, form a continuous series of solid solutions, and this has been confirmed by measurements of the

microhardness and electrical resistivity of alloys of the monocarbides [3, 19]. As the lower carbides of tantalum and vanadium
also possess the same crystal structure with similar lattice parameters, it is to be expected that they too will form a continuous
series of solid solutions, although there is no experimental evidence of this.

Continuous series of solid solutions of the monocarbides, of
the lower carbides, and of the metals are known to exist at 2000°K
from the isothermal section of the Ta−V−C system for this temperature, constructed on the basis of thermodynamic calculations
[20]. However, the diagram may not take this form at 2000°K,
since in the V−C system the eutectic V_2C_{1-x} + (V,C) melts at
1650°C, and the phase equilibria at 2000°K (1727°C) should be characterized by the presence of a liquid phase.

It was also thought to be interesting to discover, in the course
of our investigation of the Ta−V−C system, whether a ternary
compound of the η-carbide type is formed in the system. Such a
compound can be formed in Me'−Me"−C systems if compounds
of the Laves-phase or σ-phase types are formed in the binary
Me'−Me" system [12, 13].

The Ta−V−C system in the range 0-33 at.% C was investigated by means of metallography, x-ray analysis, and the determination of the temperatures at which melting began. The alloys
were prepared by melting in an arc furnace having a nonconsum-
able tungsten electrode and a water-cooled copper hearth, in an
argon atmosphere purified by melting a titanium getter. The materials used in preparing the alloys were carbothermic vanadium
containing 0.02% Al, 0.05% Fe, 0.006% Si, 0.005% S, and 0.23% C;
metalloceramic tantalum of 99.8% purity; and spectrographically
pure graphite.

Alloys were made up along the 12 at.% C, 20 at.% C, TaV_2−
V_2C, and TaV_2−Ta_2C sections. Random chemical analysis of the
alloys showed that the composition of alloys situated on the 12
at.% C section differed only slightly from that of the mix, whereas alloys that should have contained 20 at.% C had shifted to the
16 at.% C section. It was found from the micro-examination of
cast alloys that a three-phase eutectic reaction

$$\text{Liq} \rightarrow (\text{Ta, V, C}) + (\text{Ta, V})_2 C_{1-x}$$

takes place in the range investigated following the crystallization of the metallic or carbide phase, depending on the alloy composition.

The temperature at which melting began was determined from the appearance of a drop of liquid at the bottom of a blind hole which simulated the conditions of blackbody radiation, on heating the specimen by the direct passage of a current in an argon atmosphere. The temperature was measured by means of an OP-48 pyrometer calibrated against the melting points of the pure metals. The specimens had previously been annealed to equilibrium at 1600°C.

From measurements of the temperatures at which melting began on the 12 and 16 at.% C sections, we constructed a projection of the solidus surface in the vicinity of the reaction: Liq → $(Ta,V,C) + (Ta,V)_2C_{1-x}$ onto the plane of the concentration triangle (Fig. 1). The contoured surface rises from 1650°C — the eutectic temperature in the $V-C$ system — to 2850°C in the $Ta-C$ system. The monovariant lines of the solid solutions of the lower carbides of vanadium and tantalum and those of the metallic solutions were drawn on the basis of data for the lower boundaries of the homogeneity ranges of the lower carbides and of data for the solubility of carbon in Ta or V in the binary systems (the data of Gebhardt et al. [11] for the $Ta-C$ system and our own data [8] for the $V-C$ system). The course of the monovariant line of the metallic solutions is confirmed by the results of a metallographic study of cast alloys containing 5 at.% C and lying on the $TaV_2 - V_2C$ and $TaV_2 - Ta_2C$ sections; the alloy containing 5 at.% C on the $TaV_2 - V_2C$ sections has, in addition to the primary metallic phase, a small amount of the eutectic constituent, while the alloy containing 5 at.% C on the $TaV_2 - Ta_2C$ section is single-phase, in the cast

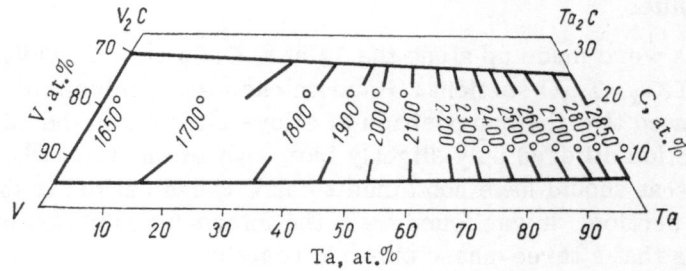

Fig. 1. Projection of solidus surface.

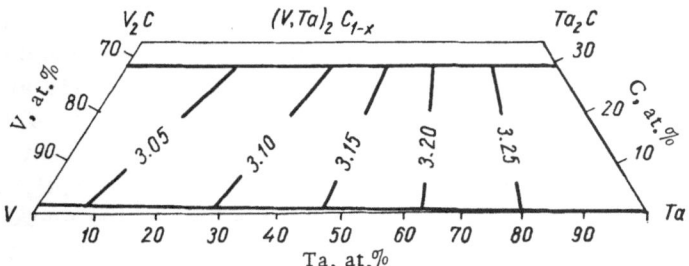

Fig. 2. Isoparametric lines of the metallic phase in the (Ta,V,C) +
(Ta,V)$_2$C$_{1-x}$ region at 1600°C.

state. The first of these alloys lies somewhat above the projection of the monovariant line of the solid solutions, whereas the
second one lies somewhat below it, in the region of the (Ta,V,C)
solid solutions.

Metallographic and x-ray diffraction studies of alloys on the
12 and 16 at.% C sections, annealed at 1600°C, revealed that all
these alloys are two-phase, consisting of a mixture of the (Ta,V)$_2$C$_{1-x}$
solid solutions and of the (Ta,V,C) metallic phase, i.e., at this
temperature equilibrium actually exists between the continuous
series of solid solutions of the carbides and the phase based on
the continuous series of metallic solutions. The lattice parameters of the phases in these alloys were determined, the x-ray
diagrams being obtained by the powder method, using CuK$_\alpha$ radiation. From the lattice parameters of the metallic phase we
constructed isoparametric lines, i.e., in this case the conodes in
the two-phase region (Ta,V,C) + (Ta,V)$_2$C$_{1-x}$ at 1600°C (Fig. 2).

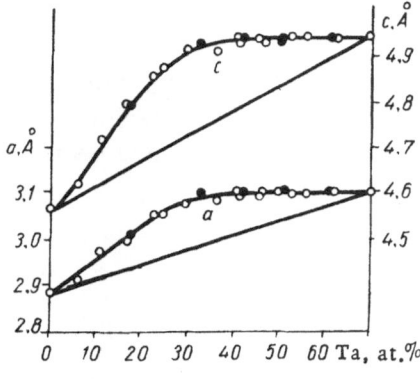

Fig. 3. Concentration dependence of the
lattice parameters of the carbide phase:
●) parameters determined graphically; O)
the composition of the carbide phase determined graphically.

From this it was possible to find the concentration dependence of the lattice constants of the metallic phase. This dependence was found to be linear. In this connection we did not take into account the solubility of carbon in the metallic phase, as this is only small at the temperature concerned. Now, knowing the composition of the alloy, the lattice parameters of the phases in it, and the concentration dependence of the lattice parameters of the metallic phase, we can determine graphically for each alloy the composition of the carbide phase that is in equilibrium with the metallic phase and can construct the curve of the concentration dependence of the lattice parameters of the carbide phase $(Ta,V)_2C_{1-x}$ (Fig. 3). For this purpose we took the carbon content of the carbide phase as constant over the whole series at 30 at.% C, which should not be far wrong at this temperature. Considerable positive deviation from the additive straight lines for a and c is found.

Investigations were also carried out on alloys annealed at 1200°C for 360 h, at which temperature a phase based on the compound TaV_2 participates in the equilibrium. The underlying structure of this phase is a cubic lattice of the $MgCu_2$ type, although there is a splitting of the lines that is probably indicative of tetragonal distortion of the lattice. The hexagonal modification of the phase based on TaV_2 and the compound of the η-carbide type were not observed at the temperatures investigated. Alloys with a large tantalum content did not attain equilibrium in the course of annealing for 360 h at 1200°C.

Conclusions

1. The projection of the solidus surface of the ternary Ta– V – C system in the region of the reaction: $Liq \rightarrow (Ta,V,C)$ + $(Ta,V)_2C_{1-x}$ has been constructed.

2. The isothermal section of the Ta–V–C system at 1600°C has been constructed for the range 0–33 at.% C, and the lattice constants of the phases present in the alloys at 1600°C have been determined.

Literature Cited

1.   A. T. Grigor'ev et al., Vestn. MGU, Ser. Khim., 4:44 (1965).
2.   L. B. Dubrovskaya, G. P. Shveikin, and P. V. Gel'd, Fiz. Metal. i Metalloved., 17:73 (1964).

3. A. E. Koval'skii and L. A. Petrova, Microhardness [in Russian], Izd. Akad. Nauk SSSR, Moscow (1951).
4. A. P. Nefedov et al., Vestn. MGU, Ser. Khim., 5:42 (1965).
5. A. P. Nefedov et al., Izv. Akad. Nauk SSSR, Neorg. Mat., 1:715 (1965).
6. V. A. Somenkov, Izv. Akad. Nauk SSSR, Neorg. Mat., 2:464 (1966).
7. M. Yu. Teslyuk, Author's Summary of Candidate's Thesis, L'vov (1965).
8. L. A. Tret'yachenko and V. N. Eremenko, Izv. Akad. Nauk SSSR, Neorg. Mat., 2:1568 (1966).
9. F. H. Ellinger, Trans. Am. Soc. Metals, 31:81 (1943).
10. E. Fromm and U. Roy, J. Less-Common Metals, 8:79 (1966).
11. E. Gebhardt, E. Fromm, and U. Roy, Z. Metallk., 57:682 (1966).
12. H. J. Goldschmidt, J. Less-Common Metals, 2:138 (1960).
13. K. Kuo, Acta Metall., 1:301 (1953).
14. M. R. Nadler and C. P. Kempter, J. Phys. Chem., 64:1458 (1960).
15. J. T. Norton and A. L. Mowry, Trans. Am. Inst. Min. Metall. Eng., 185:133 (1949).
16. H. Nowotny and R. Kieffer, Metallforschung, 2:257 (1947).
17. H. R. Ogden, F. F. Schmidt, and E. S. Bartlett, Trans. Metall. Soc., AIME, 227:1458 (1963).
18. M. L. Pochon, C. R. McKinsey, R. A. Perkins, and W. D. Forgeng, Reactive Metals, New York (1959), p. 307.
19. E. Rudy and F. Benesovsky, Planseeber. Pulvermet., 8(2):72 (1960).
20. E. Rudy, Z. Metallk., 54:112 (1963).
21. E. K. Storms and R. J. McNeal, J. Phys. Chem., 66:1401 (1962).

The NbC—Re Polythermal Section in the Niobium—Rhenium—Carbon System

Yu. N. Vil'k, S. S. Ordan'yan, A. A. Murav'ev, A. G. Miroshnichenko, and Yu. A. Omel'chenko

The structure of the NbC—Re polythermal section of the Nb—Re—C system has been investigated by the x-ray, metallographic, and chemical methods of analysis and by making microhardness and melting-point measurements. It was found that the NbC—Re system is of the eutectic type, with the eutectic point at $2225 \pm 15°C$, 70 at.% Re, and that the solubility of rhenium in NbC is small, being 1.5 at.% at 2000-2100°C. The ($\delta + \beta$) field extends to 96-98 at.% Re. The lattice parameter of the δ-phase in the δ and ($\delta + \beta$) fields as a function of its rhenium content decreases from 4.470 Å (the value given in the literature) to 4.469 Å at 2000-2100°C. The microhardness of the δ-phase is related to the solid solution of rhenium in NbC and that of the β-phase to the solid solution of carbon in rhenium.

We have investigated the NbC—Re section of the Nb—Re—C system in order to discover the most heat-resistant alloys of the cermet type based on NbC and rhenium and also in order to study the high-temperature reaction between NbC and rhenium.

The alloys were prepared by the method described in [1-3] from NbC powder (containing 11.3-11.4 wt.% C_{total}, and 0.1-0.05 wt.% C_{free}) and rhenium ("electrolytic grade" with a total impurity content of 0.1%). Ten alloys in all were prepared across the section, ranging in nominal rhenium content from 3 to 90 wt.%. The specimens were sintered for 1 h at 2000°C in a vacuum of

1×10^{-4}-1×10^{-5} mm Hg. Specimens prepared in this way were isothermally annealed for 30 min at 2000, 2100, and 2200°C. The annealing operations and the melting-point measurements were carried out in a pure argon atmosphere, the specimens being quenched after annealing. Then the alloys were investigated by the methods of metallographic and x-ray phase analysis, and the microhardness was determined in order to identify the phases present in the alloys. The experimental procedure has been described in [1-3]. The alloys were analyzed for their rhenium, carbon, and nitrogen contents. The rhenium content was determined by spectrographic and chemical analyses [5]* and the niobium content from the difference $100 - \Sigma(\text{Re} + \text{C} + \text{N})$.

The equilibrium of the alloys after heat treatment and the melting-point measurements was checked by the x-ray analysis of powders produced by grinding specimens which had been quenched after the operations described above had been carried out. As a criterion of equilibrium we chose the sharp separation of the doublets of the back-reflection lines of the NbC phase, which enabled us to measure the lattice parameters of this phase with an accuracy of not less than 0.001 Å. The photographs were taken on a URS-50-IM ionization x-ray diffractometer. In addition photographs were also taken of the surface of the specimens by using a VRS-3 camera on the URS-60 apparatus. The melting points were determined more accurately by metallographic examination of the specimens.

The NbC−Re polythermal section of the Nb−Re−C system depicted in Fig. 1 is based on the investigation of alloys situated on the NbC−Re section and on melting-point measurements.

The following point must be noted in connection with the structure of the polythermal section. We found only two phases: the δ -phase (a solid solution of rhenium in the higher carbide of niobium) and the β -phase (a solid solution of carbon in rhenium). The solubility of rhenium in NbC is small, being about 1.5 at.% at 2000-2100°C (Fig. 2). According to the results of x-ray and

* The authors wish to express their gratitude to O. V. Molchanova, N. I. Kozelkova, I. G. Kubyshkina, and A. V. Suvorova for carrying out the chemical analyses and to G. I. Kibisov and V. E. Chuchina for carrying out the spectrographic analyses of the specimens.

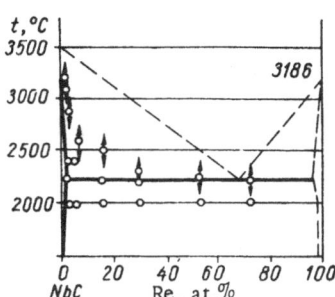

Fig. 1. Structure of the NbC—Re polythermal section in the Nb—Re—C system.

chemical analyses, alloys containing 1.32 and 1.65 at.% Re were
single phase after being held at 2000°C (from x-ray data), whereas
alloys containing 1.45 and 1.52 at.% Re were two phase after being
held at 2100°C, as was confirmed by metallographic examination.
The solution of rhenium in NbC is accompanied by a small decrease in the lattice parameter from 4.470 (as reported in the
literature [8]) to 4.469 Å. We observed the change in lattice parameter to 4.426 Å that had been found in [4] by photography in a
VRS-3 camera (diameter 143.3 mm), though only in alloys in which
the $\varkappa$-phase was retained. In specimens containing only the δ -
and β -phases the lattice parameter of the δ -phases, as determined by the VRS-3 camera, was 4.456-4.463 Å. Such a difference in lattice-parameter values measured by the two methods
is quite regular and tolerable.

It was established in [4] that the δ-phase is in equilibrium
with all the carbides of niobium. However, owing to the discrepancies found between our results on the structure of the phase
diagram of the Nb—Re—C system and those of Gorshkova et al.
[4], the complete phase diagram of the Nb—Re—C system is not
given in the present article pending our definitive determination
of it. Table 1 gives the composition of the alloys lying on the
NbC—Re section and the results of melting-point measurements
carried out on them. The specimens contained only the δ- and
β-phases.

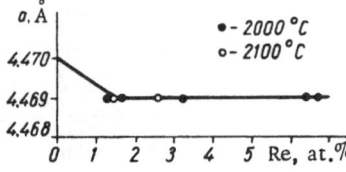

Fig. 2. Variation in the lattice parameter of the δ -phase as a function of rhenium content in the solid solution.

TABLE 1. Melting Points of Alloys
on the NbC—Re Section Determined
by the Pirani—Agte—Alterthum Method

Content of elements in the alloy, at.%			Melting point
Re	C	Nb	
0,88	49,56	49.56	3200 ± 60
1.47	49.26	49.26	3100 ± 35
3.02	48.49	48.49	2880 ± 40
6.64	46.68	46.68	2610 ± 30
15.80	42.10	42.10	2510 ± 35
29.82	35.09	35.09	2320 ± 15
52.90	23.55	23.55	2240 ± 10
71.90	14.05	14.05	2225 ± 15

The structure of the NbC—Re polythermal section (see Fig. 1) is confirmed by the results of the metallographic examination of the alloys. Figure 3a shows the structure of the alloy containing 3 at.% Re after it has been held at 2400°C for 40 min (onset of melting), and two phases are clearly visible. Figure 3b shows the alloy containing 6.6 at.% Re after it has been held at 2280°C for 1 h; two phases are plainly visible, one of them liquid. Figure 3c shows the alloy containing 15.8 at.% Re after it has been held at 2400°C for 1 h; liquid can be clearly seen. Figure 3b shows the alloy containing 52.9 at.% Re after it has been held at 2240°C for 40 min; it is a hypoeutectic alloy, with the liquid again plainly visible. Figure 3e shows the alloy containing 71.9 at.% Re after it has been held at 2240°C for 1 min; the alloy is hypereutectic and the liquid is clearly visible. A comparison of Figs. 3d and e indicates that the eutectic point must lie in the vicinity of 70 at.% Re. All the photographs are of etched specimens, the etchant being a mixture of nitric and hydrofluoric acids, used cold. Metallographic examination and photography of the structures were carried out on an MIM-8M microscope. The results reported in [4], together with our own results, point to the conclusion that the $(\delta + \beta)$ two-phase field extends as far as 96-98 at.% Re at 2000°C.

In phase identification by means of microhardness measurements made under a load of 100 g, it was found that the microhardness of the β-phase is 220-500 kg/mm^2 and that of the δ-phase 1900 kg/mm^2. The microhardness was measured after the specimens had been heat treated and quenched, a PMT-3 apparatus being used. Our microhardness results are in agreement with data given in the literature for niobium carbide [6] and for the solid solution of carbon in rhenium (β -phase) [7].

It is particularly to be noted that alloys prepared for the study of the NbC—Re section and found after vacuum sintering at 2000°C to lie in the two-phase ($\delta + \beta$) field (according to the results of all the forms of analysis employed) systematically lost rhenium and carbon after being held for 30 min in an argon atmosphere at 2100°C and were found in the three-phase ($\delta + \beta + \varkappa$) field (as shown by the x-ray and metallographic investigations); while still

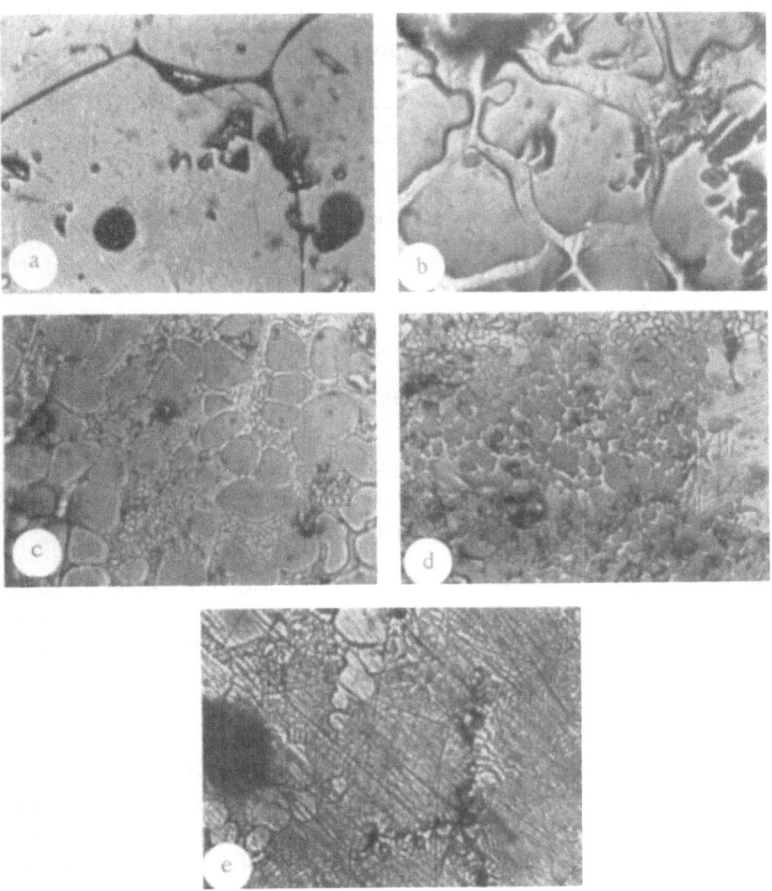

Fig. 3. Photomicrographs of the structures of alloys lying on the NbC—Re section. For explanation, see text.

greater losses of rhenium and carbon occurred after the alloys had been held at 2200°C. By this heat treatment the alloys were displaced from the $(\delta + \beta)$ field into the $(\alpha + \varkappa + \delta)$ and $(\gamma + \delta + \alpha)$ fields, where α is the solid solution of rhenium in niobium, $\varkappa$ is the intermetallic compound of niobium with rhenium, and γ is the solid solution of rhenium in Nb_2C. We observed no other phases or compounds.

Conclusions

1. The structure of the NbC−Re polythermal section in the Nb−Re−C system has been investigated by the x-ray, metallographic, and chemical methods of analysis and by making microhardness and melting-point measurements.

2. It was found that the solubility of rhenium in NbC is small, being about 1.5 at.% Re at 2000-2100°C.

3. The system NbC−Re is of the eutectic type, the temperature of the eutectic transformation being 2225 ± 15°C.

4. The eutectic point in the NbC−Re system lies in the vicinity of 70 at.% Re.

5. The $(\delta + \beta)$ field extends to 96-98 at.% Re.

6. A relationship between the lattice parameter of the δ-phase in the δ and $(\delta + \beta)$ fields and the rhenium content of the phase is given, and it is shown that during the solution of rhenium in the δ-phase its lattice parameter decreases from 4.470 Å (the value given in the literature) to 4.469 Å at 2000-2100°C.

7. The values obtained for the microhardness of the phases indicate that one of the phases (δ) consists of a solid solution of rhenium in niobium carbide (microhardness = 1900 kg/ mm²), while the second phase (β) consists of a solid solution of carbon in rhenium (220-500 kg/mm²).

8. It is established that alloys lying on the NbC−Re section, annealed in an argon atmosphere at 2100 and 2200°C, systematically lose rhenium and carbon and are displaced into the regions of ternary equilibrium: $(\delta + \beta + \varkappa)$, $(\alpha + \varkappa + \delta)$, and $(\gamma + \delta + \alpha)$.

Literature Cited

1. R. D. Avarbe et al., Zh. Priklad. Khim., 35:1967 (1962).
2. Yu. N. Vil'k et al., Teplofiz. Vys. Temp., 2:274 (1964).
3. Yu. N. Vil' et al., Zh. Priklad. Khim., 38:1500 (1965).
4. L. V. Gorshkova, T. F. Fedorov, and Yu. B. Kuz'ma, Poroshkovaya Met., No. 4 (52), 42 (1967).
5. G. M. Putvinskaya and T. G. Kibisova, Zh. Analit. Khim., No. 12, 1482 (1964).
6. G. V. Samsonov, Refractory Compounds: Handbook of Properties and Uses [in Russian], Metallurgizdat, Moscow (1963), p. 169.
7. P. Levesque, W. R. Bekebrede, and H. A. Brown, Trans. Am. Soc. Metals, 53:215 (1961).
8. E. K. Storms and N. H. Krikorian, J. Phys. Chem., 64:1471 (1960).

V. PHYSICAL PROPERTIES OF CARBIDES

An X-Ray Spectral Study of Some Compounds of Niobium and Carbon

M. I. Korsunskii, Ya. E. Genkin, and V. G. Lifshits

The emission spectra of the niobium carbides $NbC_{0.66}$, $NbC_{0.83}$, and $NbC_{0.98}$ have been investigated. An increase in the intensity of the second region of the band and a decrease in that of the short-wave region have been observed. This is due to an increase in the number of local Nb−C bonds with rise in carbon concentration. The concentration of collective electrons varies only slightly, being 2.7-3.6×10^{22} cm^{-3}. It is concluded that there is a reduction in the weight of d-states in the wave functions of the collective electrons as compared with pure niobium and that the value is 0.5-0.66.

This article reports the preliminary results of an x-ray spectral study of some niobium−carbon compounds.

The outermost emission $L\beta_2$ bands of niobium in the compounds $NbC_{0.66}$, $NbC_{0.83}$, and $NbC_{0.98}$ have been investigated. The bands were obtained on a long-wave x-ray spectrograph, using the Johann−Kapitsa method of focusing and a radius of curvature of the quartz crystal of 1 m ([$10\bar{1}1$] reflecting planes). Figure 1 shows the experimental $L\beta_2$ bands of niobium in the compounds concerned, together with bands corrected for the width of the internal level and distortions due to the apparatus [1-3]. The corrected emission band consists of two intense regions and a long-wave low-intensity third region, the form of which is not shown

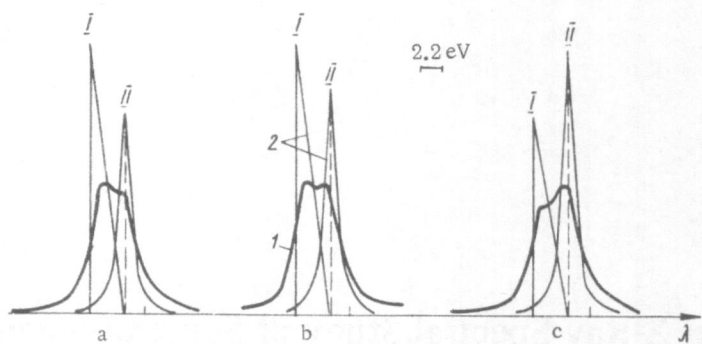

Fig. 1. Experimental (1) and corrected (2) $L\beta_2$ bands of niobium in the Nb—C system: a) $NbC_{0.66}$, b) $NbC_{0.83}$, c) $NbC_{0.98}$.

in Fig. 1, although the position of its center of gravity is indicated by a broken line. It can be seen from Fig. 1 that with rise in carbon concentration the intensity of the second region of the band increases, while the intensity of the short-wave region decreases. At the same time the width of the base of the short-wave region varies only slightly, being approximately 4.5 ± 0.5 eV. The results obtained can be interpreted as follows: the short-wave regions of the bands, which end in a sharp boundary corresponding to the Fermi energy limit, reflect the state of the collective electrons in the outer energy band. The decrease in intensity of the short-wave regions of the bands points to a reduction in the weight of the d-states in the wave functions of the collective electrons. The widths given above for the base of these regions represent a concentration of 0.6–0.8 collective electrons per atom (calculated by the approximation for almost free electrons and $m^* = m_0$).

According to [4], the second region of the band reflects the states of the electrons that participate in the local metal-metalloid bond.

The mean distance (r_1) between the niobium atoms in the three compounds with carbon is slightly greater (5-10%) than the mean distance in pure niobium (r_0); whereas the mean distance between the niobium and carbon atoms (r_2) is considerably less than r_0. This makes it very likely that a crossover transition will occur in which a vacancy from the inner L_{III} level of the niobium atom is transferred to the outer L_{II} level of the carbon atom with the emission of a quantum which is also recorded in the form of the second region of the $L\beta_2$ emission band. Hence the intensity of this region of the band increases with rise in the number of carbon atoms surrounding a given niobium atom, i.e., with increase in carbon concentration.

At the same time the intensity of the third region, which reflects the state of the electrons participating in the local niobium-niobium bonds, should decrease in comparison with the corresponding region of pure niobium, since r_1 is less than r_0 and the probability of a crossover transition taking place falls almost exponentially with increasing distance.

This is the reason for the very low intensity of the third part of the band. It must be borne in mind that it is mainly the d-states that are reflected in the intensity of the $L\beta_2$ band. The reduction in the weights of the d-states, calculated from the results of the present research, from 0.66 to 0.5 in the three compounds investigated indicates a change in the orbital moment of the collective electrons which should lead to a change in the magnetic susceptibility.

An approximate calculation of the magnetic susceptibility of $NbC_{0.98}$ [5], which takes account of the weights obtained for the various states, gives limits of variation from 10.42×10^{-6} to 19.04×10^{-6} (per mole), in satisfactory agreement with the data in [6], in which the magnetic susceptibility of a mole of NbC is 15.3×10^{-6}.

A fact that stands out is that in niobium carbide the width of the first region of the band, corresponding to the collective electrons, is less than in pure niobium and gives a conduction-electron concentration (using the free-electron approximation with $m^* = m_0$) of 0.6-0.8 electrons per nominal molecule containing a single metal atom. In pure niobium (with the same approximation) the conduction-electron concentration is 1.1-1.2 electrons per atom [3], in agreement with the change in type of conduction in niobium carbide as compared with pure niobium: in niobium carbide the conduction is predominantly of the electron type, whereas in pure niobium it is predominantly of the hole type. If the conduction of niobium carbide is assumed to be predominantly of the electron type, this gives a value for the Hall constant R of -1.73×10^{-10} m^3/C. In [6] a value of -1.32×10^{-10} m^3/C is reported for the Hall constant of NbC.

Conclusions

1. The emission L-spectra of niobium in the compounds $NbC_{0.66}$, $NbC_{0.83}$, and $NbC_{0.98}$ have been obtained.

2. An increase in the intensity of the second region of the $L\beta_2$ bands takes place wtih rise in carbon concentration, so indicating an increase in the number of local $Nb-C$ bonds.

3. The concentration of collective electrons varies only slightly $(2.7-3.6 \times 10^{22}$ cm$^{-3})$ with change in carbon concentration in the $Nb-C$ compounds investigated.

4. The magnetic susceptibility of $NbC_{0.98}$ is determined mainly by the collective electrons.

5. The weights of the d-states in the wave functions of the collective electrons of the compounds investigated is less than in pure niobium (about 1), its value being 0.5-0.66.

Literature Cited

1. M. A. Blokhin, The Physics of X-Rays [in Russian], Fizmatgiz, Moscow and Leningrad (1959).
2. Ya. E. Genkin, Transactions of Municipal Conference on Problems in General and Applied Physics [in Russian], Nauka, Alma-Ata (1966).
3. M. I. Korsunskii and Ya. E. Genkin, Dokl. Akad. Nauk SSSR, 142:1276 (1962).
4. M. I. Korsunskii and Ya. E. Genkin, Izv. Akad. Nauk Kaz.SSR, Ser. Fiz.-Mat., No. 2 (1967).
5. M. I. Korsunskii and Ya. E. Genkin, Izv. Akad. Nauk SSSR, Ser. Fiz., 28:832 (1964).
6. G. V. Samsonov, Refractory Compounds [in Russian], Metallurgizdat, Moscow (1963).

The State of the Carbon Atom in Transition-Metal Carbides

E. A. Zhurakovskii and N. N. Vasilenko

The K_α band of carbon in carbides of the transition metals of groups IV, V, and VI and in diamond and graphite has been measured by means of an ultralong-wave x-ray spectrometer with diffraction grating. The width of this band in the carbide spectra was considerably less than in pure carbon, and there was no fine structure. In all of the carbides there is a chemical shift of the CK_α band in the short-wave direction relative to its position in graphite, amounting on average to 1.5-2.0 eV. In relation to the position in diamond this shift is computed to be a few tenths of an electron volt and it takes place in the long-wave direction. From the analysis of the spectra and a comparison of them with data based on theoretical calculations and physical properties, a number of conclusions are drawn in regard to the state of the carbon atom in refractory carbides, the nature of its interaction with the valence band of the b-metals, and the magnitude and direction of charge transfer in the carbides.

Little work has been carried out on the direct experimental study of the energy spectrum of the electrons of carbon atoms in refractory carbides [1-4]. One of the present authors [5] has compared the position and shape of the K_α emission band of carbon in homogeneous titanium carbides with the corresponding K and L spectra of titanium in TiC_x. From an assessment of the relative positions of the valence bands of the two components, it proved possible to construct the total energy diagram of TiC and to account for the properties of this compound. Although data on the K and L spectra of the other transition metals and their car-

bides are incomplete, it was nevertheless of interest to extend
the x-ray spectral study of the carbon K_α band carried out for
TiC to the carbides of the other transition metals. A circumstance
favorable to this was that the K_α bands of carbon can be obtained
in various carbides, diamond, and graphite by using the same apparatus [6] and identical experimental conditions [4].

A detailed investigation may assist in the explanation of the
reasons for the appearance of a high bond strength in carbide
phases, a strength which (except in the case of WC) far exceeds
the strength characteristics of the initial pure components. In
compounds of transition metals of various groups with one particular metalloid, the nature of the interatomic bond is affected
principally by the energy characteristic of the d level of the transition metal, the degree of its noncomplexity, and the probable
existence of various types of electron configurations in which the
valence orbitals of the metalloid participate. The progress of
these investigations, together with calculations of the band structure, has enabled us to determine approximately the contribution
of the "electron" part of the strength to the total energy and
strength of the refractory compounds of the transition metals,
and possibly also to determine "whose" electrons make the greater contribution, those of the metal or those of the metalloid. In
the present communication we report preliminary results of measuring the CK_α band in the following transition-metal carbides:
TiC, VC, Cr_3C_2, ZrC, NbC, Mo_2C, HfC, TaC, and W_2C. Data for
comparing the shape and position of this band with those of the
band in graphite and diamond have been taken from [3]. The method of recording the curves has been described in part in [5]. All
the carbides were prepared from spectrographically pure raw materials by the methods given in [7], checked by phase and chemical
analyses.

The bands obtained are shown in Fig. 1, while the results of
the measurements, together with some elementary calculations
and physical properties, are contained in Table 1.

By analysis of the shifts of the K_α band of carbon (in relation
to its position in graphite and diamond), it is possible to draw
several semiquantitative conclusions about the direction and magnitude of the transferred electron charge in the carbides. As can
be seen from Table 1, the CK_α band in all the carbides undergoes

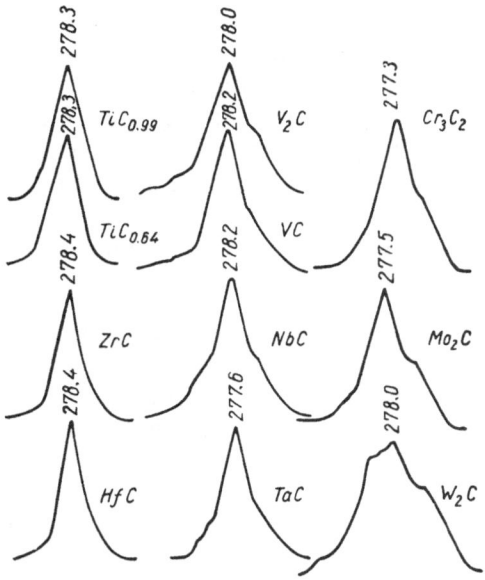

Fig. 1. X-ray emission K_α band of carbon in the carbides of some transition metals.

TABLE 1. Relative Position of the CK_α Band in Graphite, Transition-Metal Carbides, and Diamond, the Extrapolated Effective Charge Transferred to the Metallic Atom, and Some Strength Characteristics

Substance	Shift of CK_α, eV	Transferred charge in the electrons, extrapolated from TiC [9]	Melting point, °C	Modulus of elasticity · 10^{-6}, kg/cm^2 [14]
Graphite	0	—	3800	—
TiC	1.8	1.25—1.30	3250	4.6
VC	2.0	1.4	2830	4.3
Cr_3O_2	1.85	1.25	1895	3.8
ZrC	1.85	1.25	3175	5 6
NbC	1.85	1.25	3500	5.5
Mo_2C	1.35	0.94	2690	5.4
HfC	1.3	0.91	3900	—
TaC	1.2	0.84	3880	4.0
W_2C	0.8	0.56	2750	4.3
Diamond	2.3	—	1750 (converted into graphite)	—

a shift which varies in amount but is always in the higher-energy direction, its position in the graphite spectrum being taken as the reference zero. According to calculations made for heavier elements [8], this is to be interpreted as a decrease in the electron density in the valence 2s2p band of the carbon atom which evidently takes place as a result of the transfer of part of the 2p electrons of carbon to the hybrid bond orbitals. Analysis of the K and L spectra of titanium in TiC_x [5] has revealed an increase in the electron density in the band of the 3d levels of the titanium atom.

Theoretical calculations of the band structure of TiC_x have shown that there is a transfer of approximately 1.25-1.30 electrons from the carbon atom to the titanium atom [9]. From the data obtained in these researches it may be concluded that in all the carbides investigated there is also a transfer of effective charge from the carbon atom to the transition-metal atom and a transformation of the latter into a negatively charged ion. The great majority of carbides, particularly those formed with transition metals of low atomic number, exhibit a considerable shift of the CK_α band, which always takes place in the short-wave direction (see Table 1), with an accuracy of determination of the photon energy of about 0.2 eV. For TiC, ZrC, and NbC the shift obtained is in good agreement with the data reported by Holliday [2]; for Mo_2C the agreement is not so good [2]. The remaining carbides, which have been investigated for practically the first time, reveal shifts of various magnitudes. With the exception of the shift in VC (which has an anomalous nature, being the largest — 2 eV), the following regularity can be detected in the behavior of the CK_α band: the shift decreases with rise in the atomic number of the transition metal, reaching a minimum value (0.8 eV) in W_2C.

Since the metallic component of the bond, which undoubtedly exists in refractory carbides, is based on complete degeneracy of the electron wave functions and a uniform distribution of the electron gas in the interionic space of the carbide crystal, this type of bond can hardly be explained by the proposed transfer of charge to the metal. The existence of strong ionic interaction in refractory carbides is very unlikely in view of their high strength and good conductivity. It is best to attribute the charge transfer to the polarization of the covalent bond, as a result of which the density of the wave function of the covalent electrons (the pair-, single-, or sub-electron covalent bond) is not the same in the sphere of unlike atoms, and a negative charge attaches to that atom around which the life of the orbital electron is long. This very simple hypothesis enables us, albeit to a rough approximation only, to relate the magnitude of the shift of the CK_α band in carbides to the degree of polarization of the covalent bond in it. Then the reduction in the shift of the CK_α band with rise in atomic number Z ($HfC \rightarrow TaC \rightarrow W_2C$) is independent of the group to which the transition metal atom belongs and, contrary to [2], it can be related to the increase in energy of the valence nd level, which accompanies the rise in Z and which in all probability makes difficult the formation of covalent orbitals with 2ps electrons of carbon and polarization of the latter.

From the calculations in [9] and from the experimental data on the shift of the CK_α band in carbides, it is possible to determine by extrapolation the magnitude of the effective charge transferred by the carbon atom to the sphere of the transition-metal atom. The results of these calculations show (see Table 1) that the greatest effective charge induced by polarization occurs in elements of the first period, irrespective of the group to which they belong. For the same number of vacancies (as compared with their heavier analogs) the elements concerned differ in regard to the energy of the lower-lying d shell; this disproves the conclusions reached by Holliday [2], who ascribed the increased charge transfer only to carbides of elements of group IV and who attributed the supposed phenomenon to the higher covalence of these phases.

The reduction in the "electropositiveness" of the carbon atom with increase in Z of its partner is accompanied (except in the case of W_2C) by a rise in the strength characteristics of the carbides. This is indicative of the negative part played by the polarization of the covalent bond (in the general balance of the covalent—metallic bond) in regard to the strength of carbides, and is in good agreement with the general conclusions of solid-state theory about the weakening effect of ionic interactions on strength. This can also be confirmed by a study of the energy bands in transition-metal nitrides [11].

The high probability of a transfer of the effective charge from carbon to the transition-metal atom that follows from an analysis of the spectra and theoretical calculations for TiC [12] is in good agreement with the hypothesis, first advanced by Umanskii [12] and further developed by Samsonov [13], about the partial transfer of valence electrons from carbon to the 3d shell of the metal with partial filling of the defects of this band. According to the data in [9] and the results of our own investigations [3, 5], the number of 3d electrons in TiC increases from two (for the free atom) to four, taking into account the contribution of the intrinsic 4sp states of titanium. This balance is the result of a complex interaction between the electrons, which must be considered within the framework of the covalent-polarized hybrid orbitals.

The difference in shape of the CK_α band in the carbides is an outstanding feature. It differs most simply in carbides of the transition metals of group IV, where it is characterized by a small width and complete symmetry. Some widening of the band occurs with increase in atomic number of the transition metal. The simple and regular dispersion shape of the CK_α band in the carbides of the d metals of group IV is evidently associated with the almost completely filled first Brillouin zone, corresponding to approximately eight valence electrons. An increase in the electron concentration in the "quasimolecule" MeC to nine electrons, which is characteristic of carbides of d-metals of group V, causes the following energy band to be filled by the ninth, "excess" electron. This manifests itself as a complication of the shape of the CK_α band in all carbides of metals of group V without exception: a small excess, attributed by Coulson [13] to metallized π electrons, appears in the short-wave region of the CK_α band.

The appearance of a tenth valence electron in the carbides of the d-metals of group VI still further complicates the band structure; this is brought out particularly clearly in the spectra of Cr_3C_2 and W_2C.

A comparison of the CK_α band with the shape of the corresponding K and L spectra of the carbide-forming d-metal reveals little similarity. Similarities between these bands were to be expected on the assumption that the chemical bond has a predominantly metallic character in carbides of d-metals [12, 13]. The dissimilarity between the bands concerned points rather to a covalent−ionic type of bond and to a closer approximation of the nature of the hybridization in the carbides (d^2sp^3, d^4s) to that for diamond

(sp^3) than to the graphite sp^2 hybridization. It appears to us to be unjustified to attribute to the d-metal atom one type of stable electron configuration of the hybrid orbitals, and another type to the carbon atom in the same compound. Obviously there can be only a single type of hybridization common to the two shells which produces a stable crystal lattice in the given carbide.

Thus the state of the carbon atom in the various carbides of refractory transition metals, in the presence of many common features, reveals a complex dependence on the energy parameters of the electron levels of the d-metals, mainly the valence d levels, that interact with it. The relationship of the shape and position of the spectra to the crystal chemistry of the carbide phases, pointed out by Holliday, plays a subordinate part. The shift in the short-wave direction of the CK_α band, which varies in magnitude in the carbides of the d-metals of groups IV, V, and VI, provides grounds for concluding that electrons of the metalloid make the "main contribution" to the strength of metal-like carbides.

Literature Cited

1. D. Fisher and W. Baun, Ann. Chem., 37:902 (1965); J. Chem. Phys., 43:2075 (1965).
2. I. I. Holliday, Röntgenspectren und chemische Bindung, Leipzig (1966).
3. E. A. Zhurakovskii, Dokl. Akad. Nauk SSSR, Vol. 183, No. 2 (1968).
4. E. A. Zhurakovskii, Dokl. Akad. Nauk SSSR, Vol. 185, No. 4 (1968).
5. E. A. Zhurakovskii, this volume, p. 185.
6. A. P. Lukirskii and I. A. Brytov, Optika i Spektroskopiya, No. 12 (1966).
7. G. V. Samsonov and Ya. S. Umanskii, Hard Compounds of Refractory Metals [in Russian], Metallurgizdat, Moscow (1958).
8. A. T. Shuvaev, Author's Summary of Candidate's Thesis, Rostov-on-Don (1964).
9. R. C. Lye and E. M. Logothetis, Phys. Rev., Vol. 147, No. 2 (1966).
10. V. Erne and A. S. Switendick, Phys. Rev., 137:6A (1965).
11. E. A. Zhurakovskii and A. N. Vasilenko, Dokl. Akad. Nauk SSSR, Vol. 187, No. 2 (1965).
12. Ya. S. Umanskii, Carbides in Hard Alloys [in Russian], Metallurgizdat, Moscow (1947).
13. G. V. Samsonov, Izv. Sekt. Fiz. Khim. Anal., 29:97 (1956).
14. I. N. Frantsevich, E. A. Zhurakovskii, and A. B. Lyashchenko, Izv. Akad. Nauk SSSR, Neorg. Mat., No. 1 (1967).
15. C. A. Coulson and P. Taylor, Proc. Phys. Soc., 65(10):42A (1952).

The Energy Structures of Titanium Carbide

E. A. Zhurakovskii

The spectra of titanium and carbon in the homogeneity range of titanium carbide ($TiC_{0.54-1.0}$) were measured in small composition stages, under high resolution conditions. The carbide specimens were synthesized from high-purity raw materials under conditions conducive to the presence of a minimum amount of dissolved gases and free carbon in them. The combination of the K and L spectra of titanium, which are responsible for the p and d states of the metal, with the K_α band of carbon in the homogeneous carbides, which reflects mainly the 2p electron states of the carbon atom, enabled a number of considerations to be advanced in regard to the structure of the valence band of the carbide crystal and to the effect of carbon concentration on the energy spectrum.

An important role in our understanding of the electron structure of TiC_x has been played by theoretical calculations made by the method of combined plane waves [14, 16]. The results of these calculations, although somewhat conflicting, have enabled us to define with considerably more accuracy and to broaden our concepts regarding the structure of the energy spectrum of the electrons in the valence band of this carbide, and also of the nitride and monoxide that are isomorphous with it, which were previously obtained during an investigation of the emission and absorption K spectra [4, 2].

As a result of advances made by Russian instrument makers in the ultrasoft x-ray region of the spectrum, it has proved possible to obtain additional information on the details of the electron structure of TiC_x from the x-ray L spectra of titanium and also from the K_α spectra of the second component — carbon in TiC_x in which a phase with an fcc lattice of the NaCl type exists (x = 0.55-1.0), it has been possible to photograph practically all the x-ray spectra (except the absorption spectra of carbon).

In the present research an attempt has been made to construct a single energy diagram of this compound. It is important to note that the previous [4] and the present investigations were carried out on the same samples of TiC_x, in small composition stages, prepared from the purest raw materials [8]. The investigation was made on an RSM-500 ultralong-wave x-ray spectrometer, using the methods developed by A. P. Lukirskii [7]. The diffraction grating used was cut on glass which had previously been polished to a concave spherical surface of 6-m curvature, and it had 600 parallel lines per millimeter. To increase the coefficient of reflection, the diffraction grating was coated with a thin layer of gold (300 Å). The radius of the focusing mirror, which was used for high-order reflections, was 1 m, and the opening of the receiving slit was 15 μ. The entrance window of the proportional flow counter — the detector of x-radiation — was prepared from two or three layers of a very thin film made of cellulose amyl acetate and deposited in optical contact. Methylal was used as the gaseous atmosphere in the counter. The voltage on the anode was 1.5-3 kV and that on the counter was 1.1-1.4 kV; the copper anode was water cooled. The samples investigated were rubbed into the grooved surface of the anode with an agate rod. Traps containing liquid nitrogen were used to catch the oil vapors. A circuit for recording and counting the impulses was mounted on the base of the apparatus with a "Siren" preamplifier for the automatic recording of the signals on an EPP-09 recorder.

The conditions under which the K-emission and absorption spectra of titanium were obtained have been described in [4]; the K_β group of emission lines was then repeated in the second-order reflection from a quartz crystal. The results of the spectrum measurements are given in Table 1, while curves showing the variation in intensity and absorption coefficient are presented in Fig. 1. It will be seen that the shape and position of the "outermost" line of the x-ray emission spectrum of titanium in the carbides and also the fine structure of the long-wave absorption band do not alter with change in carbon concentration. There is a long-

TABLE 1. Relative Positions of the Maxima of the
K Emission and Absorption Spectra of Titanium
in the Metal and in Homogeneous Carbides

Substance	K_{β_s}	$K_{\beta}^{"}$	$\dfrac{I_{K_{\beta"}}}{I_{K_{\beta_s}}}$	Index of asym-metry	Begin-ning of edge	Absorption	
						First maximum	Principal maximum
Ti_{Me}	0.0	—	—	2.0	0.0	6.7	17.8
$TiC_{0.54}$	—1.6	—7.0	0.43	1.25	3.7	10.5	24.2
$TiC_{0.65}$	—1.6	—7.0	0.45	1.22	3.7	10.4	24.2
$TiC_{0.88}$	—1.6	—7.1	0.48	1.21	3.7	10.4	24.1
$TiC_{0.94}$	—1.6	—7.1	0.48	1.19	3.6	10.5	24.3
$TiC_{1.0}$	—1.6	—7.1	0.50	1.17	3.7	10.6	24.2

wave displacement of the K_{β_5} line in the carbides as compared
with the spectrum of the pure metal. The origin of this line is
usually associated with transfers of electrons from the hybridized
3d4sp energy band to the 1s level nearest to the nucleus. Calculation shows that in TiN there is an overlap and in TiC a complete combination of the levels of the 3d band of titanium with the
levels of the 2p band of the metalloid [14]; according to the same

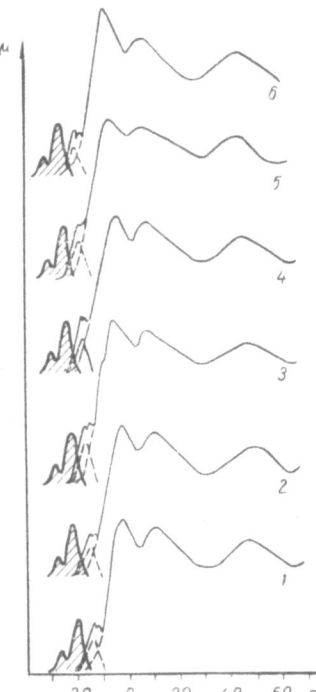

Fig. 1. K_β group of lines in the x-ray
emission spectrum and the K absorption
edge of titanium in the homogeneous
carbides: 1) $TiC_{0.54}$, 2) $TiC_{0.65}$, 3)
$TiC_{0.88}$, 4) $TiC_{0.94}$, 5, 6) $TiC_{1.0}$.

results, overlapping of the 3d and 4s bands in TiC and TiN is small, while the 4p states exist in the valence band with a very small statistical weight. As the 3d → 1s transitions are forbidden by the selection rules for the free atom and are quadruple in the spectra and weak, it is necessary, in order to explain the great brightness of the K_{β_5} band, to assume an overlap of the 3d wave functions with the p functions, which makes possible strong dipole dp → s transitions. In titanium compounds only the 2p states of the electrons of the nonmetal and in part the 4p states of titanium can possess these functions. Thus, the K_{β_5} band cannot, in principle, provide information about the "pure" d states of the electrons, which are the most interesting from the point of view of the theory of transition metals and their refractory compounds. The maximum intensity of the K_{β_5} band probably coincides with the maximum density of states in the d band and increases only as a result of the higher (by 15-20 times according to [5]) density of states in the d energy band as compared with the states of p symmetry, for which the density curve has a gentle slope. The integrated intensity, shape, and width of the K_{β_5} band of titanium in the carbide will therefore be determined by the degree of overlap of the 3d states of titanium with the 2p states of carbon and by the population of the 4p states of titanium. To make a more definite judgment from the spectra regarding the "contribution" of these states to the intensity of the K_{β_5} band ($3d_{Ti}$ or $2p_C$) is more difficult, as the selection rule for the free atom is of only limited applicability to a crystal in which dsp-hybridization of the outer energy levels undoubtedly exists. This rule is even less strictly applicable to "crossover" transitions of the $2p_C \rightarrow 1s_{Ti}$ type.

Besides the principal maximum it is possible to distinguish in the K_β spectrum a very strong long-wave maximum identified as the K_β^n satellite. This satellite has hitherto been interpreted as the result of "crossover" transitions from the valence energy levels of the second component to the K level of the metal. Calculations have shown [14, 16] that this region of the spectrum should contain a band of 2s energy levels of carbon. The energy values calculated in [14] agree satisfactorily with the experimental ones given in [2, 4]. However, the calculations in [14] failed to take into account the d_γ and d_ε subgroups of electrons in the 3d band, although the polar diagrams of their wave vectors in momentum space are completely different. Later calculations [16, 17], which took account of these electrons, confirm the possible continuance of the levels populated with d_ε electrons in the energy range lying considerably above the K_{β_5} line. Correspondingly, the electrons of d_γ symmetry, the direction of whose wave functions in K space is close to that for the p wave functions of the nonmetal, are involved as a result of the marked overlapping of their d_z^2 functions with the p functions in the localized Me−C covalent bonds, and in the x-ray spectrum they appear, with the 2p states of carbon, predominantly in the most intense maximum of the K_{β_5} band. The results of these calculations and the fine structure of the K_β emission spectrum observed experimentally are in satisfactory agreement with the additional information about the band structure of the valence band of TiC provided by the $L_{II,III}$ spectra.

In the L spectra, the electron transitions $3d \to 2p_{1/2,3/2}$ are dipolar and strong, whereas the "crossover" transition from the 2p level of the metalloid in the sphere of the titanium atoms to its $2p_{1/2,3/2}$ level is either forbidden or is possible to only a limited extent owing to the dsp-hybridization and it is obviously very weak. The $4s \to 2p$ transitions are quadrupole in the L spectra. Of the four lines present in the L spectrum of titanium, we chose for investigation those lines which are associated with the electron transitions from the high-energy levels (M levels), viz., the $L_{\alpha_{1,2}}$ and L_{β_1} lines or the so-called $L_{II,III}$ band. Figure 2 shows these spectra for a series of homogeneous titanium carbides with the same gradations of chemical composition as for the K spectra in Fig. 1. The results of the measurements made on these spectra are given in Table 2.

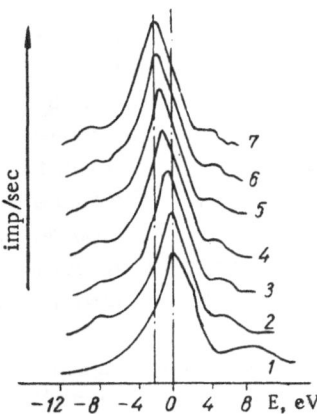

Fig. 2. X-ray emission $L_{II,III}$ spectra of titanium in the metal (1) and homogeneous carbides containing: 2) 12%, 3) 14%, 4) 16%, 5) 19%, 6) 22%, 7) 23% C.

TABLE 2. Relative Positions of the Intensity
Maxima of the Fine Structure of the $L_{II,III}$
Band of Titanium in the Metal
and in Homogeneous Carbides, eV

Substance	Position of L_{III} maximum	Position of L_{II} maximum	Position of additional long-wave maximum	$\dfrac{I_{L_{II}}}{I_{L_{III}}}$	$\dfrac{I_{L_{long}}}{I_{L_{III}}}$
Ti_{Me}	0.0	+7.5	—	0.26	—
$TiC_{0.54}$	—0.4	+6.2	—6.0	0.20	0.08
$TiC_{0.65}$	—0.6	+6.4	—5.9	0.22	0.1
$TiC_{0.84}$	—0.8	+6.0	—5.8	0.24	0.12
$TiC_{0.84}$	—1.3	+6.0	—5.9	0.18	0.12
$TiC_{\sim 1.0}$	—1.7	+5.8	—6.1	0.19	0.15
$TiC_{\sim 1.0}$	—2.0	+5.5	—6.0	0.17	0.19

As can be seen by comparing the spectrum of titanium in the carbide with that of metallic titanium, the intensity, shape, and position of the $L_{II,III}$ bands undergo substantial changes in the carbide. As has been indicated, in the case of the free atom two subbands are formed by electron transitions from the outer 3d energy band of titanium to two sublevels of p symmetry of the same atom with values of the internal quantum number J = 1/2 and J = 3/2. In a TiC crystal, as a result of the intermingling of the outer wave functions of titanium and carbon, the following more complex scheme of electron transitions should evidently be fulfilled:

$$2p_C + 3d_{Ti} + 4s_{Ti} \rightarrow 2p_{Ti}, \ J = {}^1/_2, {}^3/_2.$$

As can be seen from Fig. 2, at the same time the shape of the Ti spectra in the carbide becomes more complex, two comparatively weak intensity maxima appear at the periphery of the L band in its long- and short-wave regions. An increase in the intensity of the long-wave maximum with rise in the carbon concentration in the carbide can also be established. As the long-wave maximum has the least energy and intensity of the three subbands of the $L_{II,III}$ band and only it exhibits a concentration dependence on the carbon content, it is natural to relate its origin to transitions of 2s carbon electrons to the 2p level in titanium. According to the calculations in [14], the energy of $2s_C$ is higher than that of $2p_{Ti}$. In a similar way, the short-wave maximum can be regarded as being due to the presence in the hybridized dsp–band of the crystal of another group of levels populated mainly by 4s-electrons from titanium and in part by electrons of d_ε symmetry. Here no concentration dependence on the carbon content was found, such as occurred in the previous case; in the main its intensity increases only at the transition from metal to carbide. This is accompanied by a displacement of the short-wave maximum by 1.5-2.0 eV in the long-wave direction. Thus, the origin of the short-wave maximum in the $L_{\Pi,\Pi\Pi}$ band can be related, in contrast to the principal maximum, to the following scheme of transitions: 4s, $3d_{xy,xz,yz}-2p$, $j = 1/2, 3/2$. The population of this group of levels clearly accounts for the (Ti—Ti) bond and the electrical conductivity of TiC. In contrast, the levels forming the central part of the L band, identified as the $L_{\alpha_{1,2}}$ lines, are populated, as in the case of the K spectra, by those of the electrons which are relatively strongly bonded to the titanium atom (the Ti—C bond) and which participate in the covalent bonds with the carbon atoms (d_γ or $d_{y^2,\ x-y^2}$). The large amount of overlap between the d_γ orbitals of titanium and the p orbitals of carbon, resulting from the coincidence of the electron-density maxima of these and others with the coordinate axes of the octahedron which may represent the unit cell of TiC, accounts for the high-strength characteristics and refractoriness of TiC.

Thus, a comparison of the fine structures of the K and L spectra of titanium in the carbides reveals (bearing in mind the difference in probability of the transitions) very similar distribution patterns of the electron density in the valence band and the conduction band of this compound. Both spectra indicate a three-band structure within the boundaries of the two bands mentioned, viz., a low-energy subband with a preponderance of 2s carbon electrons, a central band ($2p_C + 3d_\gamma 4p_{Ti}$), and a high-energy band $(4s + 3d_\varepsilon)_{Ti}$. Measurements of the position of the principal maximum of $L_{\Pi,\Pi\Pi}$ band given in Table 2 shows that the maximum is progressively shifted in the long-wave direction as the carbon content of the carbide increases. As in the case of K_{β_5}, a shift of 1.5 eV in the long-wave direction takes place in the carbide, as compared with the position of the maximum in the spectrum of pure titanium. According to Shuvaev's calculations [10], this is evidence of an increase in the electron density in the part of the sphere of the titanium atom occupied by the 3d levels, although the effect of the internal screening of the 3d shell is different in the

two cases. A calculation in [16] has shown, however, that the vacancies in the 3d shell of titanium in TiC are filled not only by
electrons from the carbon atom but also that 0.75 of an electron is
transferred there from its own 4s band. On the whole the extent
of the electron charge transferred by a carbon atom to the sphere
of an atom of titanium is 1.25 electrons per atom. The results
of the measurements of the position of the principal maximum in
the L band given in Table 2 show that it is progressively displaced
in the long-wave direction as the carbon content of the carbide
increases, the greatest displacement being attained in the carbide
of stoichiometric composition. It is quite legitimate to interpret
this as a progressive transfer of 2p electrons from carbon to the
sphere of the titanium atom, where there is a very great probability of their being captured by the 3d shell, as being more favorable in terms of energy. On the other hand, the energy of the K_{β_5}
band remains practically unchanged during the filling of the vacancies in TiC_x; this is no doubt due to the different roles played
by the d and p levels of the titanium atom in the mechanism of
charge transfer. Actually, the main electron exchange between
titanium and carbon takes place between the d_{Ti} and the $2p_C$ states,
leaving out the 4p band, the maximum electron density in which
evidently also determines the maximum in the K_{β_5} band. According to the calculations in [16], the electron configuration for TiC
can be approximately represented in the following way:

$$\text{Titanium} \quad (3d_\gamma)^{9/4} (3d_g)^{7/4} (4s)^{3/4} (4p)^{1/2}$$
$$\text{Carbon} \quad (2s)^2 (2p)^{3/4}$$

These results are in excellent correlation with the x-ray spectrographic pattern for the sphere of the titanium atom. The authors of [16-18] also report a transfer of electrons from the carbon atom to the titanium atom (although there are other opinions
too [1, 29, 30]). K_α bands of carbon in homogeneous titanium carbides, graphite, and diamond, averaged from three independent re-
cords, are shown in Fig. 3. The results of measuring these spectra are

Graphite	$TiC_{0.55}$	$TiC_{0.65}$	$TiC_{0.88}$	$TiC_{0.94}$	$TiC_{1.0}$	Diamond
0	1.83	1.82	1.84	1.83	1.84	2.35

According to theory, the CK_α band is the result of the transfer of

2p carbon electrons to vacancies in the 1s shell nearest to the nucleus which are formed during excitation of the x-ray spectrum. Investigations of the CK_α band in graphite and diamond have shown that the position of its maximum is 276 eV for graphite and 278.3 eV for diamond. Thus, on the changeover from the graphite-like state to diamond, this band is displaced by 2.3 eV in the short-wave direction (the discrepancy between our results and those given in the literature lies within the limits of experimental error of 0.1 eV).

Graphite is known to possess a laminar structure with hexagonal packing of the atoms in layers, due to the sp^2 type of hybridization of the wave functions of the electrons. The length of the C—C bond in the hexagonal layers is 1.44 Å, and the distance between the layers is 3.35 Å. In graphite the CK_α band is interpreted by Coulson as the result of the superimposition of a multitude of electron levels, populated by π and σ electrons, which participate in the unlocalized quasimetallic bonds that are "spread out" along the plane of the layer (π electrons) and in the localized, directional bonds of the covalent type (σ electrons). The first are re-

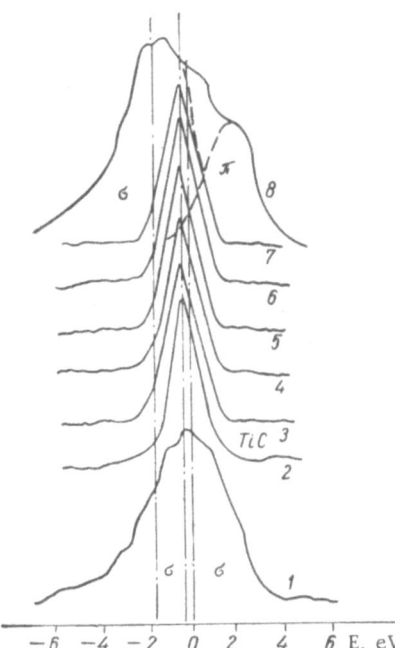

Fig. 3. The K_α emission band of carbon in diamond (1), graphite (8), and homogeneous carbides (key as for Fig. 2).

sponsible for the electrical conductivity of graphite, and the second for its high melting point (approximately 3800°C). The structure of diamond is a regular space-localized tetrahedron produced by sp^3 hybridization of the wave functions.

In the transformation from graphite to diamond, there is established between the plane "aromatic" networks of graphite layers an additional "transverse" bond of p electrons, which is now described not by the plane sp^2 but by a space-saturated hybridization of the wave functions. This evidently occurs as a result of the involvement of all the graphite electrons in the localized covalent σ bonds. There should be a complete absence of π bonds in diamond (in conformity with its dielectric properties).

It follows from a consideration of Fig. 3, in which, following Coulson, the π and σ subbands in graphite are separated (by a broken line), that the π bond lies above the σ band in terms of energy. Thus, the levels occupied by electrons that take part in the π bonds should represent the short-wave part of the CK_α band and should have their own intensity maximum. The long-wave part of the band, which is considerably broader, then represents the levels populated with electrons that take part mainly in the σ bonds. The theoretically calculated curve obtained by Coulson on the assumption that the only participants in the formation of the CK_α band are the π and σ functions of electrons having three 2p and one 2s states, is in good agreement with the shape of the band observed experimentally by Chalklin [12] and in our own research. Thus, for example, the calculated width of the π subband is 5 eV and that of the σ subband about 15 eV; according to Chalkin, the experimental width of the CK_α subband is 18 eV for graphite and 13 eV for diamond (according to our results, the width is 17.6 eV for graphite and 12.5 eV for diamond). The total width of the CK_α band is much less in the carbides.

A consideration of the shape and position of the CK_α band in the homogeneous titanium carbides indicates, first, that they display no clear dependence of the parameters on the concentration of carbon in the titanium carbide, and secondly that they are exactly similar to one another and occupy a certain intermediate position between graphite and diamond as regards energy. The shape of the CK_α band in the homogeneous carbides is much closer to that of diamond than to that of graphite. It is not so much a band as a narrow line of almost exact dispersion shape and lacking a fine structure. The width of the CK_α band in TiC is considerably smaller than that in diamond and even more so in the case of graphite. A certain broadening of the band takes place with increase in the carbon concentration (although it is so small that it might be due to distortions of the apparatus).

The smoothing out of the short-wave maximum in diamond as compared with that in graphite and its complete absence in the homogeneous carbides lead to the conclusion, by analogy with diamond, that the carbon electrons participate only in space-saturated hyyrid bonds of the covalent type; although these are no longer represented by the sp^3 electron configuration but by the d^2sp^3 configuration [15], which should produce stability of the carbide lattice. The π bonds originating in the carbon of the carbide are evidently considerably weakened, just as are the $C-C$ bonds. The comparatively high electrical conductivity, which is retained in TiC, is probably due to the 4s electrons, the degree of hybridization in which is not large according to [14]. The position of the CK_α carbon band in the carbides is in good agreement with the results of quantum-mechanical calculations. As has been mentioned, it occupies an intermediate position and is displaced by 1.8 eV in the short-wave direction relative to graphite. Correspondingly, its long-wave displacement relative to diamond is 0.5 eV, which supports the conclusion reached with regard to the transfer of 1.25 electrons from the 2p states of carbon to the sphere of the titanium atom. Actually, according to the conclusions in [10], the short-wave displacement of the CK_α band relative to its position in graphite can be attributed to just this cause. Bearing in mind that the displacement by 1.8 eV is very close to that in diamond (2.3 eV) we may assume a "diamond-like" state of the carbon atom in the carbides.

The degree of covalence of the bond in TiC is less than in diamond, as is evidenced by the intermediate position of the CK_α band. However, the covalent component of the bond should be considerably more prominent than the metallic component. Although the size of the electron charge transferred from the carbon atom to the sphere of the titanium atom is possibly not so large as the results of our measurements and of the quantum-mechanical calculations in [16] indicate, there are nevertheless, bearing in mind the small interatomic distances in TiC, grounds for assuming the existence of a considerable Coulomb interaction in TiC. However, the ionic bond in TiC appears not as a purely Coulomb interaction induced, like that in NaCl, by an irreversible transfer of charge from one atom to another and by the resulting interaction between them, but as a definite degree of polarization of the covalent bond. In our simplified concept it is based on the inequality of dwell times of the covalently bonding electrons, represented by the dsp-hybrid wave function of the bond, in the sphere of one atom and the other. However, the true charge distribution in TiC is evidently more complex. A combination of the measured x-ray spectra of titanium and carbon in their pure form and in TiC of stoichiometric composition makes it possible to determine the position of the bottom of the valence band and the Fermi surface in relation to the k level. The energy of the k level and that of the bottom of the band relative to that of the unexcited crystal (which was adopted as the energy zero) were determined by taking account of the work function. According to [32], the work functions for titanium and carbon are 4.4 and 4.17 eV, respectively. These data are reproduced on the proposed energy diagram for TiC (Fig. 4). The bottom of the conduction band in this diagram was determined relative to the vacuum potential, i.e., to the electron affinity. According to the results of various researches, this characteristic is for carbon 1.12 ± 0.6 [34], 2.1 ± 0.9 [33], and 1.24 eV [35]. The most reliable value is probably 1.2 eV. There are no corresponding data available in the literature for titanium, the closest "transition" metal for which this value has been determined being copper (2.4

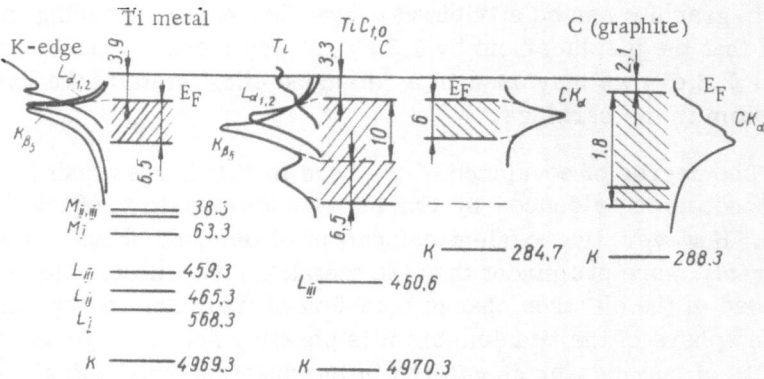

Fig. 4. Energy diagram for TiC.

eV). Hence the electron affinity for titanium can be roughly estimated as 2 eV. In constructing the proposed energy diagram we found that the energy structures in the vicinities of the titanium and carbon atoms in TiC differ among themselves considerably more than they do in the original pure elements. It is significant that the band of valence electrons close to the titanium atom is approximately three times narrower than it is in graphite; this may be due to the greater density of the d states as compared with the s and p states [5]. The difference in shape and extent of the valence band near each of the atoms may also be due to the complex superposition of the metallic, covalent, and ionic bonds, which, according to the results of numerous researches [8, 9, 11, 14–30], occurs in TiC_x.

Literature Cited

1. M. P. Arbuzov and B. V. Khaenko, Poroshkovaya Met., No. 8 (1968).
2. M. A. Blokhin and A. T. Shuvaev, Izv. Akad. Nauk SSSR, Fiz., 26:429 (1962).
3. E. E. Vainshtein, X-Ray Spectra of Atoms in Molecules of Chemical Compounds and Alloys [in Russian], Nauka, Moscow and Leningrad (1966).
4. E. A. Zhurakovskii and E. E. Vainshtein, Dokl. Akad. Nauk SSSR, Vol. 122, No. 3 (1958); Vol. 127, No. 3 (1959); Vol. 129, No. 6 (1959); Vol. 140, No. 3 (1961).
5. Yu. P. Irkhin, Fiz. Metal. i Metalloved., 11:10 (1961).
6. A. P. Lukirskii and I. A. Brytov, Optika i Spektroskopiya, No. 12 (1966).
7. A. P. Lukirskii, Author's Summary of Doctor's Thesis, LGU, Leningrad (1963).
8. G. V. Samsonov and Ya. S. Umanskii, Hard Compounds of Refractory Metals [in Russian], Metallurgizdat, Moscow (1958).
9. Ya. S. Umanskii, Carbides in Hard Alloys [in Russian], Metallurgizdat, Moscow (1947).
10. A. T. Shuvaev, Author's Summary of Candidate's Thesis, RGU, Rostov-on-Don (1964).
11. H. Biltz, Z. Physik, 153:338 (1958).
12. F. G. Chalklin, Proc. Roy. Soc., 194:1036 (1948).
13. C. A. Coulson and P. Taylor, Proc. Phys. Soc., 65(10):42A (1952).
14. V. Erne and A. C. Switendick, Phys. Rev., 137:6A (1965).
15. C. T. Kimball, J. Chem. Phys., 8:188 (1940).
16. R. C. Lye and E. M. Logothetis, Phys. Rev., 147:2 (1966).
17. P. Costa and R. R. Conte, International Symposium on Compounds of Interest in Nuclear Reactor Technology (1964), pp. 3-27.
18. D. A. Robins, Powder Metallurgy, 2:172 (1958).
19. R. Kiessling, Powder Metallurgy, 3:177 (1959).
20. E. Dempsey, Phil. Mag., 8:285 (1963).
21. N. Engel, Trans. Am. Soc. Metals, 57:610 (1964).

22. Yu. N. Surovoi, L. A. Shvartsman, and V. I. Alekseev, Fiz. Metal. Metalloved.,
     20:80 (1965).
23. W. H. Phillipp, NASA, TN D-3533 (1966).
24. W. S. Williams, Science, 152:34 (1966).
25. E. K. Storms, The Refractory Carbides, Academic Press, New York and London
     (1967).
26. H. J. Goldschmidt, Interstitial Alloys, Butterworths, London (1967).
27. J. E. Holliday, J. Appl. Phys., 38:4720 (1967).
28. D. W. Fisher and W. L. Baun, J. Appl. Phys., 39:4757 (1968).
29. S. P. Denker, J. Less-Common Metals, 14:4 (1968).
30. L. Ramquist, J. Anal., Vol. 153 (1969).
31. L. N. Dobretsov, Electronic and Ionic Emission [in Russian], Moscow and Lenin-
     grad (1952).
32. N. S. Burel'nikova, Uspekhi Fiz. Nauk, 65:351 (1958).
33. L. N. Branscomb and D. S. Burch, Phys. Rev., 111:504 (1958).
34. B. Edlen, J. Chem. Phys., 33:98 (1960).
35. V. I. Vedeneev et al., Ionization Potentials and Electron Affinity [in Russian],
     Izd. Akad. Nauk SSSR, Moscow (1962).

Soft X-Ray Spectra of Vanadium Carbides
in the Homogeneity Range

E. A. Zhurakovskii, V. P. Dzeganovskii,
and T. N. Bondarenko

The L spectra of vanadium and the K_α band of carbon in the homogeneity ranges of the vanadium carbides V_2C and VC have been measured on an RSM-500 ultralong-wave spectrometer. The results are compared with previously obtained emission and absorption K spectra of vanadium in carbides, and the results of quantum-mechanical calculations on isomorphous titanium carbides are included. The existence of a three-band energy structure of the valence band is shown, reflecting the metallic—covalent—ionic nature of the chemical bond in the carbides. Conclusions are drawn regarding the direction of charge transfer, and an energy diagram is proposed for $VC_{0.688}$.

The physical nature of the nonstoichiometric compounds of variable composition that transition metals form with hydrogen, carbon, nitrogen, and oxygen has not yet been adequately elucidated. A sharp change has been discovered in various properties of these compounds in the homogeneity range [1-4], and these changes have been related to the electron structure [5]; the specific role played by the metalloid vacancies in the shaping of the N(E) curve of the density of states of the valence band has been revealed [2, 5, 6], and the inadequacy of the energy state of the transition-metal atoms in the lattice has been demonstrated [6].

The fullest information about the structure of the energy spectrum of the electrons in a solid is given by quantum-mechanical

calculations and x-ray spectra. X-ray spectral studies of the vanadium carbides V_2C and VC [7, 8] based on the emission and absorption K spectra of vanadium have revealed a clear dependence
of the spectra on the concentration of carbon. However, the K
spectra of the metal do not completely reflect the structure of the
valence bond and provide only indirect information about the behavior of the wave functions of the d symmetry of vanadium and the
sp symmetry of carbon [9], which are just the states responsible
for the formation of the high bonding forces in the carbides [10, 11].
Consequently we have measured directly the reflecting d states of
the $L_{II,III}$ spectrum of vanadium and of the K spectrum of carbon
(the sp state) in a series of homogeneous carbides: $VC_{0.39}$, $VC_{0.50}$,
$VC_{0.76}$, $VC_{0.85}$, $VC_{0.88}$. An energy diagram has been constructed
from the results obtained and from data available in the literature.
The methods of preparing the carbides, studying the spectra, and
constructing the diagram have been described in [12-14], and here
we give only the particular conditions under which the carbon spectra were photographed with oil-vapor evacuation on the RSM-500
apparatus [14].

In spite of the presence of two freezing nitrogen traps at the
entries to the x-ray camera and additional cooling of the body of
the tube by liquid nitrogen, it proved impossible completely to
protect the target from oil vapor falling on it. In order to assess
the distortions that might be introduced into the shape of the carbon K_α line in the carbide by oil vapors condensed in a thin layer
on the anode, we investigated the effect of the degree of evacuation
and exposure time on the intensity of the carbon K_α line from the
oil. It emerged that during the first 20 minutes' recording in a
vacuum of $1-5 \times 10^{-6}$ mm Hg the CK_α line from the oil does not
exceed 5% of the maximum intensity of the CK_α line from the carbide and does not in any way distort the latter (Fig. 1). With more
prolonged periods of recording its effect on the shape of the CK_α
band in the carbide could no longer be neglected. Hence only one
carbide of a particular composition was rubbed into the grooved
surfaces of the copper anode, and the first recording lasted 20
min. The reproducibility of the shape of the CK_α line in a given
carbide was checked by recording this line for another sample of
the same carbide, freshly rubbed into the anode. No change in the
chemical composition of the specimen owing to carburization by
the oil vapors could take place, since in none of the experiments

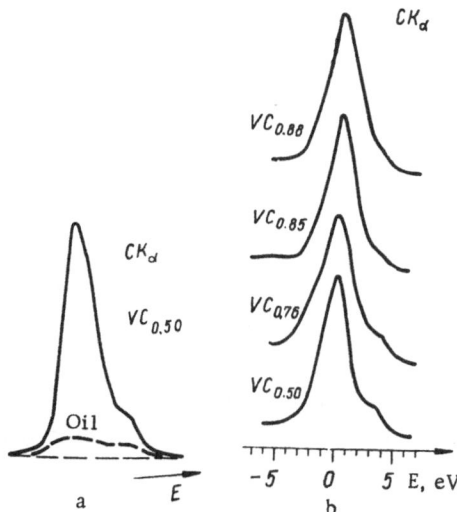

Fig. 1. Experimental method (a); CK$_\alpha$ band in the homogeneity range of the vanadium carbides VC$_{0.50-0.88}$ (b).

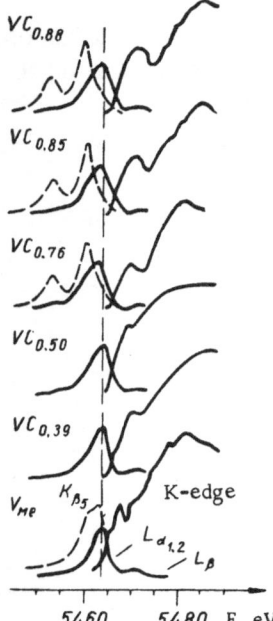

Fig. 2. Combination of K and L emission spectra and K absorption edges in the homogeneous vanadium carbides V$_2$C and VC.

did the target temperature exceed 800°C, whereas the carbide-formation range in vanadium carbides lies above 1400°C. The intensity of the oxygen K_α line in VC_x was found to be very small. X-ray phase analysis of the carbide samples made after the spectra had been recorded also confirmed that there was no change in the chemical composition of the carbides and that no impurity phases were present.

Figure 2 shows the K_α bands of carbon in $VC_{0.50-0.88}$ and the $L_{II,III}$ spectra of vanadium in pure vanadium and in the homogeneity ranges, combined with the K_α emission [7, 8] and absorption spectra. The combination was made through the photon energy of the K_α line [13]. The results of the measurements are given in Table 1 and depicted in the energy diagram shown in Fig. 3. The individual bands of the K and L spectra have already been interpreted (see this volume, previous article).

Consideration of the results obtained shows that the principal $L_{\alpha_{1,2}}$ band in the carbides is displaced by 1 eV in the long-wave direction relative to the spectrum of pure vanadium, the displacement increasing with the carbon concentration. According to [7, 8], the K_{β_5} band is also displaced by 2.5-3.0 eV in the long-wave direction on the formation of the carbide, and in this case too the displacement increases with the carbon concentration. In contrast to the $L_{\alpha_{1,2}}$ band, the shape of which is constant, on the short-

TABLE 1. Principal Parameters of X-Ray Spectra

Substance	Displacement, eV			Width, eV	
	K_{β_5} (relative to the metal)	$L_{\alpha_{1,2}}$	CK_α (relative to graphite)	$L_{\alpha_{1,2}}$	CK_α
Vanadium	0	0	—	3.2	—
$VC_{0.39}$	—	0	—	4.4	—
$VC_{0.50}$	−2.5	0	+1.3	5.2	2.8
$VC_{0.76}$	—	−0.8	—	5.7	3.5
$VC_{0.85}$	—	−0.8	—	5.8	3.0
$VC_{0.88}$	−3.0	−1.0	+1.6	5.9	3.1
				±0.1	±0.2

Note: The L spectra of vanadium and the K bands of carbon were obtained with a resolution of 0.3-0.4 eV.

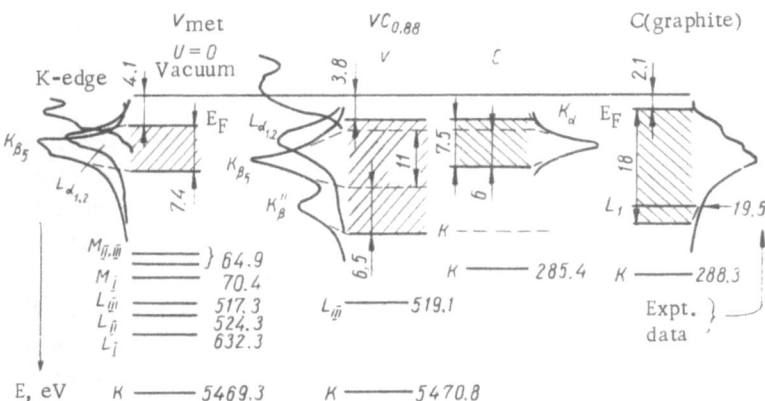

Fig. 3. Proposed energy diagram for VC (the principle of construction is the same as for TiC).

wave side of the K_{β_5} band the excess separates and increases with reduction in the carbon content [8]. With rise in its intensity, that of the principal K_{β_5} band and its satellite K_{β}'' diminishes. The CK_{α} bands in V_2C and VC are displaced relative to the corresponding band in the graphite spectrum by 1.3 and 1.8 eV (with an error in the determination of 0.3 eV). On the short-wave side of the CK_{α} band the excess separates and increases with reduction in the carbon content, just like the K_{β_5} band in vanadium. At the composition $VC_{0.88}$, which is closest to the stoichiometric composition, the short-wave excesses at the foot of the CK_{α} and K_{β_5} lines become barely perceptible, so providing grounds for the following provisional conclusions:

1. The character of the emission bands of vanadium and carbon, their relative positions, and their change with carbon concentration leads us to assume the existence of a three-band structure of the energy spectrum in vanadium carbides. The first of these bands (I), occupying the central position in the spectrum, contains the energy states of the electrons localized in the covalent Me$-$C bonds, and corresponds to the region of marked overlapping of the $3d_{\gamma}$ orbitals of vanadium with the $2p_{\sigma}$ orbitals of carbon. In the spectra it corresponds to the main parts of the K_{β_5} and $L_{\alpha_{1,2}}$ emission bands and the CK_{α} band of carbon.

The second (II) band is the conduction band and includes energy states of the antibonding type, corresponding to the 4s and

3d states of vanadium and the $2p_\pi$ states of carbon. In the spectra it corresponds to the short-wave excesses (subbands) of the K_{β_5} and CK_α bands.

The third (III), low-energy, band represents the states of the electrons which are localized in the vicinity of the ionic nuclei of the carbon atoms and which have a predominantly 2s character with a small [10] admixture of the 2p states of carbon and also of the 4p and 3d states of vanadium, as a result of which it may appear in the K spectrum (K_β^{π} satellite). The existence of marked overlapping of the $2p_\sigma$ and $3d_\gamma$ orbitals follows from a consideration of the relative positions of the K and L spectra and of the results of theoretical calculations [10, 11] for the isostructural carbides of titanium. Thus it is possible to distinguish in the spectrum covalent $Me-C$, metallic $Me-Me$, and ionic components of the bond which appear in the I, II, and III energy bands, respectively.

2. With the transition from the higher carbide to the lower, as the carbon vacancies in the homogeneity ranges of the two carbides become filled, there is a substantial redistribution of the contributions of the various electron states, no doubt as a result of the change in the character of the chemical bond from being predominantly metallic in V_2C_{1-x} to being covalent−metallic in VC_{1-x}. With increase in the electron concentration the contribution of the 3d electrons to the $Me-Me$ bonds gradually diminishes, while their participation in the $Me-C$ bonds increases, i.e., there is a gradual change in the symmetry of the d wave function from the thrice-degenerate d_ε ($d_{xz,xy,yz}$) to the twice-degenerate d_γ (d_{z^2,x^2-y^2}), which in turn may be the result of a strengthening of the p character of the hybrid $Me-C$ bond.

3. The long-wave displacement of the K and L bands and the short-wave shift of the K_{α_1} line [13] of vanadium on the formation of the carbides, and likewise the short-wave displacement of the CK_α bands in the two carbides relative to graphite, point to only a slight transfer of charge from carbon to vanadium, which takes place within the framework of the polarized covalent bond.

4. It follows from the combination of the K and L spectra that the changes in electron structure that accompany the changes in carbon content in the carbides (or in the number of vacancies) occur mainly in the region of enhanced density of the d states below

the Fermi surface, which in carbides of group V metals lies, according to [15-17], outside the boundaries of the principal band of the bonding states with a capacity of about six electrons [16, 17] and is evidently completely populated. This surface cuts off approximately one third (the total capacity is approximately three electrons) [10, 15-17] of the following additional band of the antibonding states with a predominantly d_ε character, which overlaps the first. If, in accordance with [10], we take the capacity of the low-energy one third of the band as two electrons, then the nine valence electrons of vanadium carbide are distributed in the three-band diagram as 2 + 6 + 1. A change in the number of carbon vacancies will affect mainly the electron density of the antibonding states. The increasing number of vacancies in VC_{1-x} corresponds to a defective lattice, which is more favorable in terms of energy, since vacancy formation takes place as a result of the liberation of the higher-energy orbitals of the antibonding d_ε band and is accompanied by a reduction in the number of valence electrons and by a lowering of the Fermi level. The energy instability of the defect-free VC lattice is evidently due to the absence of vanadium carbide of the stoichiometric composition $VC_{1.0}$ (the maximum concentration was $VC_{0.88}$) [12]. The existence of the typically metallic d_ε band (conduction), a considerable contribution to which is made by carbon electrons, may, in principle, account for the existence of wide concentration ranges of the metalloid within whose limits compounds of variable composition remain stable.

Literature Cited

1. E. A. Zhurakovskii and V. P. Dzeganovskii, Poroshkovaya Met., No. 5, 57 (1964).
2. V. I. Chirkov, Author's Summary of Candidate's Thesis, Rostov-on-Don (1964).
3. V. S. Neshpor, S. V. Airapetyants, and S. S. Ordan'yan, Izv. Akad. Nauk SSSR, Neorg. Mat., 11:855 (1966).
4. H. Bittner and H. Goetzki, Monatsh. Chem., 93:1000 (1962).
5. L. Ramqvist, J. Ann., Vol. 153 (1969).
6. C. Froidevau and D. Rossier, J. Phys. Chem. Solids, 28:1197 (1967).
7. V. P. Dzeganovskii and E. A. Zhurakovskii, Dokl. Akad. Nauk SSSR, 167:1283 (1966).
8. E. Z. Kurmaev, S. A. Nemnonov, and A. Z. Men'shikov, Izv. Akad. Nauk SSSR, Ser. Fiz., 31:6 (1967).
9. M. A. Blokhin, The Physics of X-Rays [in Russian], Fizmatgiz, Moscow and Leningrad (1959).
10. V. Eme and P. Switendick, Phys. Rev., 137:6A (1965).
11. R. Lye and E. Logothetis, Phys. Rev., 147:1 (1966).

12. G. V. Samsonov and Ya. S. Umanskii, Hard Compounds of Refractory Metals [in Russian], Metallurgizdat, Moscow (1957).
13. E. A. Zhurakovskii and V. P. Dzeganovskii, Ukrainsk. Fiz. Zh. (1969).
14. A. P. Lukirskii, I. A. Brytov, and N. I. Komyak, in: Apparatus and Methods for X-Ray Analysis [in Russian], Vol. 2, SKB RA, Leningrad (1967).
15. S. D. Denker, J. Phys. Chem. Solids, 25:1397 (1969).
16. S. A. Nemnonov, E. Z. Kurmaev, K. M. Kolobova, and A. Z. Men'shikov, Fiz. Metal. i Metalloved., 34:1066 (1968).
17. E. A. Zhurakovskii, Dokl. Akad. Nauk SSSR, 184:1317 (1969).

The Effect of Neutron Irradiation on the Structure of Titanium and Chromium Carbides

M. S. Koval'chenko, V. V. Ogorodnikov, A. G. Krainii, and L. F. Ochkas

An x-ray study has been made of the change in lattice parameters of TiC and Cr_7C_3 irradiated in a reactor with integrated neutron fluxes of 10^{16}, 10^{18}, and 10^{20} neutrons/cm^2. The broadening of the lines on the diffraction patterns of annealed, hot-compacted, and irradiated specimens has been investigated by means of harmonic analysis. The effect of compression on the change in lattice parameter of the irradiated carbides was also investigated. Analysis of the possible positions of the displaced atoms showed that the stable defect configuration in TiC is a dumbbell in the <111> direction. The energy stored in the form of static lattice distortions has been calculated from the change in volume of the unit cell on irradiation.

Neutron irradiation of titanium carbide TiC and chromium carbide Cr_7C_3 leads to a significant change in their lattice parameters and to an increase in hardness, brittleness, and electrical resistivity [6, 7, 12]. These changes are due to the accumulation of lattice defects during irradiation. The object of the present research was to determine which types of defect can give rise to the observed changes in properties.

An x-ray study has been made of the change in lattice parameter, and a harmonic analysis has been made of the broad-

TABLE 1. Volume Changes in the Lattice and Calculated Values
of the Effective Temperature, Mean Square Displacements,
and Stored Energy in Irradiated Titanium
and Chromium Carbides

Com- pound	Neutron dose, neutrons/ cm2	Lattice parameters, A		$\dfrac{\Delta V}{V}$, %	ΔT, deg	$\sqrt{\overline{u}}$, A	$\sqrt{\overline{\Delta u^2}}$, A	Stored ener- gy, ΔU	
		a	c					J/mole	J/g
TiC	Not ir- radiated	4.3177	—	—	—	0.084	—	—	—
	10^{16}	4,3230	—	0.39	200	0.101	0,056	4110	68.7
	10^{18}	4,3255	—	0.54	310	0.112	0,074	6660	111.0
	10^{20}	4.3257	—	0.57	320	0,113	0.076	6870	115.0
Cr_7C_3	Not ir- radiated	13.9965	4.5244	—	—	0.112	—	—	—
	10^{18}	14,0138	4.5245	0,24	90	0.126	0,057	10 050	25,2
	10^{20}	14.0172	4.5344	0.52	180	0,138	0,081	21 100	52.7

ening of the (111) and (222) diffraction reflections from hot-compacted and irradiated specimens of TiC by using the template
method [3]. A specimen of TiC hot compacted and then annealed
at 2400°C acted as a standard in the line-broadening analysis.

The lattice parameters were calculated from the position of
the integrated center of the diffraction line, which divides the area
of the diffraction peak into two equal parts [15]. The area was
found by means of a planimeter. The average was taken of the results from three specimens, the diffraction peaks being recorded
twice for each specimen (in the direct and reverse directions). To
calculate the lattice parameter of TiC, we recorded the reflections
from the {600} plane, using CuK_β-radiation, and to determine the
parameters of Cr_7C_3 we recorded the reflections from the (6.4.10.2)
and (10.0.10.1) planes, using CrK_β radiation. Table 1 contains the
values of the lattice parameters of the carbides investigated before and after irradiation and also data on the relative changes in
the parameters and volumes of the unit cells.

Analysis of the line broadening in the diffraction patterns of
irradiated specimens revealed that the broadening is due not to
the effect of irradiation but to deformation during the hot compaction of TiC powder since with hot-compacted specimens the
line broadening is practically the same before and after irradiation. The mean size of the blocks D and the values of the mean
square deformation $(\overline{\Delta L^2})^{1/2}$ were determined from the shape of

the curve for the dependence of the coefficients of true broadening A_n on the atomic number n with subsequent separation of the effects of the degree of dispersion and the microdeformation. The mean size of the blocks in hot-compacted and irradiated specimens was 325 Å. It was found from the magnitude of the microdeformation $\overline{(\Delta L^2)}^{1/2}$ which characterizes the mean displacements of the atoms from the lattice points that in the TiC investigated there are regions of uniform deformation having linear dimensions of 250–300 Å. The relative microdeformation $\varepsilon = \overline{(\Delta L^2)}^{1/2}/L$ is approximately 1.5×10^{-3}, corresponding to microstresses $\sigma = E\varepsilon \approx 0.7$ kN/mm^2. Similar results have been obtained for refractory metals in [10].

The effect of annealing on the change in lattice parameters of the irradiated carbides of titanium and chromium has been investigated. For this purpose four specimens of each of the carbides were chosen for irradiation with a dose of 10^{20} neutrons/cm^2. One of the four specimens was not annealed, but was investigated with the others after each stage of annealing. The scatter of the parameter values for this specimen was used to determine the confidence limits of the measurements made on the remaining specimens. The lattice parameter of irradiated TiC remained practically constant on annealing at temperatures up to 800°C (see Fig. 1). The lattice parameter gradually recovers with further rise in temperature. Annealing at 1200°C causes a reduction in

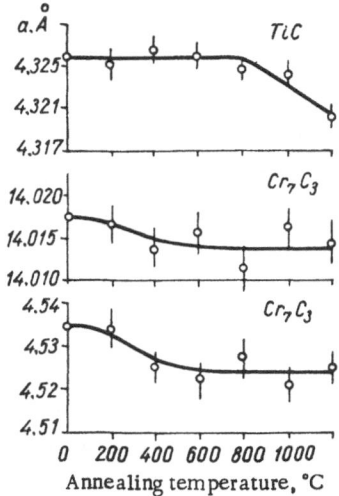

Fig. 1. Lattice parameters of carbides irradiated with a dose of 10^{20} neutrons/cm^2 as a function of annealing temperature. Annealing time 1 h.

the parameter a to 4.322 Å, which is 0.004 Å greater than the initial value.

In the case of chromium carbide there is a partial recovery of the parameter a and a complete recovery of the parameter c on annealing. The principal change takes place at 400°C, and further increase in the annealing temperature to 1200°C does not alter the lattice parameters (the variations in the parameters not exceeding the confidence limits).

The results obtained show that the change in properties of the carbides on irradiation is due mainly to the accumulation of point defects and possibly also to small agglomerations of them. If the agglomerations were fairly large, this would produce an increase in the width of the diffraction reflections. It has already been found that the latter is not affected by irradiation. The change observed in the lattice parameters is due to interstitial or introduced atoms which cause static displacements of the lattice planes.

It is important to determine the locations of the interstitial atoms in the various crystal structures. As has been shown by Vineyard [1] on a mathematical model of a copper crystal, using an electronic computer with a large memory, many of the positions of the interstitial atoms in the lattice that had previously been proposed were unstable. The only stable situation was an arrangement of two atoms in the form of a dumbbell, in which the interstitial atom displaced one of the neighboring atoms from its lattice site and the two of them located themselves along one of the equivalent crystallographic directions (for copper $\langle 100 \rangle$). A similar conclusion may also be reached on the basis of simple and obvious crystallochemical calculations based on the condition of minimum relaxation of the lattice around the intrusive atom and on the consideration of several positions of this atom.

1. Cube center. This is the freest position in the copper lattice. If we place the excess atom here, the distance from its nearest neighbors, assuming they remain in their original sites, is $0.5a$, where a is the lattice parameter.

2. Dumbbell along the $\langle 110 \rangle$ direction. In this case the shortest distance between the atoms, as we have not distributed them freely (assuming that the atoms surrounding the dumbbell retain their former positions), is not more than $0.47a$.

3. Dumbbell along the ⟨111⟩ direction. With this arrangement of the atoms the shortest distance that can be attained between them is 0.54a.

4. Dumbbell along the ⟨100⟩ direction. In this case we have the greatest distance that can be reached between adjacent atoms, viz., 0.55a.

If the lattice relaxation around the air of atoms in the dumbbell is taken into account, it is to be expected that it will be least for position 4, since less separation of them around the dumbbell is required in this case for equilibrium distances to be established between the atoms. In copper, according to Vineyard, the distances between the atoms increase from 0.55 to 0.62a.

Similar considerations can be used in discussing the radiation breakdowns in carbides. In the case of TiC the cube centers are occupied by carbon atoms, and although the closeness of the packing of the atoms in the TiC lattice is the same as in copper, the distances between the atoms are smaller, and so it is reasonable to consider only the dumbbell configurations. Table 2 gives the calculated distances between the atoms in the region of the interstitial atom. The arrangement of the atoms in the dumbbell was chosen in such a way that the distances between them and the neighboring atoms were as large as possible. In consequence, the greater the minimum distance between the atoms, the more favorable is the given configuration as regards the condition of minimum relaxation of the surrounding atoms. It can be seen from Table 2 that in the TiC structure the most stable dumbbell con-

TABLE 2. Minimum Interatomic Distances
and Values of Young's Modulus in
Relation to Crystallographic Direction
in Titanium Carbide

Direction of arrangement of dumbbell	Minimum distance between the atoms				Young's modulus E_{hkl}, gN/m²
	Ti — Ti		Ti — C		
	Å	$\dfrac{d_{min}}{2r_{Ti}}$	Å	$\dfrac{d_{min}}{r_{Ti} + r_C}$	
⟨100⟩	1.44	0.49	1.44	0.65	459
⟨110⟩	2.04	0.70	1.61	0.72	248
⟨111⟩	2.19	0.75	1.77	0.79	220

figuration should be that consisting of Ti atoms arranged along the $\langle 111 \rangle$ direction.

Since the carbon atoms are arranged in a similar way to the atoms of titanium, an analogous dumbbell configuration should be stable for them too. The accumulation of radiation defects in a solid increases its internal energy, i.e., stored energy appears as a result of the static displacements of the atoms during irradiation [8]. The magnitude of the static displacements and the corresponding stored energy for deformed metals are usually estimated from the reduction in intensity of the lines on the x-ray pattern. Because of the considerable expansion of the lattice of irradiated materials, the amount of stored energy can be determined from the change in volume of the unit cell. There is a well-established similarity between the lattice expansion of irradiated materials and the expansion on heating due to dynamic displacements of the atoms by thermal vibrations. The atomic displacements due to third-order distortions correspond to some degree to the average state of the lattice disturbed by thermal agitation. Lattice expansion on heating is represented by the formula

$$V = V_0 \exp \left(\int_{T_0}^{T} \beta(T) \, dT \right), \tag{1}$$

where V_0 is the initial volume at temperature T_0, V is the volume at temperature T, and $\beta(T)$ is the coefficient of volume expansion in relation to temperature. For a small temperature range $\Delta T = T - T_0$, the volume change is expressed by the formula

$$V = V_0 (1 + \beta \Delta T). \tag{2}$$

By using Eqs. (1) and (2), we can determine the effective temperature ΔT from the change in lattice parameter of the irradiated material.

The temperature dependence of the coefficient of volume thermal expansion is, for hexagonal crystals,

$$\beta = 2\alpha_a + \alpha_c, \tag{3}$$

where α_a and α_c, the coefficients of linear thermal expansion along the a and c axes, respectively, are in the form [19]

$$\left. \begin{array}{l} \alpha_a = AC_{V_x} + BC_{V_z} + CT \\ \alpha_c = LC_{V_x} + MC_{V_z} + NT \end{array} \right\}, \tag{4}$$

where

$$A = \frac{2}{3} \cdot \frac{\gamma_x}{V} (s_{11} + s_{12}),$$

$$B = \frac{1}{3} \frac{\gamma_z}{V} s_{13},$$

$$L = \frac{4}{3} \frac{\gamma_x}{V} s_{13},$$

$$M = \frac{1}{3} \frac{\gamma_z}{V} s_{33},$$

$$C = \frac{1}{3} [2G_1 (s_{11} + s_{22}) + G_2 s_{13}],$$

$$N = \frac{1}{3} [4G_1 s_{13} + G_2 s_{33}].$$

s_{11}, s_{12}, s_{13}, and s_{33} being the elastic constants of a hexagonal crystal, $G = a^2/\varkappa$, $\varkappa$ the compressibility, γ_x and γ_y Grüneisen's constants in the directions of the a and c axes, V the volume of 1 g-mole, and C_{V_x} and C_{V_y} the components of the specific heat at constant volume in the directions of the a and c axes. For compounds

$$C_V = 3pRD\left(\frac{\Theta}{T}\right), \tag{5}$$

where R is the gas constant, p is the number of atoms in a molecule of the compound, $D(\Theta/T)$ is the Debye specific heat function

$$D\left(\frac{\Theta}{T}\right) = 3\left(\frac{T}{\Theta}\right)^3 \int_0^{\frac{\Theta}{T}} \frac{\xi^4 e^\xi}{(e^\xi - 1)^2} d\xi, \tag{6}$$

values for which are tabulated in [16], and Θ is the characteristic Debye temperature.

For an isotropic body Eq. (4), bearing in mind that s_{13} and s_{33} are zero, while $\gamma_x = \gamma_y$ and $C_{V_x} = C_{V_y}$ and also that

$$s_{11} = \frac{1}{E} \quad \text{and} \quad s_{12} = -\frac{\nu}{E},$$

where E is the modulus of elasticity and ν is Poisson's ratio, are transformed into the expression

$$\alpha = \frac{2(1-\nu)}{3E}\left(\frac{\gamma}{V} C_V + GT\right), \tag{7}$$

and the temperature dependence $\beta = 3\alpha$, can be written in the form

$$\beta(T) = K'C_V + N'T,\tag{8}$$

which is suitable for calculating the temperature dependence of the coefficient of thermal expansion of refractory compounds. For TiC the constants K' and N', determined from the passage of the $\beta(T)$ curve through the experimental points given in [16, 11], are K' = 4.91×10^{-7} mole/J and N' = 3.0×10^{-9} deg^{-2}. It follows from Eq. (1) that

$$\ln\left(\frac{V}{V_0}\right) = \ln\left(1 + \frac{\Delta V}{V}\right) = \int_{T_0}^{T} \beta(T)\, dT.\tag{9}$$

By graphical integration of the dependence $\beta(T)$, it is easy to calculate the values of $\ln(V/V_0)$ for various temperatures and, by comparing the data derived from the graph with the values of $\ln(V/V_0)$ obtained on irradiation, to determine the effective temperature. Owing to the inadequate data for the coefficient of thermal expansion of Cr_7C_3, the effective temperature for this compound was determined from Eq. (2) without taking into account the temperature dependence $\beta(T)$.

The energy stored in the form of static displacements of the atoms can be determined from a formula similar to that for the change in enthalpy of a body on heating:

$$\Delta U = \frac{1}{2}\int_{T_0}^{T} C_p(T)\, dT.\tag{10}$$

The factor 1/2 is used because only the potential energy is being considered. As the temperature dependence C (T) = A + BT + CT^{-2}, Eq. (10) on integration takes the form

$$\Delta U = \frac{1}{2}\left[A + \frac{1}{2}B(T + T_0) + \frac{C}{T_0 T}\right]\Delta T.\tag{11}$$

Values of the stored energy calculated by the method described above are given in Table 1. From the effective temperature ΔT it is also possible to determine the mean square displacements of the atoms by means of the well-known formula

$$\sqrt{\overline{u^2}} = \sqrt{\frac{9h^2 T}{Mk\Theta}\left[\Phi(x) + \frac{x}{4}\right]},\tag{12}$$

where $(\overline{u^2})^{1/2}$ is the mean square displacement, $\hbar = h/2\pi$ is Planck's constant, M is the atomic mass, K is Boltzmann's constant, $x = \Theta/T$, and $\Phi(x)$ is the Debye function

$$\Phi(x) = \frac{1}{x}\int_0^x (e^\xi - 1)^{-1}\,\xi d\xi. \tag{13}$$

The Debye temperature for Cr_7C_3 is determined from the formula [16]

$$\Theta \approx 137\sqrt{\frac{T_m}{MV_0^{2/3}}},$$

where T_m is the melting point and M and V_0 are, respectively, the atomic weight and specific volume per gram-atom. The results of calculating the mean square displacements for unirradiated and irradiated carbides in relation to the effective temperature are given in Table 1. For the unirradiated compounds $\Delta T = 0$ and $T = T_0 \approx 300°K$. The static displacements caused by irradiation are $(u^2_T - u^2_{T_0})^{1/2}$ for the effective temperatures T and T_0.

The concentration of interstitial atoms can be estimated from the static displacements Δu^2. The authors of [5] have derived a formula for the mean square static displacements in substitutional solid solutions which, in the case of irradiated materials, can be modified in the following way:

$$\overline{\Delta u^2} = \gamma c_i (\Delta R)^2 + c_i U_i^2, \tag{14}$$

where c_i is the concentration of interstitial atoms, ΔR is the distance by which the atoms of the first coordination sphere around the interstitial atom should be separated for normal distances between them to be established, U_i is the distance from the interstitial atom to the nearest lattice point, and γ is a numerical factor determined by the coefficient of closeness of packing of the atoms in the lattice $\gamma = 3.56q$. For the dumbbell arrangement of the displaced atoms described above, $u_i^2 = u_1^2 + u_2^2$, where u_1 and u_2 are the displacements of the atoms in the dumbbell. Relaxation of the lattice around the dumbbell should be very anisotropic, since ΔR is a maximum in the direction of the dumbbell axis and a minimum in the direction perpendicular to it, so that $(\Delta R)_{max} = d - d_{min}$, where d is the equilibrium distance between the atoms and d_{min} is the minimum distance between the atoms forming the dumbbell (see Table 2). For making approximate calculations we can use the value $(\Delta R)_{av} = (\Delta R)_{max}/2$, since $(\Delta R)_{min} = 0$.

An approximate calculation of the value of c_i for TiC showed that the concentration of "excess" atoms or dumbbells in TiC irradiated with a dose of 10^{20} neutrons/cm^2 is 0.2 at.%. In making the calculation we bore in mind that the static displacements are due mainly to interstitial Ti atoms, since the C atoms are considerably smaller in size and are evidently present in smaller quantity owing to their greater mobility. The displacements of the atoms around the vacancies should be even smaller. An estimate of the extreme limits of c_i gives 0.12–0.23 at.% of interstitials in the metal sublattice.

By using the conclusions of the continuous defects theory [17], we obtain a value for the interstitial concentration

$$c_i = \frac{\Delta U\,(c)}{\dfrac{6\mu\Omega}{\gamma}\,\varepsilon^2}, \tag{15}$$

where ΔU is the stored energy, μ is the shear modulus, Ω is the atomic volume, $\gamma = 3(1 - \nu)/(1 + \nu)$, ν is Poisson's ratio, and ε is the deformation in the region of the interstitial atom.

Because the interstitial atoms can be combined in more or less large complexes, the actual concentration of the dumbbells will be somewhat higher, and the concentration of the complexes themselves will be somewhat lower than the calculated value of c_i.

The above concentration of interstitial atoms formed as a result of the interaction of fast neutrons with carbides enables us to make a rough estimate of the "productivity" of a single neutron. The number of neutrons that collide with atoms of the material is determined by the expression [16]

$$N = \Phi S\,(1 - e^{-n\sigma}), \tag{16}$$

where Φ is the integrated fast-neutron flux (10^{19} neutrons/cm^2), S is the surface area of the specimen (3.5 cm^2), σ is the scattering cross section for fast neutrons at the target atoms, n is the thickness of the target, which in nuclear physics is expressed by the number of atoms n in a column of section 1 cm^2 and height equal to the thickness of the target (cm). For TiC $\sigma = 1.2$ barn (found additively from the data on the developed section of the elements (13)), and n $= 10^{23}$ cm^{-2}.

Calculation by means of Eq. (16) gives N = 4 × 10^{18} neutrons. As the light carbon atoms lose a considerable part of their energy by the ionization of the surrounding atoms [9], the main part of the displaced atoms will be produced by titanium atoms that have obtained their energy from the neutrons (2 × 10^{18} atoms, which corresponds to 4 × 10^{-5} of the total number of titanium atoms). Thus the productivity of a single neutron

$$P = \frac{N_i}{N} = \frac{2 \cdot 10^{-3}}{4 \cdot 10^{-5}} = 50 \; i/n, \qquad (17)$$

i.e., one scattered neutron creates 50 interstitial atoms which are retained after irradiation.

The value obtained is an order of magnitude less than the theoretical estimate made by Seitz and Keller [4], according to which one copper atom which has obtained an impulse from a neutron with an energy of 1 MeV produces 380 Frenkel pairs (this figure will be of the same order for TiC). From this estimate it follows that approximately 90% of the interstitials are annealed out of TiC during irradiation.

The reduction in the lattice parameters of irradiated carbides on annealing should be accompanied by the liberation of stored energy.

Conclusions

1. Neutron irradiation leads to an increase in the lattice parameters of the carbides of titanium and chromium. Annealing irradiated Cr$_7$C$_3$ at temperatures above 400°C causes a recovery of the parameters. Such a recovery takes place in TiC above 1200°C.

2. It is shown that the block size and microstresses in hot-compacted TiC specimens do not change on irradiation with a dose of up to 10^{20} neutrons/cm^2.

3. A method is proposed, and results given, for the determination of the amount of stored energy and concentration of point defects in irradiated carbides from data on the change in lattice parameter. A rough estimate of the "productivity" of neutrons based on the experimental results showed that one neutron creates

approximately 50 interstitial atoms which do not undergo radiation annealing. Of the order of 90% of the interstitial atoms are annealed out in the course of irradiation with a dose of 10^{20} neutrons/cm^2.

Literature Cited

1. J. Vineyard, Uspekhi Fiz. Nauk, 74:435 (1961).
2. V. I. Gol'danskii and E. M. Leikin, Transformations in Atomic Nuclei [in Russian], Izd. Akad. Nauk SSSR, Moscow (1958).
3. S. S. Gorelik; L. N. Rastorguev, and Yu. A. Chakov, Radiographic and Electron-Diffraction Analyses of Metals [in Russian], Metallurgizdat, Moscow (1964).
4. F. Seitz and I. S. Keller, in: Proceedings of an International Conference on the Peaceful Uses of Atomic Energy, Geneva (1955).
5. V. I. Iveronova and A. P. Zvyagina, Izv. Akad. Nauk SSSR, Ser. Fiz., 20:729 (1956).
6. M. S. Koval'chenko and V. V. Ogorodnikov, Poroshkovaya Met., No. 10, 48 (1966).
7. M. S. Koval'chenko and V. V. Ogorodnikov, At. Energ., 21:302 (1966).
8. M. S. Koval'chenko and V. V. Ogorodnikov, At. Energ., 22:138 (1967).
9. S. G. Konobeevskii, in: The Action of Nuclear Radiation on Materials [in Russian], Izd. Akad. Nauk SSSR, Moscow (1962), p. 5.
10. B. M. Levitskii and L. D. Panteleev, in: The Action of Nuclear Radiation on Materials [in Russian], Izd. Akad. Nauk SSSR, Moscow (1962), p. 209.
11. J. Ney, The Physical Properties of Crystals and Their Description in Terms of Tensors and Matrices [Russian translation], IL, Moscow (1960).
12. V. V. Ogorodnikov et al., Fiz.-Khim. Mekhan. Mat., 2:532 (1966).
13. B. Price, K. Horton, and K. Spinney, Protection from Nuclear Radiation [in Russian translation], IL, Moscow (1959).
14. G. V. Samsonov and K. I. Pomoi, Alloys Based on Refractory Compounds [in Russian], Oborongiz, Moscow (1961).
15. D. M. Kheiker and A. S. Zevin, X-Ray Diffractometry [in Russian], Fizmatgiz, Moscow (1963).
16. Ch'ien Hsüeh-sen, Physical Mechanics [in Russian], Mir, Moscow (1965), p. 220.
17. J. Eshelby, The Continuous Theory of Dislocations [Russian translation], IL, Moscow (1963).
18. A. Lieberman and W. Grandall, J. Am. Ceram. Soc., 35:304 (1952).
19. D. Riley, Proc. Phys. Soc., 57:486 (1945).

Some Physical Properties of Pyrolytic Boron Carbide

V. S. Neshpor, V. P. Nikitin, and V. V. Rabotnov

The specific electrical resistivity, absolute differential thermo-emf, Hall coefficient, thermal conductivity, and microhardness have been measured on specimens of pyrolytic boron carbide having theoretical density and a compact laminar-columnar structure. It is suggested that the p-conduction in boron carbide is due to the partial replacement of carbon atoms in boron carbide of the limiting composition B_4C by boron atoms which have a lower valence.

Boron carbide, the stoichiometric composition of which corresponds to the formula B_4C is a high-temperature, semiconducting material (melting point = 2470°C) which possesses great hardness and comparatively good resistance to scaling [7]. According to [14], boron carbide may be regarded as a unique interstitial phase with a group of carbon atoms in voids approximating to close packing of boron icosahedra (B_{12}), so that the chemical formula of boron carbide should be written as (B_{12})C_3. The true symmetry of the elementary cell of boron carbide is rhombohedral, although its structure is more often considered in terms of hexagonal axes. A group of three carbon atoms is arranged in the form of a linear chain in the direction of the greater diagonal of the rhombohedral unit cell [7, 1]. Some of the carbon atoms in this chain may be replaced by boron with the formation of boron-richer phases of variable composition based on B_4C = (B_{12})C_3 [4-6].

The theoretical calculation of the parameters of the energy bands in boron carbide made in [12] showed that the spacing between the valence band and the conduction band in the center of the Brillouin zone is 2.5 eV while at the zone boundary, in the direction of the (111) wave vector, it is about 1 eV. These results are not in bad agreement with the width of the forbidden band in B_4C (about 1.64 eV) in the intrinsic conduction region (above 1600°C) determined experimentally in [8].

Investigations of the electrical properties of boron carbide [6-8, 11, 13, 17] have been carried out mainly on specimens produced by the sintering or hot compaction of B_4C powder and having a certain residual porosity. Samsonov and his coworkers [7, 6] have shown that the electrical, and in particular the thermoelectric, properties of B_4C specimens exhibit a marked dependence on the porosity of the specimens, although it is impossible to approximate this by any analytical relationship.

In the present research we have investigated the electrical and thermal conductivities, the thermo-emf, the Hall coefficient, and the thermal expansion of B_4C specimens obtained by deposition from a gaseous phase containing boron trichloride vapor, methane, and hydrogen at atmospheric pressure, on to a graphite substrate heated to 1100-1300°C.

The boron carbide specimens for the investigations were in the form of platelets or semicylinders 0.8-1.2 mm thick produced by deposition on the inner surface of hollow graphite molds of appropriate shape. Before the specimens were deposited the inner surface of the graphite mold was coated with a thin layer of pyrographite which restricted the effect of the substrate on the composition of the carbide and ensured good separation of the specimens from the substrate without breaking. The ratios by volume of the gaseous BCl_3 and CH_4 corresponded to the stoichiometry of B_4C, whereas the flow rate of hydrogen was 20 times greater than that required by the equation

$$4BCl_3 + 4H_2 + CH_4 = B_4C + 12HCl$$

in order to eliminate processes which cause a substantial deterioration in the structure of the deposited carbide. Specimens of the desired shape were cut from the deposited tubes or platelets by the electroerosion method followed by polishing with diamond paste. The chemical composition of the B_4C specimens was (wt.%):

B	C	Fe	Mn	Mg	Si	Al	Ca	Cu	Cr	Ni	Ti
78.5	21.3	0.2	0.004	0.002	0.001	0.002	0.002	0.002	0.004	0.002	0.002

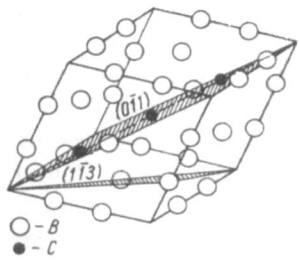

Fig. 1. Position of the $(0\bar{1}1)$ and $(1\bar{1}3)$ planes in the unit cell of B_4C.

The specimens contained a slight excess of boron over the stoichiometric composition (21.72% C and 78.28% B), i.e., they corresponded to phases in which the carbon atoms in the chain had been partially replaced by boron. X-ray phase analysis revealed that the specimens contained a single phase B_4C with lattice parameters (in hexagonal axes) a = 5.61 and c = 12.00 Å.

Comparison of the intensities of the diffraction maxima in the x-ray patterns* taken from the surface of tubular specimens of pyrolytic B_4C with the diffractogram of fine powder (approximately 10 μ) obtained from the same specimens (with the same geometry of the reflecting surface), and also with the intensity of the diffraction maxima of a standard powder diffractogram of B_4C [18], shows that in pyrolytic B_4C specimens there is no clearly defined growth texture, such as, for example, that which characterizes pyrolytic graphite having marked anisotropy of the crystals with (001) cleavage planes. Nevertheless, the diffractogram of the surface of the compact pyrolytic material reveals a substantial increase in the intensity of the reflections from the (011) planes and to a smaller extent those from the (113) planes and possibly the (203) planes (in rhombohedral axes). This leads to the conclusion that these crystallographic planes have a definite preferred orientation parallel to the surface of the specimens of pyrolytic boron carbide. The position of the $(01\bar{1})$ and $(1\bar{1}3)$ planes in the rhombohedral cell of B_4C is shown in Fig. 1. It may be noted that the chains of carbon atoms in the unit cell are situated just in the $(0\bar{1}1)$ plane which shows the clearest signs of preferred orientation.

The density of the specimens was 2.51 g/cm³, which is close to the theoretical density of B_4C (2.52 ± 0.01 g/cm³) [7].

---

*The x-ray structural analysis was made by A. G. Miroshnichenko and V. D. Novozhilova.

Metallographic analysis of microsections of the B_4C specimens cut perpendicular to the deposition surface showed that the specimens possess the columnar-laminar structure which is characteristic of pyrolytic materials, with close packing of the crystal-growth columns. There was no sign of any inclusions of impurity phases, cracks, or voids on the microsections. The microhardness of the specimens was 4750 ± 250 kg/mm^2 (under a load of 200 g), which is in good agreement with the value determined in [15] on B_4C microcrystals (4210 ± 280 kg/mm^2).

The thermal conductivity of the specimens was measured by a comparative stationary-heat-flow method, which is a modification of the method described in [3].

The thermal conductivity $\varkappa$ and the absolute differential thermo-emf α in the direction parallel to the deposition surface, for specimens of B_4C immediately after deposition and after annealing at 1000°C for 2 h in a vacuum of 5×10^{-5} mm Hg, are as follows:

	After deposition	After annealing
$\varkappa$, W/m · deg	8.5	18 ± 3
α, μV/deg	245 ± 6	216 ± 15

Annealing substantially increases the thermal conductivity, but the thermo-emf and the electrical resistivity remain practically unchanged. The latter can evidently be explained by the fact that as a result of the comparatively low electrical conductivity of B_4C ($\sigma \sim 1 \times 10^3$ Ω^{-1} · cm^{-1}) its thermal conductivity is due mainly to the scattering of phonons (the electron contribution to the thermal conductivity is approximately 0.8 W/m · deg, i.e., about 4% of the measured value). Annealing without substantially affecting the structure of the material, removes the internal stresses set up in it after deposition as a result of the difference in the thermal expansion coefficients of the deposit and the substrate and of the existence of a temperature gradient through the thickness of the specimen. The reduction in internal stresses, which are clearly local in character as a result of the compact columnar-laminar structure of the material, causes a decrease in the contribution of phonon scattering to the thermal resistance of the specimen and to an increase in its thermal conductivity.

The Hall coefficient R was measured on the platelet specimens of 0.8-mm thickness using direct current and a field strength of

9870 Oe, the current density through the specimen being 17.5 A/cm^2. The Hall coefficient at room temperature had a positive sign and its value was $(+2.19 \pm 0.01) \times 10^{-2}$ cm^3/C.*

The positive sign of the Hall coefficient and the thermo-emf point to hole conduction in the B$_4$C specimens, which may evidently be attributed to the tendency of this compound to form phases of variable composition (with a limiting stoichiometric composition of B$_4$C), in which the component with the lower valence (boron) replaces the component with the higher valence (carbon), thus leading to the creation of acceptor levels or stray mobile holes in the system of B$-$C valence bonds. As will be shown below, conduction is maintained in boron carbide even at elevated temperatures. As the positive thermo-emf has been observed in boron carbide by many investigators [7, 8, 13, 6, 11], it may be assumed that the ideal stoichiometric composition corresponding to the atomic ratio B:C = 4:1 is difficult to attain. On the assumption that the conduction in the specimens investigated is unary, we obtain for the current-carrier concentration in specimens of pyrolytic boron carbide

$$P = \frac{3\pi}{8} \cdot \frac{1}{Re} = 3.4 \cdot 10^{20} \text{ cm}^{-3},$$

and for the Hall mobility

$$U_n = \frac{8}{3\pi} \cdot \sigma R = 0.13 \text{ cm}^2/\text{V} \cdot \text{sec} \cdot$$

The temperature dependence of the specific electrical resistivity ρ and that of α were investigated by means of the apparatus previously described [4, 5].

The temperature dependence of the electrical resistivity of specimens of pyrolytic boron carbide (the average of three independent measurements on heating and cooling), plotted on semilogarithmic coordinates, is shown in Fig. 2 (curve 1). In the range from room temperature to 600°C, the temperature dependence of ρ can be represented by the exponential function $\rho = \rho_0 \exp(-\Delta E/kT)$ and is due to the thermal activation of impurity current carriers. The activation energy ΔE, calculated from the slope of the linear

---

* The Hall-coefficient measurements were made by laboratory assistant N. A. Skaletskaya.

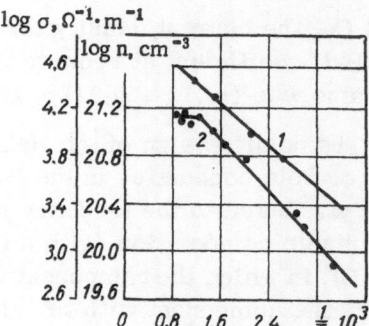

Fig. 2. Temperature dependence of the specific electrical resistivity (1) and current-carrier concentration (2) for pyrolytic boron carbide.

portion of curve 1 in Fig. 2, is 0.22 eV. Changes in the character of the temperature dependence of ρ above 600°C are due to the depletion of impurity levels, while the temperature is still too low for the appearance of significant intrinsic conduction.

The thermo-emf increases very slowly with temperature in the range 20–500°C, but thereafter its growth becomes more marked. In character, it approximates to the relationship found in [11] for B_4C specimens produced by hot compaction. The significant increase in α with temperature in the low-temperature region observed in [6] probably indicates a different type of impurity center.

It was not possible to make measurements at temperatures above 1000°C owing to the chemical reaction of the chromel/alumel thermocouple junctions with the material of the specimens, which led to the corrosion penetration of the thin-walled tubular specimens.

In the region of impurity conduction there is a linear relationship between α and the logarithm of the electrical conductivity. This shows that, in spite of the relatively high current-carrier concentration for a semiconductor, the following simple equation [15] can be used to represent the temperature dependence of α for the specimens investigated:

$$\alpha = \frac{K}{e}\left(2 - \frac{\mu}{kT}\right), \tag{1}$$

where μ is the chemical potential level, assuming the mechanism of scattering of current carriers at acoustic vibrations of the lattice which is typical of purely covalent semiconductors.

By using the experimental values of α, we obtain μ = 0.1 eV, i.e., the chemical potential level is situated in the middle of the energy range between the ceiling of the valence band and the acceptor level, as required by theory. This enables us to use N. L. Pisarenko's simple formula to calculate α:

$$\alpha = \frac{K}{e}\left[A + \ln \frac{2(2\pi m^* kT)^{3/2}}{h^3 p} \right], \tag{2}$$

where m* is the effective mass and A = 2 for atomic lattices. This gives for m*, using the experimental values of α and p, the ratio m*/m = 0.44, which is typical of many semiconductors of this type. The temperature dependence of the current-carrier concentrations, calculated from Eq. (2), using the established value of m* and the experimental values of α, is shown in Fig. 2 (curve 2). It will be seen that in the temperature range investigated the temperature dependence of the current-carrier concentration can be represented by the exponential function $P = P_0 \cdot \exp(-\Delta E/kT)$, while the value of ΔE calculated from the slope of the straight line is 0.18 eV, in satisfactory agreement with the value of ΔE calculated from the temperature dependence of the electrical resistivity.

The thermal expansion of pyrolytic boron carbide in the range 800-1200°C has been measured by an optical method, using the heating and optical systems of the MP-301V testing machine.*

The results obtained are shown in Fig. 3. The coefficient of thermal expansion increases comparatively slowly with temper-

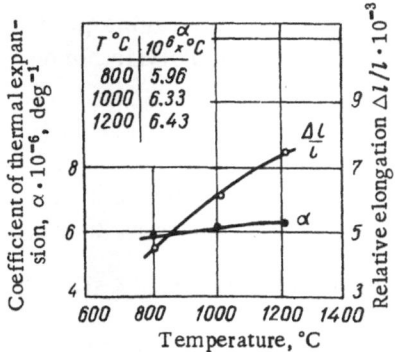

Fig. 3. Temperature dependence of the relative elongation and coefficient of thermal expansion of pyrolytic boron carbide (parallel to the deposition surface).

* The thermal-expansion measurements were made by B. B. Mogilevich.

ature in the range investigated and is higher than the values given in the literature (4.5×10^{-6} deg^{-1}), for which the temperature range was not reported [7]. This difference is due to the effect of the preferred orientation of the crystals in the specimens of pyrolytic boron carbide, although it is also possible that the value reported in the literature relates to lower temperatures.

Conclusions

1. The specific electrical resistivity, the absolute differential thermo-emf in the temperature range 20-100°C, the Hall coefficient, the thermal conductivity, and the microhardness have been measured on monolithic specimens of pyrolytic boron carbide having the theoretical density and a compact laminar-columnar structure.

2. In the temperature range investigated boron carbide is an impurity semiconductor of the p-type with a current carrier concentration of 3.4×10^{20} cm^{-3}, a Hall mobility of 0.13 cm^2/V · sec, and an effective mass $m^*/m_0 = 0.44$.

3. The activation energy of the impurity current carriers is 0.22 eV; the chemical potential level is situated in the middle of the energy range between the ceiling of the valence band and the acceptor level.

4. The p-conduction in boron carbide is evidently due to the partial replacement of carbon atoms in boron carbide of the limiting composition $B_4C = (B_{12})C_3$ by boron atoms which have a lower valence.

The authors wish to thank Yu. D. Kondrashev for taking part in the discussion of the structure of specimens of pyrolytic boron carbide.

Literature Cited

1. G. S. Zhdanov and N. G. Sevast'yanov, Zh. Fiz. Khim., 52:326 (1943).
2. G. S. Zhdanov, G. A. Meerson, N. N. Zhuravlev, and G. V. Samsonov, Zh. Fiz. Khim., 28:1046 (1954).
3. A. F. Ioffe, The Physics of Semiconductors [in Russian] (1957), pp. 377, 165.
4. N. V. Kolomeets et al., Zh. Tekh. Fiz., 28:2382 (1958).
5. V. S. Neshpor, in: Apparatus for Investigating the Physicomechanical Properties and Structure of Materials [in Russian], Vol. 7, GOSINTI, Moscow (1962), p. 3.

6.   G. V. Samsonov, G. N. Makarenko, and G. G. Tsebulya, Izv. Akad. Nauk SSSR, Otd. Tekh. Nauk, No. 4 (1960).

7.   G. V. Samsonov et al., Boron, Its Compounds and Alloys [in Russian], Izd. Akad. Nauk UkrSSR, Kiev (1960).

8.   G. V. Samsonov and V. S. Sinel'nikova, Ukrainsk. Fiz. Zh., No. 6 (1961).

9.   M. I. Sokhor and V. I. Kudryavtsev, Abrazivy, No. 2 (1963).

10. H. Clark and I. Noard, J. Am. Chem. Soc., 65:2115 (1943).

11. V. Gug, Compt. Rend. Acad. Sci. Paris, 260:5282 (1965).

12. M. Zamazaki, J. Chem. Phys., 27:746 (1957).

13. I. Lagrenaudie, J. Chim. Phys. et Phys.-Chim. Biol., 50:352 (1953).

14. V. Matkovich, R. Giese, and I. Economy, Z. Krist., 122:116 (1965).

15. S. Mierzejewska and T. Niemyski, J. Less-Common Metals, 8:368 (1965).

16. C. F. Powell, I. E. Campbell, and B. W. Gonser, Vapor Plating (1955).

17. R. Ridgway, Trans. Am. Electrochem. Soc., 63:369 (1933); 66:117 (1934).

18. X-Ray Diffraction Data for Analysis Cards, ASTM, Philadelphia (1955), Card 6-05555.

Emissive Power and Specific Electrical Resistivity
of Zirconium and Niobium Carbides
at High Temperatures

V. A. Petrov, V. Ya. Chekhovskoi,
A. E. Sheindlin, I. I. Kashekhlebova,
V. A. Nikolaeva, and L. P. Fomina

The emissive power and specific electrical resistivity of zirconium and niobium carbides have been investigated in vacuum and in argon at temperatures up to 3300-3500°K. The existence of an "initial" section on the curve for the $E_t(T)$ relationship is demonstrated in the case of both carbides. It is suggested that this is due to the oxidized condition of the surface. The deviation of the $\rho(T)$ relationship from a straight line above 2750°K in the case of niobium carbide is attributed to the incongruent evaporation of this carbide.

The integrated hemispherical emissive power, the monochromatic emissive power at a wavelength of 0.65 μ, and the specific electrical resistivity of zirconium and niobium carbides have been investigated by the method previously described in [4]. The test specimens were heated by passing an electric current through them. From measurements of the current strength I, the potential drop over the isothermal section of the test specimen U, the true temperature T, and the geometrical dimensions of the test section of the specimen, we determined the integrated hemispherical emissive power E_t and the specific electrical resistivity ρ from the formulas

$$E_t = \frac{UI}{\sigma \cdot F_0 T^4},\tag{1}$$

$$\rho = \frac{U}{I} \cdot \frac{S_0}{l_0},\tag{2}$$

where $\sigma = 5.67 \times 10^{-8}$ W/m$^2 \cdot$ deg^4, the Stefan−Boltzmann constant; F_0 is the area of the emitting surface at room temperature; S_0 is the cross section of the specimen at room temperature, and l_0 is the length of the test section of the specimen. The isothermal character of the specimen over a length of about 60 mm was produced by the special design of the current leads in which contact was made through the graphite packing. The length of the test section was 30 mm.

The monochromatic emissive power for the wavelength of 0.65 μ was determined from measurements of the true and brightness temperatures of the emission surface using the formula

$$\ln E_{\lambda=0.65\mu} = \frac{c_2}{\lambda_e}\left(\frac{1}{T} - \frac{1}{T_{br}}\right),\tag{3}$$

where $c_2 = 1.438\ \mu \cdot$ deg, the second constant in Planck's emission law; $\lambda_e = 0.65\ \mu$, the effective wavelength of the pyrometer; and T_{br} (°K) is the brightness temperature of the emitting surface.

Thest test specimens were in the form of rods 5.3 mm in diameter and about 120 mm long, prepared by cold compaction followed by sintering twice in an argon atmosphere at a maximum temperature of 2500°C. The properties of the test specimens are given in Table 1. The quality of the surface finish of the specimens before sintering was of the sixth class.

The true temperature was measured with an optical micropyrometer calibrated against a standard-temperature lamp of the second class. The pyrometer was focused on the bottom of a radial hole 0.4 mm in diameter and about 3 mm long. A correction was made to the measured temperature to take account of the absorption and reflection in the observation glass and also of the temperature drop between the bottom of the pyrometer hole and the emission surface. In calculating the latter, use was made of data from [13] for the coefficient of thermal conductivity of ZrC and data from [10] for that of NbC.

TABLE 1. Characteristics of Specimens Investigated

Specimen	Chemical composition, wt.%					Lattice parameter, Å	Total porosity, %	Microhardness, kg/mm²	Mean grain size, μ
	C_{total}	C_{free}	N	O	Me				
Before experiment									
ZrC	13.28	1.14	0.72	0.41	85.33	4.695	12.6	1900	10.3
NbC	11.64	0.63	0.26	0.07	87.85	4.470	15.1	1600	8.6
After experiment									
ZcC	11.70	0.48	0.23	0.06	86.20	—	13.3	2290	20.3
NbC	10.88	0.35	—	0.21	89.61	—	16.9	2000	20

The maximum limiting error in determining E_t in the present research is 6% at 1200°K and 10% at 3300°K; the error in determining $E_{0.65\mu}$ does not exceed 12 and 7%, respectively, at these temperatures, while the error in determining ρ is not more than 3% over the whole temperature range.

To determine the emissive power and specific electrical resistivity of ZrC, three experiments were carried out on each of two specimens. The first specimen was tested twice, with an interval of three months between the two tests; the second specimen was tested only once. The whole temperature range was covered in the following way in a single test:

1) test in vacuum of 2-4 × 10⁻⁴ mm Hg from 1300 to 2400°K and back again;
2) test in argon from 2100 to 3000°K and back again;
3) test in argon from 2900 to 3300-3500°K.

The results of all three tests agreed well among themselves, and so the experimental points relating to only one of them are given in Fig. 1. All three tests were characterized by an "initial" section of the temperature dependence of the emissive power, where the values for the emissive power lie above the experimental results as represented by the basic mean curve, which is reproduced with both rising and falling temperature. According to this curve the integrated emissive power increases uniformly from 0.4 to 0.6 over the range 1300-3300°K. The higher values of E_t found on the "initial" section of the curve are evidently due to the presence of an oxide film formed on the surface of the specimen during cooling after sintering in the argon atmosphere. In

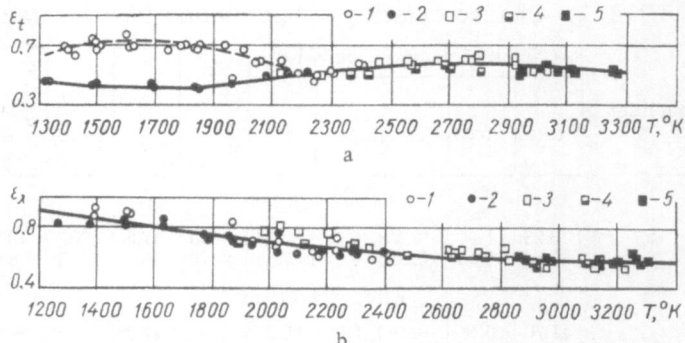

Fig. 1. Experimental results for the determination of the integrated (a) and spectral (b) emissive power of zirconium carbide: 1) rising temperature in vacuum; 2) falling temperature in vacuum; 3, 5) rising temperature in argon; 4) falling temperature in argon.

addition, some part may be played by mechanical inclusions of graphite in surface of the specimens, as the latter were sintered in a covering of graphite. Above 2200°K the oxide film evaporated and at the same time the graphite inclusions, if there were any to begin with, were eliminated. Some increase in the emissive power at temperatures below 1700°K on the stabilized curve may be attributed to the appearance of an oxide film as a result of diffusion of oxygen from the body of the specimen to the surface or to insufficient vacuum. The agreement between the results on the "initial section" and those on the stabilized curve in the second test on the same specimen, carried out three months after the first one, showed that the surface properties, which govern the emission characteristics, are completely retained.

Experiments made on another batch of ZrC [5] likewise showed the existence of an "initial" section of the curve and a "stabilized" surface condition, though the integrated emissive power of ZrC was higher; the difference was most marked in the range 1200-2400°K, the results being in satisfactory agreement at 2400-3100°K. This was probably due to the considerable effect of oxygen on the emissive power, since in contrast to the present work, the oxygen content of the specimens investigated in [5] had not changed after the experiment and was 0.41 wt.%.

Rather more data are available on the monochromatic emissive power at a wavelength of 0.65 μ. However, the discrepancies

between the results of the various authors are also large and can
be attributed only to experimental errors. Hysteresis in the
$E_{0.65\mu}$ (T) relationship, observed in [12], should be considered separately. Shaffer [12] ascribed the hysteresis to the existence of a
phase change in ZrC at 2000-2200°C. His conclusions were crit-
icized by Grossman [8], who obtained a constant value of 0.63 for
$E_{0.65\mu}$ and observed no hysteresis. It was noted in [8] that the
sharp fall in emissive power reported in [12] and the hysteresis
in the $E_{0.65\mu}$ (T) relationship are probably due to the fact that there
takes place on the ZrC surface a reaction with oxygen and possibly
with other impurities present in the argon with which the apparatus
was filled; when the temperature rose, the reaction products decomposed and evaporated.

The suggestion regarding the effect of oxygen on the emissive
power seems to us to be feasible, although it cannot yet be taken
as firmly established fact. Moreover, it is not possible to say
anything about the source of the oxygen for the surface oxidation,
since the oxygen content was very considerable in the specimens
investigated. The "initial" section of the E_t(T) relationship observed in the present research and a certain increase in the emissive power below 1700°C confirm once again how important it is
that further investigations should be made at once into the reasons
for the change in surface properties.

The experimental results obtained in the present work for the
monochromatic emissive power show good agreement with one another in all cases within the limits of accuracy of measurement,
and in the range 1300-3300°K $E_{0.65\mu}$ falls from 0.9 to 0.6. Hysteresis such as was reported in [12] was not observed, but all the experimental points obtained with a falling temperature in the range
2400-1200°K lie below the points obtained with a rising temperature (see Fig. 1).

Figure 2 shows the results of various authors for the specific
electrical resistivity of ZrC having a composition close to the
stoichiometric (all the results have been converted to porosity-
free material by means of Odelevskii's formula [2]). The temperature dependence of the specific electrical resistivity obtained
in the present investigation is not in bad agreement with that derived in our previous work [5]; it is almost linear up to temperatures of the order of 2400°K, but departs significantly from linear-

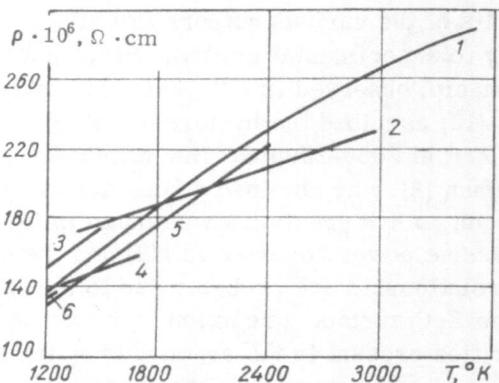

Fig. 2. Specific electrical resistivity of ZrC based
on data from: 1) present work, 2) [10], 3) [11], 4)
[12], 5) [5], 6) [2].

ity at higher temperatures. The agreement of the available results up to 2400°K can be regarded as satisfactory, bearing in mind
the different methods used to prepare the specimens.

The emissive power and specific electrical resistivity of NbC
were determined in vacuum in the range 1350-2200°K and in the
opposite direction down to 1350°K; then in argon from 2300 to
3100°K and back again, and finally from 3200 to 3500°K. The ex-

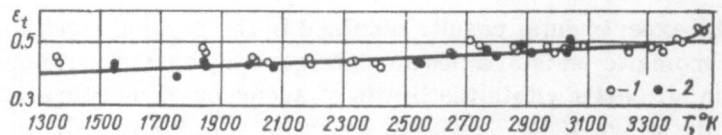

Fig. 3. Results of experiments to determine the integrated emissive
power of NbC: 1) experimental points obtained with rising temperature;
2) experimental points obtained with falling temperature.

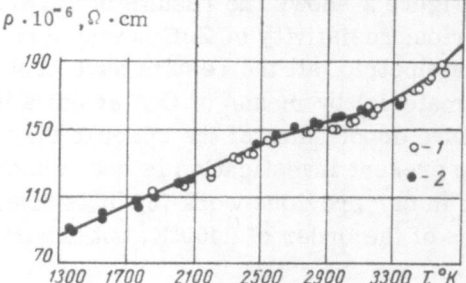

Fig. 4. Specific electrical
resistivity of NbC. Key as
for Fig. 3.

perimental points for determining the integrated emissive power
of NbC are shown in Fig. 3. The "initial" section of the $E_t(T)$ relationship ends in the case of NbC at approximately 1800°K, i.e.,
earlier than in ZrC. It is not shown in Fig. 3. E_t for NbC increases linearly from 0.40 to 0.49 over the range 1300-3300°K.
The steep rise in E_t at higher temperatures is due to the high rate
of evaporation of NbC.

The results obtained determining the monochromatic emissive
power of NbC can be represented by a linear $E_{0.65\,\mu}(T)$ relationship.
Up till now the emissive power of NbC has been studied even less
than that of ZrC. The results reported in [7], according to which
the emissive power of NbC has a constant value of 0.85 over the
range 800-1800°C, are probably somewhat too high.

The temperature dependence of the specific electrical resistivity of NbC determined in the present work is shown in Fig. 4.
From 1300 to 2700°K the experimental points lie close to the mean
straight line, but at about 2750°K the points begin to deviate below
the straight-line relationship. The experimental points obtained
on subsequently lowering the temperature from 3150 to 2500°K
lie somewhat above the points obtained with a rising temperature
over the same range. This departure of the $\rho(T)$ relationship
from a straight line and the subsequent increase in the resistivity
when the temperature change is in the opposite direction is probably due to incongruent evaporation of NbC and predominantly to
loss of carbon; this is fully confirmed by chemical analysis of the
specimens after the experiments have been carried out.

An increase in the electrical resistivity of the carbides with
diminishing carbon content up to 1200°C has been found in [1, 6].
The electrical resistivity of NbC above 1200°C has been determined in [3]. The high results reported in [3], as compared with
those of the present investigation, are probably due to the use of
specimens having a carbon content much less than the stoichiometric.

The change in carbon content in the surface layer of NbC
during evaporation, observed in the present work, may affect
the emissive power, though it may be assumed that at each temperature there is formed on the surface of the specimen a certain stable composition determined only by the temperature, and
that the results obtained for E_t and E_λ do in fact represent the
emissive power of the surface of NbC.

Literature Cited

1. V. S. Neshpor et al., Izv. Akad. Nauk SSSR, Neorg. Mat., 2:855 (1966).
2. V. I. Odelevskii, Zh. Tekh. Fiz., 21:678 (1951).
3. Yu. B. Paderno, I. G. Barantseva, and V. A. Yupko, in: High-Temperature Inorganic Compounds [in Russian], Naukova Dumka, Kiev (1965), p. 199.
4. V. A. Petrov, V. Ya. Chekhovskoi, and A. E. Sheindlin, Teplofiz. Vys. Temp., 1:24 (1963).
5. V. A. Petrov, V. Ya. Chekhovskoi, and A. E. Sheindlin, in: Electricity from MHD, Proc. Symposium on Magnetohydrodynamic Electrical Power Generation, Salzburg, July 4-8, 1966, Vol. 2, IAEA, Vienna (1966).
6. G. V. Samsonov and A. D. Panasyuk, Teplofiz. Vys. Temp., 4:207 (1966).
7. V. S. Fomenko, Yu. B. Paderno, and G. V. Samsonov, Ogneupory, No. 1, 40 (1962).
8. L. N. Grossman, J. Am. Ceram. Soc., 48:236 (1965).
9. F. H. Morgan, J. Appl. Phys., 22:108 (1951).
10. D. S. Neel and C. D. Pears, "Progress in international research on thermodynamic and transport properties," Second Symposium on Thermophysical Properties (1952), p. 500.
11. T. Riethof, B. D. Accione, and E. R. Branyan, Temperature — Its Measurement and Control in Science and Industry, Vol. 3, 2 (1962), p. 515.
12. P. T. B. Shaffer, J. Am. Ceram. Soc., 46:177 (1963).
13. R. E. Taylor, J. Am. Ceram. Soc., 45:353 (1962).

Variation in the Lattice Parameters and Static Distortions in the Homogeneity Range of Transition-Metal Monocarbides

I. I. Timofeeva and L. A. Klochkov

The variation in the lattice parameters of carbides of metals of groups IV and V with reduction in their carbon content has been analyzed on the basis of the authors' own results and of data available in the literature. The static lattice distortions in the nonstoichiometric carbides of titanium, zirconium, and hafnium have been measured.

The monocarbides of transition metals of groups IV and V are known to consist of a fcc phase of the NaCl type having a more or less broad homogeneity range.

In order completely to characterize the carbides, it is necessary to know the relationship between the lattice parameters and the chemical composition. The results of numerous measurements of the lattice parameters of monocarbides are somewhat conflicting; this is due not so much to the inadequate precision of the x-ray methods used to determine the parameters as to the inadequate purity of the carbides investigated and to the incomplete chemical analysis of them. Nevertheless, if we trace the variation of the lattice parameter in the homogeneity range from the results of recent carefully conducted researches [1-3, 12-20], we can see the following features.

In all the monocarbides of transition metals of groups IV and V there is a tendency for the lattice parameter to decrease as the

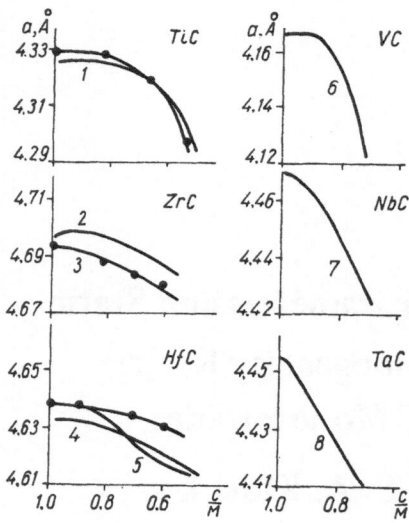

Fig. 1. The crystal lattice parameters of monocarbides of transition metals of groups IV and V as a function of the carbon content, according to the results of various authors: 1, 5) [16]; 2) [14]; 3, 4) [13]; 6) [20]; 7) [19]; 8) [12]; ●) present results.

carbide becomes more deficient in carbon. However, the variations in the lattice parameter differ in the carbides of the metals of the two groups: as can be seen from Fig. 1, the rate of change of the lattice parameter in all carbides of group V transition metals is greater than that of carbides of group IV transition metals. The observed difference can be explained on the assumption that the bonds in the carbides consist of two components [2, 4]: one is responsible for the Me−C bond and the other for the Me−Me bond. In stoichiometric carbides the bonds are strongly screened by the electron clouds of the carbon atoms [4, 11], and they are so far from one another that their influence is small. However, the lattice becomes more deficient in carbon, the screening of the Me−Me bonds becomes less, while the contribution of the Me−Me interaction increases correspondingly. This marks the beginning of a steeper fall in the lattice parameter as the distance from the stoichiometric composition increases. Then the course of the curves is determined mainly by the nature of the Me−Me interaction. The steep slope of the curves for the carbides of the group V transition metals is evidently due to the strengthening of the covalent bonds in the group V metals and in their quasimetallic compounds that arise from the increase in the number of electrons in the bonding part of the d-band on passing from group IV metals to group V metals. This corresponds to the higher statistical weight of the atoms with stable d^5 configurations in the localized part of the valence electrons (which are the bearers of the covalent properties) in group V metals as compared with group IV metals [9, 10].

The reduction in the interplanar distances with impoverishment of the carbide in carbon, which is common to all the carbides, may be regarded as a consequence of the defects formed in the lattice — unoccupied sites in the carbon sublattice. These defects must give rise to considerable static displacements of the atoms in the carbide lattice in the homogeneity range, and this is in fact observed in titanium carbide containing 30 at.% C [7].

Many physical properties of solids are known to be determined not only by the arrangement of the atoms at the points of the ideal crystal lattice, but often also by the deviations from this arrangement. Evidently the variations in the lattice parameters of the transition-metal carbides in the homogeneity range may also be due to the static distortions of the lattice.

We have determined the lattice parameters and static displacements of the atoms in the lattices of the carbides of titanium, zirconium, and hafnium in the homogeneity range. The carbide specimens were produced by direct synthesis of the metal concerned and carefully heated carbon black in a vacuum furnace under a pressure of 10^{-3} mm Hg. The temperature used in producing the titanium carbides was 1700°C, that for zirconium carbides 1800°C, and that for hafnium carbides 2100°C. For taking the x-ray photographs, the specimens were crushed in a metal mortar; then they were annealed in a zirconium carbide crucible in a vacuum furnace at 900-1000°C (pressure 10^{-3} mm Hg) and the resolution of the $K_{\alpha_1 \alpha_2}$ doublet of the outermost lines was checked.

The photographs were taken on a URS-50IM diffractometer with scintillation recording of the x-rays in filtered CuK_{α} radiation. The lattice parameter was determined from the center of gravity of the diffraction lines. The photographs were calibrated against the lines from single-crystal quartz. The accuracy of determination of the lattice parameters was ±0.0005 Å.

In determining the static distortions in the crystal lattice of low-carbon carbides of titanium, zirconium, and hafnium, we took as specimens having an undistorted lattice the respective carbides having a composition close to the stoichiometric ($TiC_{0.97}$, $ZrC_{0.96}$, $HfC_{1.0}$).

It follows from the theory of x-ray scattering [5] that static displacements of the atoms lead to a reduction in the intensity of the corresponding interference maxima. A photograph with a standard was used to eliminate the influence of extinction on the intensity of the first lines [7]. In the specimens investigated, we compared the intensity ratios of the (511) lines of titanium carbide, those of the (531) lines of zirconium carbide, and those of the (440) lines of hafnium carbide — of stoichiometric composition and in the homogeneity range —with the (422) lines of aluminum, which was adopted as standard for the photograph. In calculating the static distortions, we took into account differences in the values of a number of line-intensity factors in the x-ray diagrams of the carbides in the homogeneity range.

It should be noted that there actually exist in the lattice not only static but also dynamic distortions due to the thermal vibrations of the atoms. However, because the carbides investigated differ in their refractoriness and the photographs were taken at room temperature, the main contribution to the change in intensity with reduction of the carbon content is made by the static distortions.

As can be seen from the results obtained (Table 1), a considerable increase in the static lattice distortions with rise in the deficiency of the carbide in carbon is characteristic of all the specimens. The greatest static displacements of the atoms in the series of transition-metal carbides of group IV occur in titanium carbide, diminishing as we pass to the carbides of zirconium and hafnium. This is evidently due to the fact that the energy stability of the d^5 configurations of the localized part of the valence electrons of the metal atoms increases with rise in the principal quantum number; the covalency increases and so does the "tightness" of the Me—Me interaction in the carbide lattice, which prevents displacement of the atoms in it.

The results obtained in the present research for the variation of the lattice parameters of the carbides of titanium, zirconium, and hafnium in the homogeneity range are in agreement with the results reported in [13, 16].

From a comparison of the variation in the lattice parameters of the carbides of titanium, zirconium, and hafnium in the homogeneity range with the size of the static displacements of the atoms

TABLE 1. Static Lattice Distortions $(\overline{u^2})^{\frac{1}{2}}$
in the Homogeneity Range of
TiC, ZrC, and HfC

| Composition | Content, wt.% | | Lattice parameter, Å | $\sqrt{\overline{u^2}}$ |
	metal	C_{comb}		
$TiC_{0.97}$	80.6	19.4	4.3304	—
$TiC_{0.82}$	82.7	17.2	4.3309	0.160
$TiC_{0.66}$	85.6	14.4	4.3211	0.184
$TiC_{0.50}$	88.0	12.0	4.2975	0.199
$ZrC_{0.96}$	89.0	11.3	4.6929	—
$ZrC_{0.8}$	90.3	9.8	4.6886	0.066
$ZrC_{0.7}$	91.2	8.4	4.6838	0.120
$ZrC_{0.6}$	91.9	7.4	4.6799	0.138
$HfC_{1.0}$	93.8	6.4	4.6392	—
$HfC_{0.9}$	94.0	6.0	4.6378	0.080
$HfC_{0.7}$	95.1	4.9	4.6338	0.091
$HfC_{0.6}$	96.3	3.8	4.6318	0.095

in the lattice, it may be concluded that a greater change in the parameter is accompanied by larger static distortions in the lattice
of the corresponding carbide. It should be noted that the static
distortions are produced as a result of random displacements
of the atoms from their equilibrium positions in the lattice and
that they come into equilibrium in volumes of the same order of
magnitude as the unit cell, while the lattice parameter gives the
mean value of the interatomic distances.

In order to obtain information about the joint effect of the static
distortions and the change in parameter on the state of the crystal lattice of the carbides as they become impoverished in car-
bon, it is interesting to consider the ratio $(\overline{u^2})^{\frac{1}{2}}/a$, where $(\overline{u^2})^{\frac{1}{2}}$
represents the static displacements of the atoms from their equi-
librium positions (the static distortions of the crystal lattice) and
a is the lattice parameter of the carbide concerned. The rela-
tionship between $(\overline{u^2})^{\frac{1}{2}}/a$ and the carbon content of the carbide is
shown in Fig. 2.

Noteworthy is the sharper increase in the ratio $(\overline{u^2})^{\frac{1}{2}}/a$ (and
also in $(\overline{u^2})^{\frac{1}{2}}$) with reduction in carbon content of the zirconium
carbide. According to [9, 10, 6, 8], this may be provisionally interpreted in the following way. In this carbide it is evidently pos-

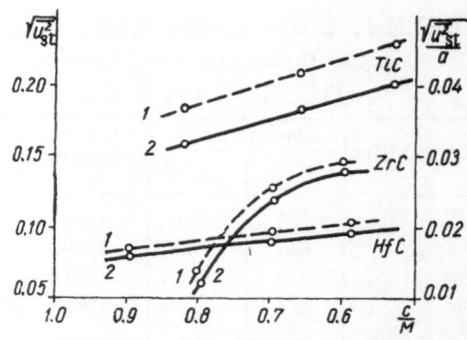

Fig. 2. Static distortions $\overline{(u^2)}^{\frac{1}{2}}$ (1) and $\overline{(u^2)}^{\frac{1}{2}}/a$ ($a =$ lattice parameter) as a function of the carbon content of carbides of transition metals of group IV.

sible for the valence electrons to be partially localized in the completely unoccupied $4f$-shell of zirconium, thereby reducing the statistical weight of the stable d^5 states and consequently the degree of covalency of the bond. In the case of hafnium carbide, the effect of the corresponding transfers to the $5f$ level is offset by the increasing energy stability of the d^5 configurations of the hafnium atoms in HfC.

Literature Cited

1. R. G. Avarbe et al., Zh. Priklad. Khim., 35:1976 (1963).
2. A. I. Avgustinik et al., Izv. Akad. Nauk SSSR, Neorg. Mat., 3:286 (1967).
3. S. I. Alyamovskii, P. V. Gel'd, and I. I. Matvienko, Zh. Strukt. Khim., 2:445 (1961).
4. P. V. Gel'd and V. A. Tskhai, Fiz. Metal. i Metalloved., 16:493 (1963).
5. R. James, Optical Principles of X-Ray Diffraction [Russian translation], IL, Moscow (1950).
6. I. R. Kozlova, V. N. Gurin, and A. P. Obukhov, Poroshkovaya Met., No. 12, 68 (1966).
7. S. M. Nikolaeva and Ya. S. Umanskii, Izv. Akad. Nauk SSSR, Ser. Fiz., 20:631 (1956).
8. I. F. Pryadko, Poroshkovaya Met., No. 12, 61 (1966).
9. G. V. Samsonov, Ukrainsk. Khim. Zh., 31:1233 (1965).
10. G. V. Samsonov, Poroshkovaya Met., No. 12, 49 (1966).
11. V. A. Tskhai and P. V. Gel'd, Trudy Inst. Khimii UFAM, Akad. Nauk SSSR, 9 (1966).
12. L. Bowman Allen, J. Phys. Chem., 65:97 (1961).
13. F. Benesovsky and E. Rudy, Planseeber. Pulvermet., 8:66 (1960).
14. H. Bittner and H. Goretzki, Monatsch. Chem., 93:1000 (1962).

15. R. Lesser Brauer, Z. Metallk., 49:12 (1958).
16. H. Goretzki, Untersuchung der magnetischen, elektrischen und termoelek-
     trischen Eigenschaften der Karbide und Nitride der 4-5a Übergangsmetalle
     (1963).
17. R. V. Sara, J. Am. Ceram. Soc., 48:243 (1965).
18. R. V. Sara, J. Met. Soc. AIME, 233:1683 (1965).
19. E. K. Storms and H. Krikorian, J. Phys. Chem., 63:10 (1959).
20. E. K. Storms and R. S. McNeal, J. Phys. Chem., 66:1401 (1962).

Electron-Fractographic Study of the Fractures in Zirconium Carbide after Tensile Testing in the Range 20-2000°C

V. N. Turchin, G. A. Rymashevskii, and A. G. Lanin

A systematic study has been made of the fractures in ZrC specimens tested in tension in the range 20-2000°C, using optical magnification and the electron microscope. It was found that failure at all temperatures was of the macrobrittle type. It was concluded from the tests that the development of microplastic processes, which adsorb considerable energy during fracture, begins long before the region of the brittle−ductile transition (2000-2200°C) is reached.

The fractures of ZrC specimens tested in tension in the range 20-2000°C have been systematically studied by means of optical magnification and the electron microscope, using the ordinary method of two-stage carbon replicas [3].

Analysis shows that fracture at all temperatures of testing is of the macrobrittle type. The fracture surface at 20°C is finely crystalline and consists of typical brittle relief elements (river pattern, "terraces," Walner lines) (Fig. 1). Failure takes place both at the grain boundaries and through the body of the grains [2]. At 1000°C, besides the brittle elements, there are regions in which the well-known microplasticity [1] begins to appear during the development of fracture. Brittle relief elements are entirely absent from the fractures of specimens tested at 1500 and 2000°C.

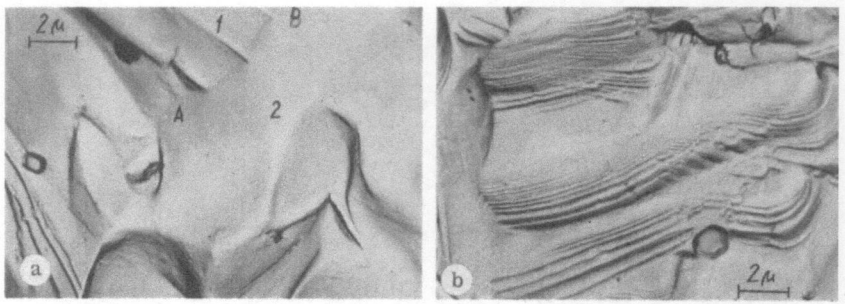

Fig. 1. Fracture in ZrC at 20°C.

The existence of microplasticity during slip gives to the river
patterns the appearance of "ragged ridges" formed by shear [1]
(Fig. 2). In the fracture at 2000°C there are characteristic bands
with a relatively high degree of deformation. A rise in the test
temperature is accompanied by a reduction in the mean size of the
facets on the fracture of the specimen. A relationship of this kind
is well known in carbon steel tested in the region of the brittle—
ductile transition [5]. At elevated temperatures the facets present
a distorted surface, which evidently results from the additional
plastic deformation that occurs even after the formation of local
discontinuities [4].

From what has been said above, it may be concluded that
microplastic processes, which absorb considerable energy, begin to develop long before the region of the brittle—ductile transition (2000-2200°C), as determined by tensile tests, is reached,
and the well-known analogy with the behavior of steels tested in
the brittle range appears.

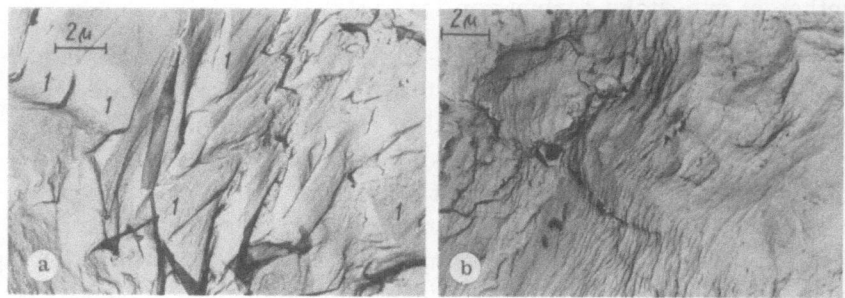

Fig. 2. Fracture in ZrC at 1500°C.

Literature Cited

1. R. G. Arons et al., Zavod. Lab., No. 11 (1965).
2. R. John et al., Review of the Microstructural Features in Shear Failure. The Atomic Mechanism of Failure [Russian translation], Metallurgizdat, Moscow (1963).
3. A. N. Pilyankevich, The Practice of Electron Microscopy [in Russian], Mashgiz, Moscow (1961).
4. N. P. Shapov, Zavod. Lab., Vol. 17, No. 2 (1951).
5. E. M. Shevandin and I. A. Razov, Cold Brittleness and the Limiting Plasticity of Metals in Shipbuilding [in Russian], Nauka, Leningrad (1965).

Physical Properties of Carbides of
Rare-Earth Metals

V. L. Yupko, G. N. Makarenko, and
Yu. B. Paderno

The results are reported of a complex study of the physical properties
of a large group of dicarbides and some sesquicarbides of rare-earth
metals carried out on the same specimens. Regular features were
found in the changes in properties of the carbides at the transition
from one phase to another. A diagram of the band structure of the
rare-earth metal carbides is proposed.

The present article is concerned with a detailed study of the
physical properties of rare-earth-metal carbides having the compositions MeC_2 and Me_2C_3.

Powders of all the dicarbides except SmC_2 were produced by
reduction of the oxides of the respective metals by carbon black at
1800-1900°C for 1 h in a vacuum of 10^{-1}-10^{-2} mm Hg. Compact
specimens were obtained by sintering the powders by hot compaction in graphite dies in an argon atmosphere at 1700-1800°C for
15 min under a pressure of 100 kg/cm^2. The variation in chemical composition of the specimens lay within the limits of analytical error (±0.1%). The porosity of the specimens was 3-20%.
The sesquicarbides and the dicarbide SmC_2 were synthesized from
the elements by arc melting.

All the dicarbides had a tetragonal structure of the CaC_2 type,
while the sesquicarbides had a cubic structure of the Pu_2C_3 type.

TABLE 1. Chemical Composition and Crystal Lattice Parameters of Rare-Earth Metal Carbides Investigated

Carbide	Content, wt.%				Lattice parameter, Å	
	Me	C_{comb}	C_{free}	$Me + C_{com}$	a	c
YC_2	78.8	20.8	1.00	99.6	3.66	6.10
LaC_2	85.5	14.5	0.20	100.0	3.94	6.57
CeC_2	85.2	14.2	0.15	99.4	3.88	6.46
PrC_2	85.2	14.6	0.50	99.8	3.86	6.40
NdC_2	85.3	14.3	0.30	99.6	3.83	6.40
SmC_2	86.1	14.0	not found	100.1	—	—
GdC_2	87.0	13.0	» »	100.0	3.71	6.26
TbC_2	86.7	13.3	» »	100.0	—	—
DyC_2	87.1	12.5	0.30	99.6	3.66	6.15
ErC_2	87.3	12.2	0.40	99.5	3.62	6.07
TuC_2	87.2	11.9	1.10	99.1	3.60	6.04
Y_2C_3	84.0	15.8	not found	99.8	—	—
La_2C_3	88.6	11.6	» »	100.2	8.82	—
Ce_2C_3	84.4	11.7	» »	100.1	8.44	—
Nd_2C_3	88.6	11.1	» »	99.7	8.54	—

TABLE 2. Some Physical Properties of Rare-Earth Metal Carbides

Carbide	Hall coefficient $R \cdot 10^4$, cm^3/C (20°C)	Specific electrical resistivity ρ, $\mu\Omega \cdot$cm (20°C)	Thermal coefficient of electrical resistivity $\alpha_{20-800} \cdot 10^3$, 1/deg	Coefficient of thermo-emf α, μV/deg (20°C)	Coefficient of thermal expansion $\alpha_{100-900}$ 10^6, deg^{-1}	Melting point, °C	Temperature range of phase transformation, °C
YC_2	-11 ± 1	30 ± 2	3.0	-4.9 ± 0.5	9.2 ± 0.3	2210 ± 60	1000—1170
LaC_2	-5.3 ± 0.2	45 ± 2	3.9 ± 0.4	-1.8 ± 0.5	11.9 ± 0.3	2360 ± 60	900—1070
CeC_2	-2.8 ± 0.1	60 ± 2	2.5 ± 0.3	-0.2 ± 0.5	13.5 ± 0.3	2290 ± 60	850—1050
PrC_2	-5.40 ± 0.2	36 ± 3	3.9	-2.4 ± 0.5	12.1 ± 0.3	2160 ± 60	940—1080
NdC_2	-4.5 ± 0.2	49 ± 5	2.4 ± 0.3	-2.1 ± 0.5	9.3 ± 0.3	2260 ± 60	900—1140
SmC_2	-4.4 ± 0.2	34 ± 1	3.1 ± 0.3	-0.3 ± 0.5	11.4 ± 0.3	—	1100—1220
GdC_2	-3.0 ± 0.1	35 ± 5	2.8 ± 0.3	-1.8 ± 0.5	13.5 ± 0.3	2260 ± 60	1050—1250
TbC_2	-2.6 ± 0.1	35 ± 1	2.8 ± 0.3	-1.8 ± 0.5	10.2 ± 0.3	2100 ± 60	1100—1220
DyC_2	-3.0 ± 0.1	32 ± 2	2.9 ± 0.3	-2.1 ± 0.5	11.4 ± 0.3	2250 ± 60	1150—1300
ErC_2	-3.2 ± 0.1	57 ± 10	2.2 ± 0.3	-2.8 ± 0.5	12.4 ± 0.3	2280 ± 60	1050—1270
TuC_2	$+136 \pm 4$	451 ± 15	0.30 ± 0.03	-4.2 ± 0.5	10.7 ± 0.3	2180 ± 60	1120—1370
Y_2C_3	-34 ± 1	446 ± 13	0.45 ± 0.04	-11.1 ± 0.0	10.3 ± 0.3	1800 ± 60	—
La_2C_3	-0.8 ± 0.1	340 ± 10	—	-1.9 ± 0.5	14.6 ± 0.3	1430 ± 50	—
Ce_2C_3	-1.4 ± 0.04	398 ± 12	—	-2.2 ± 0.5	14.7 ± 0.3	1530 ± 50	—
Nd_2C_3	$+12.8 \pm 0.4$	322 ± 10	—	-2.2 ± 0.5	13.5 ± 0.3	1620 ± 50	—

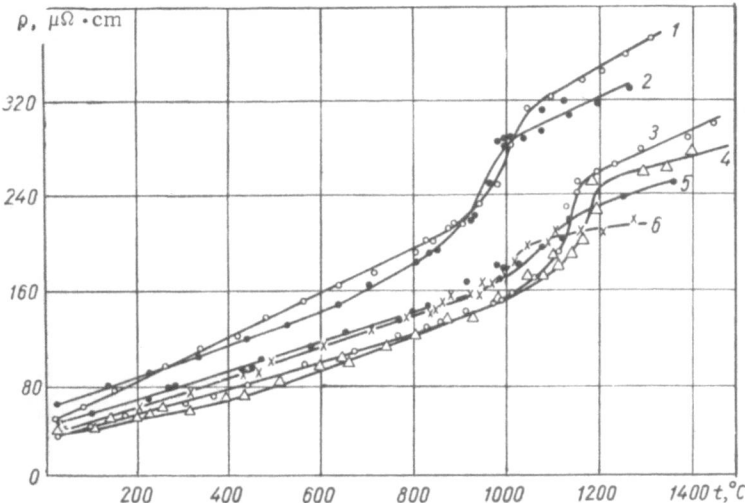

Fig. 1. Temperature dependence of the specific electrical resistivity of the dicarbides LaC$_2$ (1), CeC$_2$ (2), YC$_2$ (3), SmC$_2$ (4), NdC$_2$ (5), and PrC$_2$ (6).

The lattice parameters were in satisfactory agreement with data in the literature [10, 11], and the ratio of the components in the compounds was close to the stoichiometric ratio. Before measurements were made, the specimens were homogenized for 8 h, at 1600°C in the case of the dicarbides and at 1100°C in that of the sesquicarbides. The compositions of the carbide phases investigated in the present research are given in Table 1.

The Hall effect at room temperature, the specific electrical resistivity in the range 20-1500°C, the coefficient of the thermo-emf at 40-1300°C, and the coefficient of thermal expansion in the range 100-900°C were investigated, and the melting points of the respective carbides were also determined. The measurements were made by methods previously described in [7, 9, 6], except in the case of the coefficient of the thermo-emf, which was measured on an apparatus we developed with controlled heating of the thermocouple.

As has been noted in [5, 8], the small values of the specific electrical resistivity and the coefficient of the thermo-emf in dicarbides (Table 2) and the nature of their temperature dependence (Figs. 1 and 2) indicate the existence of a considerable proportion of metallic bonding in these compounds. Calculation

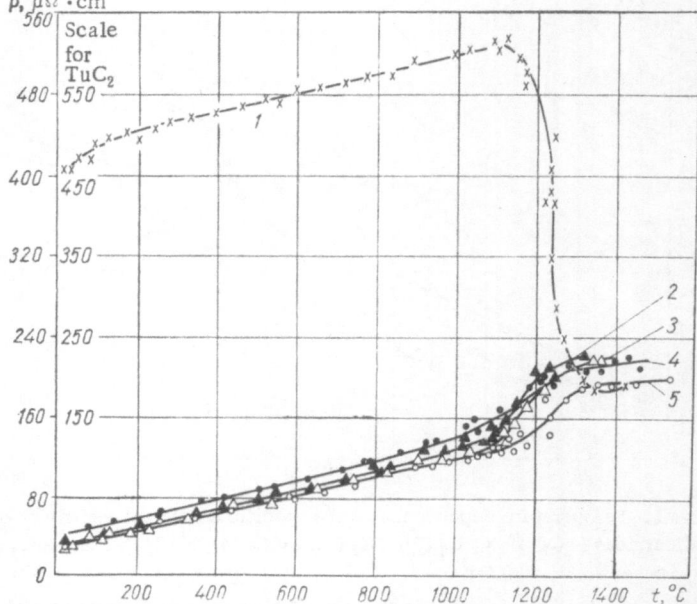

Fig. 2. Temperature dependence of the specific electrical resistivity of the dicarbides TuC_2 (1), TbC_2 (2), GdC_2 (3), ErC_2 (4), and DyC_2 (5).

TABLE 3. Physical Characteristics of Rare-Earth Metal Carbides

Carbide	Effective carrier concentration $n^* \cdot 10^{-22}$, deg^{-1}	Effective carrier concentration per metal atom n_a^*, 1/Me atom	$1/\rho m\theta^2 \cdot 10^9$, $\mu\Omega^{-1} \cdot cm^{-1} \cdot g^{-1} \cdot deg^{-2}$	$\delta \cdot 10^{-23}$ cm/V$^2 \cdot$ sec^2	R_{mc}/R_{mm}	Characteristic temperature θ, °K	Mean square amplitude $\sqrt{\overline{u^2}}$, Å
YC_2	0.60	0.25	2.30	—77.0	1.017	381	0.17
LaC_2	1.20	0.60	1.40	—16.3	1.053	316	0.16
CeC_2	2.23	1.10	1.00	—4.9	1.065	310	0.16
PrC_2	1.17	0.55	1.90	—25.8	1.055	298	0.17
NdC_2	1.36	0.64	1.30	—11.7	1.050	311	0.16
SmC_2	1.44	0.65	—	—23.5	1.047	—	—
GdC_2	2.07	0.90	1.40	—15.4	1.032	337	0.15
TbC_2	2.25	1.04	1.80	—13.0	1.036	294	0.16
DyC_2	2.11	0.86	1.80	—18.0	1.035	304	0.15
ErC_2	1.98	0.79	0.97	—6.1	1.032	308	0.15
TuC_2	0.04	0.02	0.13	+4.2	1.031	296	0.15
Y_2C_3	0.18	0.06	—	—1.1	—	—	—
La_2C_3	7.80	3.30	0.26	—0.04	0,968	189	0.20
CeC_3	4.62	1.80	0.20	—0.05	0,995	199	0.19
Nd_2C_3	0.51	0.20	0,23	+0.65	—	203	0,19

of the values of the effective carrier concentration n*, using the
Hall-effect data, in the single-band-model approximation also
gives values that are typical of compounds of the metallic type
(Table 3). To judge from the signs of the Hall effect and the thermo-emf, transfer in the dicarbides takes place mainly by electrons.

From the nature of the temperature dependence of the electrical resistivity, thulium dicarbide is closer to the semimetals
(see Fig. 2). The appearance in the sign of the Hall effect of this
compound of positive current carriers at a value of the parameter
$1/\rho m\odot$ which is small in comparison with other dicarbides makes
it possible to relate the anomalous nature of its electrical properties with a reduction in the concentration of conduction electrons.

The rare-earth metal sesquicarbides, which are similar to
TuC_2 as regards the high level of their electrical resistivity and
the small values of the thermal coefficient of the electrical resistivity and of the parameter $1/\rho\, m\odot$, are also characterized by a
reduction in the concentration of conduction electrons as compared
with the quasimetallic dicarbides; this is confirmed by the small
value of n* for the carbide Y_2C_3 and the positive value of the Hall
effect in the carbide Nd_2C_3 (see Tables 2 and 3). High values of n*
for the conduction electrons in the carbides La_2C_3 and Ce_2C_3 as
compared with the dicarbides may be due to the calculated results being too great owing to the fact that the single-band model
is applicable to only a small extent to these compounds; this follows
from the smallness of their δ parameter (see Table 3), which can
be used as a criterion of the applicability of the single-band model
[4]. The least excessive estimate of the value of n* for the conduction electrons from this point of view is for the carbide Y_2C_3,
0.18×10^{22} electrons/cm^3, which is evidently nearest to the true
concentration of conduction electrons in carbides of the Me_2C_3
type.

Analysis of the results obtained in the present research leads
to certain conclusions regarding the energy structure of the compounds studied.

A consideration of the values of the parameter δ for the quasimetallic rare-earth metal carbides show that the most accurate
estimate of the effective concentration of conduction electrons per
metal atom n_a^* for the dicarbide YC_2 is 0.25 electrons/Me atom.
This value is evidently the nearest to the true concentration of conduction electrons in dicarbides of trivalent rare-earth metals.

An x-ray spectral investigation has shown [11] that conduction in rare-earth-metal dicarbides takes place in the d band of the metal. It may therefore be assumed that transfer in it by means of electrons is due to the very small filling capacity of the latter. The smallness of the estimated value of n_a^* for the conduction electrons in dicarbides as compared with unity indicates that a large part of the 5d electrons in the rare-earth metal atoms enter the antibonding orbitals of the C_2 complexes, where the great stability of the length of the $C-C$ bond in the series of dicarbides is due in all probability to the approximately identical filling capacity of the antibonding orbitals by electrons. The appearance of metallic properties and the increase in length of the $C-C$ bonds of the C_2 complexes as we pass from CaC_2 to the rare-earth-metal dicarbides are evidence of the simultaneous filling of the conduction band of the metal and of the band of the antibonding states of the C_2 complexes by 5d electrons of the atoms as a result of their energy overlap.

The appreciable contribution made by holes to the electrophysical properties of these dicarbides is evidently connected with the unoccupied states in the band of the antibonding states of the C_2 complexes, since, by analogy with the dicarbide CaC_2, it is to be expected that the band corresponding to the $Me-C$ bonds is filled in this class of compound. The smallness of n_a^* in the case of TuC_2, in which holes determine the sign of the Hall effect, points to the almost complete filling of the band of the antibonding states of the C_2 complexes in the dicarbides; this is a result of its splitting under the influence of the crystalline field into two subbands, each with a capacity of one electron from the metal atom, as has been suggested for the carbide VC [1].

Thus, it may be concluded that the concentration of electrons and holes in rare-earth metal dicarbides is considerably less than unity, and that the comparatively small contribution made by holes to the electrical resistivity, Hall effect, and thermo-emf coefficient is due to their low mobility; this is a consequence of the great density of the band of the antibonding states resulting from the great distance of the C_2 complexes from one another (3.5 Å). This leads to the conclusion that its degree of spread in the series of rare-earth metal dicarbides does not vary so much as the conduction bands of the metal; this conclusion is supported by the ex-

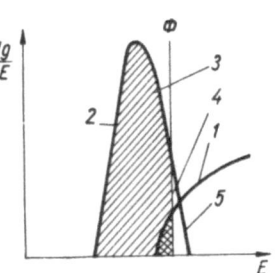

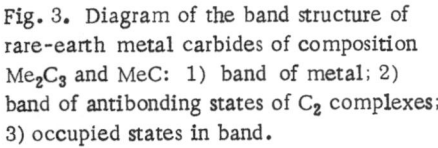

Fig. 3. Diagram of the band structure of rare-earth metal carbides of composition Me_2C_3 and MeC: 1) band of metal; 2) band of antibonding states of C_2 complexes; 3) occupied states in band.

istence of a correlation between the electrophysical properties and the degree of overlap of the wave functions of neighboring metal atoms in the dicarbides.* In YC_2 in which, in accordance with the value of the ratio R_{mm}/R_{mc} (see Table 3), we find the strongest Me−Me interaction among all the dicarbides investigated, the conduction band of the metal spreads out and the conduction electrons have the greatest mobility; this leads to comparatively large (in absolute terms) negative values of the Hall effect and of the coefficient of the thermo-emf, and to a small value of the specific electrical resistivity. The dicarbide CeC_2, in which the Me−Me interaction is weakest, has high values of the electrical resistivity and small values of the Hall effect and the coefficient of the thermo-emf; while the other dicarbides (which have intermediate values of the ratio R_{mc}/R_{mm}) have intermediate values of these properties.

The higher values of n_a^* for dicarbides of the second subgroup of the rare-earth metals as compared with those of the first evidently reflect qualitatively a certain increase in the carrier concentration in the band of the metal with rise in the atomic number of the rare-earth element, in agreement with the greater probability of a $f \rightarrow d$ transition in the second subgroup [2, 3].

The analysis we have made leads us to put forward the following diagram of the band structure of quasimetallic carbides of rare-earth metals (Fig. 3). The low-density conduction band of the metal and the high-density band of the antibonding states of the C_2 complexes overlap to only a slight extent, and this leads to lit-

---

*The extent of Me−Me interaction in the carbides can be assessed from the value of the ratio R_{mm}/R_{mc}, where R_{mm} and R_{mc} are the radii of the metal atoms in the metals and the corresponding carbides (see Table 3).

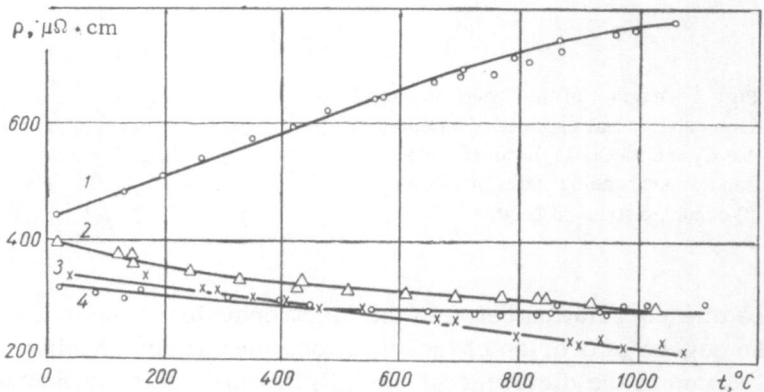

Fig. 4. Temperature dependence of the specific electrical resistivity of the rare-earth metal sesquicarbides: Y_2C_3 (1), Ce_2C_3 (2), LaC_2C_3 (3), and Nd_2C_3 (4).

tle filling of the former and to almost complete filling of the latter by 5d electrons of the rare-earth metal atoms. The diagram assumes a smaller dependence of the degree of filling of the band of antibonding states of the C_2 complexes on the valence state of the metal than in the case of the conduction band of the metal, which is important for our understanding of the electrical properties of the dicarbides.

The semimetallic properties of TuC_2 and the appearance of holes in its Hall effect in accordance with the proposed diagram can be attributed to a reduction in the effective valence of the metal in this compound as compared with the other dicarbides; this results in a fall in the concentration of conduction electrons and to an increase in the hole concentration with a shift of the Fermi level in the lower-energy direction.

The similarity of the electrophysical properties of rare-earth-metal carbides of the Me_2C_3 type with those of TuC_2 enables us to extend the band-structure diagram to this class of compounds. The reduction in the contribution made by electrons to the electrical transfer in carbides of the Me_2C_3 type may be due to the breakdown of part of the C_2^{-2} complexes into C^{-4} ions. An approximate calculation shows that to explain the electrophysical properties of Me_2C_3-type carbides it is essential that not more than 30% of the carbon shall be present in the form of C^{-4} ions, and this is confirmed by the results of analysis of the products of decomposition of the carbides by water [3].

The reduction in the electrical resistivity of the sesquicarbides with rise in temperature (Fig. 4) is evidence of an increase in the filling capacity of the conduction band of the metal resulting from the transfer to it of electrons from the overlapping high–density band of the antibonding states of the C_2 complexes, after which they participate more effectively in the electrical transfer.

Estimates of the Debye temperatures and mean square amplitudes of the elastic vibrations of the structural complexes of the carbides calculated from their melting points (see Table 3) indicate the weakness of the atomic interaction in carbides of the Me_2C_3 type as compared with dicarbides. The same conclusion is reached by comparing the coefficients of thermal expansion of carbides of the Me_2C_3 and MeC_2 types of the same rare-earth metals.

Literature Cited

1.   A. I. Avgustinik et al., Izv. Akad. Nauk SSSR, Neorg. Mat., 3:286 (1967).
2.   M. A. El'yashevich, Rare-Earth Spectra [in Russian], Gostekhizdat, Moscow (1953).
3.   T. Ya. Kosolapova, G. N. Makarenko, and L. T. Domasevich, Zh. Priklad. Khim. (1970).
4.   S. N. L'vov, V. F. Nemchenko, and G. V. Samsonov, Dokl. Akad. Nauk SSSR, 135:34 (1960).
5.   G. N. Makarenko et al., Izv. Akad. Nauk SSSR, Neorg. Mat., 1:1787 (1965).
6.   V. N. Paderno, I. G. Barantseva, and V. L. Yupko, High-Temperature Inorganic Materials [in Russian], Naukova Dumka, Kiev (1965), p. 88.
7.   Yu. B. Paderno, B. M. Rud, and V. A. Virovtseva, Informatsionnoe Pis'mo PIM, Akad. Nauk UkrSSR, 37 (1966).
8.   Yu. B. Paderno and V. L. Yupko, Izv. Akad. Nauk SSSR, Neorg. Mat., 3:398 (1967).
9.   V. S. Sinel'nikova and V. A. Virovtseva, Zavod. Lab., 8:37 (1961).
10. F. Spedding, K. Gschneider, and A. Daane, J. Am. Chem. Soc., 80:4499 (1959).
11. R. Vickery, R. Sedlacek, and A. Rubens, J. Chem. Soc., No. 2, 498 (1958).

Regularities in the Variation of the Electrophysical Properties of Solid-Solution Alloys of Some Carbides of Transition Metals of Groups IV and V

A. I. Avgustinik, O. A. Golikova, L. V. Kudryasheva, and S. S. Ordan'yan

Solid-solution alloys of the systems ZrC–NbC, ZrC–TaC, TiC–NbC, and TiC–TaC have been prepared, and their electrical resistivity, thermo-emf, and Hall effect have been studied at room temperature. Complete similarity was found in the variation of these electrophysical parameters in stoichiometric solid solutions in the systems Me^{IV}– $Me^{V}C$. The concentration dependence of the above parameters has been analyzed.

The isostructural nature and the closeness of the physical properties of the monocarbides of transition metals of groups IV and V does not result in complete similarity between their thermodynamic [1], crystallochemical [7], and electro- and thermophysical properties. There is a substantial difference between the effective concentrations of free current carriers in $Me^{IV}C$ and $Me^{V}C$ carbides, which may be attributed to differences in their electron structure [16]. An investigation of the electrophysical properties of TiC–VC solid solutions has shown [23] that in monocarbides of group IV metals having eight valence electrons, there is about 0.1 electron per formula unit. It is possible that carbides of group IV transition metals, which are characterized by a very small density of states in the vicinity of the Fermi level, have semiconducting properties [12, 18] when produced in strictly stoichiometric form (with the chemical bond entirely of the covalent type and all electrons localized in the Me − C bonds). However, the presence of even a single vacancy in the metalloid sublattice results in the formation of one or two unscreened Me − Me bonds [8, 17] based on Me − Me interaction, and a conduction band is created.

On the other hand, monocarbides of group V transition metals (having nine valence electrons) have one or more electrons per formula unit, four of the electrons of the metal atom being used to create a localized Me − C bond, while a fifth passes into the conduction band.

A mechanism for conduction in monocarbides of metals of groups IV and V has been proposed. The authors of [21, 28] have suggested that the current-carrier concentration in these carbides is determined by the total number of valence electrons in the metal and carbon. In this case defective carbides of group V metals of composition $Me^V C_{0.75}$ should approximate in current-carrier concentration to carbides of group IV metals, i.e., the current-carrier concentration should fall continuously with decrease in carbon content. Experimentally, it is found that the current-carrier concentration in NbC_x, assuming unary electron conduction in the homogeneity range, passes through a minimum at $x \approx 0.80$–0.85.

Thus, the nature of the variation of the current-carrier concentration, electrical resistivity, and thermo-emf in monocarbides of group V metals can be accounted for by assuming that, in addition to the Me − C valence band, there is a band overlapping it which is responsible for the Me − Me bond and which broadens as the number of carbon vacancies increases [9].

With a knowledge of the nature of the variation of some of the electrophysical properties of monocarbides of transition metals of groups IV and V, it is interesting to trace the effect of gradually substituting in the carbide group V metal atoms for group IV metal atoms on the electrical properties (resistivity, thermo-emf,

Hall effect), i.e., to study the change in properties of stoichiometric $Me^{IV}C - Me^{V}C$ solid solutions.

With this object in view, solid-solution alloys of the systems ZrC−NbC, ZrC−TaC, TiC−NbC, and TiC−TaC were prepared from powdered zirconium, niobium, and titanium of 99.8% purity, tantalum (99.2%), and acetylene soot (99.99%) heated at 1000°C in a stream of argon in a Tammann furnace.

The alloys were synthesized by holding them for 4-6 h at 2000°C in a vacuum of the order of 5×10^{-5} mm Hg.

The composition of the alloys was checked by chemical and x-ray structural analysis. Chemical analyses were carried out to determine the contents of the three components of the $Me^{IV}C - Me^{V}C$ solid solutions investigated. The nitrogen content did not exceed 0.05 wt.%.

The results of the x-ray analyses showed that all the alloys were single phase and had an fcc lattice of the NaCl type. The lattice parameters of the solid-solution alloys varied in accordance with Vegard's law with a small deviation in the negative di-

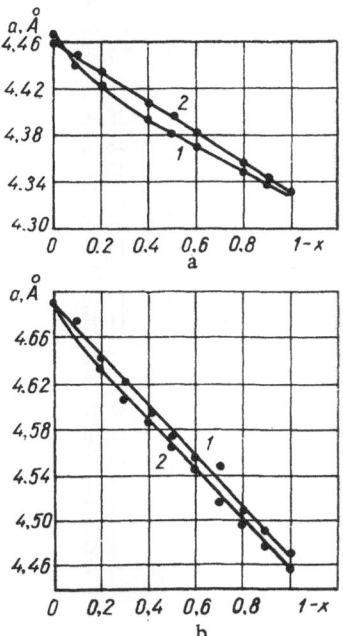

Fig. 1. Variation of lattice parameters with composition for: a) (NbC,TaC)−TiC: 1) $Nb_xTi_{1-x}C_{1.0}$; 2) $Ta_xTi_{1-x}C_{1.0}$; b) ZrC−(NbC,TaC): 1) $Zr_xNb_{1-x}C_{1.0}$; 2) $Zr_x \cdot Ta_{1-x}C_{1.0}$.

TABLE 1. Carbon Content in Alloys Investigated

Composi-tion	Calculated composition					Carbon content from chemical analysis, wt.%
	Me^{IV}		Me^V		C^*, at.%	
	at.%	wt.%	at.%	wt.%		
ZrC — NbC	47	82.98	3	5.39	11.63	11.17
	45	79.40	5	8.99	11.61	11.29
	40	70.45	10	17.94	11.61	11.54
	35	61.55	15	26.87	11.58	11.56
	30	52.67	20	35.77	11.56	11.40
	25	43.82	25	44.64	11.54	11.27
	20	35.00	30	53.48	11.52	11.49
	15	26.21	35	62.29	11.50	11.50
	10	17.46	40	71.07	11.47	11.45
	5	8.71	45	79.83	11.46	11.29
ZrC — TaC	47	78.95	3	9.99	11.06	11.12
	45	73.17	5	16.30	10.53	—
	40	60.24	10	29.86	9.90	9.86
	35	49.09	15	41.70	9.21	8.98
	30	39.35	20	52.03	8.62	8.42
	25	30.81	25	61.08	8.11	8.25
	20	23.24	30	69.14	7.62	7.50
	15	16.49	35	76.40	7.11	7.18
	10	10.43	40	82.71	6.86	6.76
	5	4.96	45	88.51	6.53	6.70
TiC — NbC	50	80.00	—	—	20.00	19.80
	45	66.98	5	14.42	18.60	18.95
	40	55.65	10	26.96	17.39	17.60
	30	36.30	20	47.80	15.52	15.80
	25	29.16	25	56.12	14.60	14.58
	18	19.46	32	67.03	13.51	13.65
	10	10.00	40	77.50	12.50	12.78
	5	4.78	45	83.28	11.94	12.00
	—	—	50	88.55	11.47	11.35
TiC — TaC	50	80.00	—	—	20.00	19.75
	45	58.94	5	24.63	16.37	16.63
	40	44.34	10	41.80	13.86	14.10
	30	25.44	20	63.96	10.61	10.40
	25	19.28	25	71.34	9.49	9.37
	18	11.91	32	79.82	8.31	8.62
	10	5.78	40	87.02	7.21	7.45
	5	2.69	45	90.67	6.68	6.90
	—	—	50	93.68	6.22	6.62

*Carbon content = 50 at.%.

rection and were in good agreement with data published in the literature [2, 11, 19].

The variation in the lattice parameters of the solid solutions in the systems $TiC-NbC$, $TiC-TaC$, $ZrC-NbC$, and $ZrC-TaC$ is shown in Fig. 1. As we synthesized alloys containing the minimum number of vacancies in the carbon sublattice (see Table 1), the variation in lattice parameter may be regarded as being the result of a change in the forces of interatomic action that occurs on substituting one kind of metal atom for another, since equality of the lattice parameters was to be expected on geometrical grounds ($r_{Ti} = 1.45$ Å; $r_{Ta} = 1.45$ Å; $r_{Nb} = 1.45$ Å; $r_{Zr} = 1.46$ Å). The closeness of the lattice-parameter curves to Vegard's straight line and the results of chemical analysis provide confirmation that the concentration of defects in the carbon sublattice is minimal, and this enabled us to trace the effect of metal atoms of differing valence.

From the synthesized alloys we prepared specimens in the form of cylinders (10 mm in diameter and 10-12 mm high) and rods (65 × 5 × 6 mm); these were sintered for 2 h at 2200°C in a TVV-4 furnace in a vacuum of 10^{-4}-10^{-5} mm Hg. Residual porosity in the sintered specimens was 20-25%.

The specimens so obtained were used to determine the electrical resistivity ρ, the thermo-emf α, and the Hall constant R at room temperature by the methods already described in [12].

Corrections for the porosity of the specimens were introduced for the electrical resistivity and Hall constant in accordance with the methods given in [19, 25], respectively.

In the $Zr_x Nb_{1-x} C_{1.00}$ and $Zr_x Ta_{1-x} C_{1.00}$ solid solutions the thermo-emf falls approximately to the alloy with an equiatomic ratio of zirconium and niobium, and then becomes practically independent of composition. The decrease in the thermo-emf may be due to an increase in the concentration of free electrons, in accordance with ordinary theory. Nevertheless, there is no proportionality between the thermo-emf and the metal concentration in the range 20-50 at.% Nb, the thermo-emf remaining practically constant with continuous rise in the concentration of niobium. This is evidently the result of a gradual change in the band structure, i.e., with the gradual changeover from ZrC to NbC. In addition to the rise in the concentration of the current carriers, their effective mass may also increase, and it is this which gives rise to a very small change in the thermo-emf in the $Me_x^{IV}Me_{1-x}^V C_{1.0}$ solid solutions investigated.

The variation in the electrical resistivity in the system TiC−NbC and TiC−TaC corresponds to the change in the current-carrier concentration and the thermo-emf, and is determined by the change in metal (Nb or Ta) content in the solid solution.

The nature of the change in the three electrical properties investigated in the TiC−NbC and TiC−TaC solid solutions is similar to that in the ZrC−NbC and ZrC−TaC solid solutions. The absolute values of the electrical resistivity of TiC, NbC, ZrC, and TaC are in good agreement with the data available for the very pure substances, viz., $\rho_{TiC} \approx 240 \, \mu\Omega \cdot cm$ [15]; ρ_{NbC} and $\rho_{TaC} \approx 50 \, \mu\Omega \cdot cm$ [22]; and $\rho_{ZrC} \approx 220 \, \mu\Omega \cdot cm$ [10]; this is in contrast with the results in [3, 4], in which the small value given for the specific resistivity is no doubt due to the presence of the impurities oxygen and nitrogen and to the uncontrolled vacancy content, which leads, according to [6, 13], to a change in the corresponding electrophysical properties. The gradual decrease in electrical resistivity with rise in Me^V content may be due only to the effect of the additional valence electrons introduced by the group V metal.

The concentration of free electrons rises with increase in the content of niobium (or tantalum) atoms, these metals having a higher valence than zirconium (Fig. 2). Similar results have been obtained by the authors of [5] in an investigation of the electrical properties of TiC−VC alloys, and evidently such results are general among the $Me^{IV}−C−Me^V C$ systems.

In all the solid solutions investigated, the Hall emf (and also the thermo-emf) had a negative sign, indicating that the current carriers are electrons.

Fig. 2. Current-carrier concentration as a function of niobium and tantalum content $(1-x)$ at room temperature: +) $Zr_x Nb_{1-x} \cdot C_{1.0}$; O) $Zr_x Ta_{1-x} C_{1.0}$.

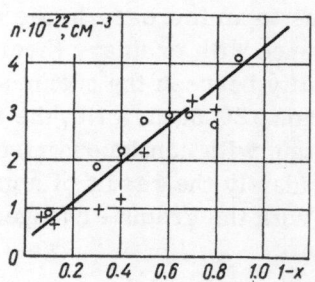

The current-carrier concentration varies within wide limits in the TiC–NbC (TaC) solid solutions from 1.6×10^{21} cm^{-3} for TiC to 7×10^{22} cm^{-3} for NbC (TaC).

The similarity of the current-carrier concentrations in alloys of the various Me^{IV}–C–Me^{V}C systems having the same molar concentration indicates that in stoichiometric solid solutions, i.e., in alloys containing carbon in approximately the theoretical amount, of the sum total of the valence electrons, which varied from eight per molecule in Me^{IV}C to nine for Me^{V}C, some part, which is approximately the same for all the alloys, is localized in the Me–C bonds, while the rest – which increases continuously as we pass from Me^{IV}C to Me^{V}C – is collectivized, as is shown by those properties that are dependent on the concentration of these electrons.

Conclusions

1. Alloys situated on the sections TiC–NbC, TiC–TaC, ZrC–NbC, and ZrC–TaC of the ternary Me^{IV}–C–Me^{V} systems have been synthesized from the high-purity elements, and their electrical resistivity (ρ), thermo-emf (α), and Hall emf have been investigated at room temperature.

2. There is complete similarity between the changes in these electrical parameters for the stoichiometric solid solutions in the Me^{IV}C–Me^{V}C systems.

3. The decisive factor in the change in the electrophysical properties of the alloys is the change in current-carrier concentration due to "excess" electrons introduced by the atoms of group V metals.

4. The band structure of carbides of group IV metals is retained in $Me^{IV}_x Me^{V}_{1-x}C_{1.0}$ alloys up to the equiatomic concentration of the metal atoms; thereafter the characteristic features of the band structure of carbides of group V metals gradually begin to appear.

Literature Cited

1. R. G. Averbe, Poroshkovaya Met., No. 1, 41 (1965).
2. A. I. Avgustinik, in: Studies in the Field of the Chemistry of Silicates and Oxides [in Russian], Vol. 4, Nauka, Moscow (1965).
3. A. I. Avgustinik et al., Izv. Akad. Nauk SSSR, Neorg. Mat., 3:286 (1967).
4. A. I. Avgustinik et al., Izv. Akad. Nauk SSSR, Neorg. Mat., 2:1439 (1966).

5.   A. I. Avgustinik, in: Studies in the Field of the Chemistry of the Silicates and Oxides [in Russian], Nauka, Moscow (1965), p. 244.

6.   I. G. Barantseva, V. N. Paderno, and Yu. B. Paderno, Poroshkovaya Met., No. 2, 70 (1967).

7.   P. V. Gel'd and V. A. Tskhai, Fiz. Metal. i Metalloved., 16:493 (1963).

8.   P. V. Gel'd and V. A. Tskhai, Zh. Strukt. Khim., 5:4235 (1963).

9.   P. V. Gel'd and V. A. Tskhai, Zh. Strukt. Khim., 5:576 (1963).

10.   O. A. Golikova et al., Fiz. Tverd. Tela, 7:3698 (1965).

11.   A. E. Koval'skii and Ya. S. Umanskii, Zh. Fiz. Khim., 20:769 (1946).

12.   Kieffer and Schwartzkopf, Hard Alloys

13.   A. Ya. Kuchma and G. V. Samsonov, Izv. Akad. Nauk SSSR, Neorg. Mat., 2:1970 (1966).

14.   V. S. Neshpor et al., Izv. Akad. Nauk SSSR, Neorg. Mat., 2:855 (1966).

15.   V. I. Odelevskii, Zh. Tekh. Fiz., 21:682 (1951).

16.   V. A. Tskhai and P. V. Gel'd, in: Physicochemical Studies of Compounds of Rare Refractory Elements [in Russian], Vol. 1, Ural'sk. Filial Akad. Nauk SSSR, Sverdlovsk (1966), p. 9.

17.   V. A. Tskhai and P. V. Gel'd, Zh. Strukt. Khim., 5:2275 (1964).

18.   C. Agte and K. Moers, Z. Anorg. Chem., 198:233 (1931).

19.   C. Agte and K. Alterthum, Z. techn. Physik, 6:130 (1930).

20.   H. Biltz, Z. Physik, 153:338 (1958).

21.   H. Bittner et al., Monatsh. Chem., 94:518 (1963).

22.   H. Juretschke, J. Phys. Chem. Solids, 4:118 (1958).

23.   S. Noguchi and T. Sato, J. Phys. Soc. Japan, 15:2359 (1960).

24.   S. Noguchi and T. Sato, J. Phys. Soc. Japan, 19:136 (1964).

25.   J. T. Norton and A. L. Mowry, Trans. Am. Inst. Min. Metal. Eng., 185:133 (1949).

26.   A. Munster, K. Sagel, and G. Schlamp, Nature, 174:1154 (1954).

27.   J. Piper, J. Appl. Phys., 33:2394 (1962).

28.   I. Zoth et al., Acta Metall., 13:379 (1965).

Thermal Coefficient of the Electrical Resistivity of Carbides and Their Solid Solutions

V. G. Grebenkina and E. N. Denbnovetskaya

In order to discover materials with a very low temperature coefficient of the electrical resistivity, a study has been made of the temperature dependence of the electrical resistivity of carbides of the transition metals of groups IV to VI, of the carbides TiC, ZrC, and NbC in their homogeneity ranges, and of the complex carbides of the systems ZrC—WC and VC—WC. It is shown that the coefficient diminishes with increase in the proportion of covalent bonding in the Me—Me bond in the compounds investigated.

The results of a study of the electrical properties of refractory compounds have been given in [7], the authors of which discuss the regular features of the change in the thermal coefficient of the electrical resistivity of transition-metal carbides and borides. However, the results in [7] are limited, and it is not possible to generalize about large groups of refractory compounds or to analyze broadly the physical essentials of the change in the temperature coefficient of the refractory compounds.

Consequently, in the present research we have determined experimentally the temperature coefficient of electrical resistivity of carbides of transition metals of groups IV to VI, of the carbides TiC, ZrC, and NbC in their homogeneity ranges, and of the systems ZrC—WC and VC—WC, which consist of limited solid solutions. Carbide specimens of stoichiometric composition and the homogeneity ranges were prepared by sintering powders produced by the methods previously described in [8] and by hot compaction. Specimens of the complex systems were obtained by the method described in [1]. The results of x-ray phase analysis showed that all the specimens of individual carbides were single phase, while the components of the complex systems had the following maximum solubilities:

System	Solubility of WC, mol.%	Temperature at which carbide alloys were produced, °C
ZrC — WC	30	1800
VC — WC	70	2400

The temperature dependence of the electrical resistivity was investigated over the range from room temperature to 1000°C. The temperature coefficient was calculated by means of the expression

$$\alpha = \left(\frac{\rho_t - \rho_\Theta}{\rho_\Theta T} \right)_\Theta$$

and the differential temperature coefficient $\left(\frac{d\rho}{dT} \right)_\Theta$ was also calculated. The specific electrical resistivity of the compounds investigated corresponding to the dense state were calculated from Odelevskii's formula:

$$\rho_0 = \rho \frac{2 - 3P}{2}.$$

The results of calculating the temperature coefficients and the differential coefficients are given in Table 1.

The change in the thermal coefficient of electrical resistivity of the carbides TiC, ZrC, and NbC in their homogeneity range is shown in Fig. 1.

TABLE 1. Thermal Coefficients of Electrical
Resistivity and Differential Coefficients
of Individual Carbides

Phase	$\frac{d\rho}{dT} \cdot 10^2$, $\mu\Omega \cdot$ cm/ deg	$\alpha \cdot 10^3$, deg^{-1}	Phase	$\frac{d\rho}{dT} \cdot 10^2$, $\mu\Omega \cdot$ cm/ deg	$\alpha \cdot 10^3$, deg^{-1}
TiC	8.3	0.80	TaC	3.3	0.79
ZrC	6.6	0.88	Cr_3C_2	2.2	0.19
HfC	5.0	0.94	Mo_2C	1.6	0.20
VC	3.3	0.38			
NbC	3.5	0.52	WC	5.5	1.63

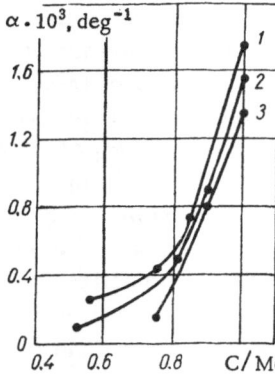

Fig. 1. Concentration dependence of the thermal coefficient of electrical resistivity of the carbides TiC_x (1), ZrC (2), and NbC (3) in the homogeneity range.

It will be seen that the differential coefficient decreases as we pass from the carbides of transition metals of group IV to those of group V and then to those of group VI; the coefficient (and the differential coefficient) fall with increase in the degree of defectiveness of the crystal lattice of the carbides in the homogeneity range. The results obtained can be explained in terms of current ideas regarding the electron structure of transition-metal carbides. Two bands can be distinguished in the energy spectrum of carbides having an NaCl-type structure [10]: (1) a band of electron states producing the Me−Me bonding, which determine the magnetic and electrical properties of the carbides; and (2) a band of electron states producing the Me−C bonding, which determine the mechanical properties; i.e., refractory carbides are regarded as compounds having covalent-metallic bonds. It has been found in [3] that there is a considerable proportion of the covalent component of the interatomic bond in the case of the d−elements and that this increases as we pass from metals of group IV to those of group V and then to those of group VI with rise in the number of electrons in the bonding part of the d band, which leads to a strengthening of the covalent bond in the same direction in quasimetallic compounds [6]. The strengthening of the covalent bond in the Me−Me band of the carbides corresponds to an increase in the statistical weight of the atoms having stable d^5 configurations [9].

The electrical resistivity is shown to be made up of two parts: $\rho = \rho_0 + \rho_t$; here ρ_0 is the temperature-independent part of the resistivity due to scattering of electrons at static imperfections in the crystal lattice, and ρ_t is the part of the resistivity due to scattering of conduction electrons at thermal vibrations of the atoms.

From the results for the electrical resistivity and Hall constant
of the carbides, it has been found from the single-band-model approximation (as the Hall coefficients and thermo-emf of these compounds are negative at room temperature) that the concentration
of carriers falls, while their mobility increases, as we pass from
titanium carbides to zirconium carbides and then to hafnium carbides, this leading to a reduction in the electrical resistivity on
passing from TiC to ZrC and then to HfC. The number of carriers
does not vary with rise in temperature, but their mobility decreases, as was shown in the case of ZrC and NbC in [5]; with increase in the proportion of the covalent bond in the Me — Me bond,
the mobility changes less with rise in temperature; consequently
the electrical resistivity varies less, which leads to a reduction
in the differential thermal coefficient of the resistance. The temperature coefficient calculated at the Debye temperature does not
as a rule reveal the regular features observed in the differential
coefficient, a fact that can be accounted for by the considerable influence which the value of ρ_θ exerts on the thermal coefficient.

It is known from [4] that the monocarbides TiC, ZrC, and NbC
are phases of variable composition. The electrical resistivity
increases with the degree of deficiency of the carbide lattice in
carbon, carbides of stoichiometric composition having the least
resistance; on the other hand the thermal coefficient and the differential coefficient increase as we pass from carbide phases deficient in carbon to the phases of limiting composition. A reduction in the carbon content of the carbide phase increases the probability of a direct exchange of electrons between the metal atoms;
this leads to their localization at the covalent type of bonds, and
the number of unlocalized electrons that act as current carriers
increases somewhat. The electrical resistivity increases with
reduction in the carbon content of the phase; this is evidently due
to a simultaneous considerable reduction in the mobility of the
conduction electrons due to their intensive scattering at vacancies
in the carbon sublattice, which act as scattering centers. The
temperature component of the electrical resistivity (ρ_t) varies
less with increase in the defectiveness of the carbon sublattice
than it does in carbides of limiting composition. Clearly, the effect of the thermal vibrations of the atoms on the electrical resistivity is greater than that of the thermal vibrations of the electrons [2]. Hence a certain rise in the concentration of the un-

localized part of the electrons with increase in the deficiency of carbon in the lattice has little effect on the variation of ρ_t; it is mainly the electrons in the carbide phases most deficient in carbon which are localized at the Me−Me bonds, thus leading to a slower rise in ρ_t and a reduction in the differential coefficients of resistivity in the homogeneity range as compared with carbides of stoichiometric composition.

The existence of mutual solubility in transition-metal carbides makes it possible to trace the effect of the electron structure, which varies continuously with composition, on the thermal coefficient of resistivity. The results of an investigation of the thermal coefficient and Hall constant at room temperature in the complex ZrC−WC and VC−WC carbides are shown in Fig. 2.

It is clear that the thermal coefficient and the Hall coefficient vary additively in the two-phase region in all the systems. The following regularities can be seen in the solid-solution regions in all the systems: the thermal coefficient decreases with rise in the WC content in the solid solutions (the differential coefficient varies similarly); the reduction in the thermal coefficient is accompanied by an increase in the Hall constant; this constant changes sign from negative to positive at a definite WC content in all the complex carbides investigated, thus indicating a transition from predominantly electron conduction to predominantly hole-type conduction. The proportion of covalent bonding in the Me−Me band

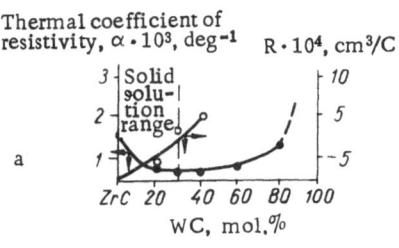

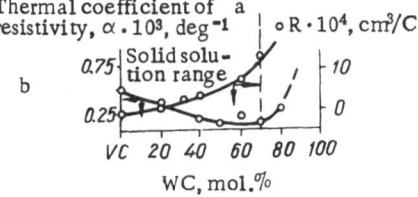

Fig. 2. Concentration dependence of the thermal coefficient of electrical resistivity (a) and the Hall coefficient (b) in complex carbides.

increases when atoms of metals of groups IV and V are replaced by tungsten atoms; this is due to an increase in the statistical weight of the atoms having stable d^5 configurations in tungsten as compared with the transition metals of groups IV and V. Conduction electrons of the Me − Me band are utilized in the saturation of the strengthened covalent bond of the Me − Me band, and as a result their concentration falls and the metallic nature of the alloys also diminishes. The nature of the change in the thermal coefficient of electrical resistivity and in the differential coefficient of the complex carbides shows that they decrease with the formation of the electron structure with the greatest localization of electrons into stable configurations.

Literature Cited

1. E. N. Denbnovetskaya, Poroshkovaya Met., No. 3, 32 (1967).
2. J. Simon, Electrons and Phonons [Russian translation], IL, Moscow (1962).
3. M. I. Korsunskii and Ya. E. Genkin, Dokl. Akad. Nauk SSSR, 142:1276 (1962).
4. A. Ya. Kuchma and G. V. Samsonov, Izv. Akad. Nauk SSSR, Neorg. Mat., 2:1970 (1966).
5. Yu. B. Paderno et al., Poroshkovaya Met., No. 2, 60 (1967).
6. P. A. Reznitskii, Izv. Akad. Nauk SSSR, Neorg. Mat., 2:1966 (1966).
7. G. V. Samsonov and V. S. Sinel'nikova, Poroshkovaya Met., No. 4, 59 (1962).
8. G. V. Samsonov and Ya. S. Umanskii, Hard Compounds of Refractory Metals [in Russian], Metallurgizdat, Moscow (1957).
9. G. V. Samsonov, Poroshkovaya Met., No. 12, 49 (1966).
10. H. Biltz, Z. Physik, 133:338 (1958).

A Study of Some Physical Properties of
Alloys of the System TaC–WC

E. N. Denbnovetskaya and S. N. L'vov

A study has been made of the concentration dependence of the specific electrical resistivity, Hall coefficient, absolute thermo-emf,
magnetic susceptibility, microhardness, and crystal lattice constant
at room temperature, and also of the temperature dependence of the
specific electrical resistivity and the coefficient of the thermo-emf
in the range 77-300°K, of alloys in the system TaC–WC in the solid-
solution region. The regular features observed in the changes in physical properties and the qualitative explanation of them are in accordance with current ideas regarding the electron structure of transition-
metal carbides. The authors have put forward a suggestion concerning
the strengthening of the covalent bond in the Me–Me band of the
solid solutions.

From a study of the physical properties of solid solutions of
transition-metal carbides it is possible to obtain valuable information about their electron structure. The physical properties
of complex carbides formed from continuous mixture of individual
carbides having a crystal lattice of the NaCl type have been extensively studied [10, 14, 15, 16, 18, 20], but the same is not true
of systems formed from carbides of limited natural solubility.

The hexagonal tungsten carbide WC is known to have considerable solubility in carbides with an NaCl-type lattice (for example,
in TiC, ZrC, HfC, NbC, TaC, and VC), whereas the solubility of
the latter in WC is negligible.

In the present research we have investigated some physical properties of alloys of the TaC−WC system. The superconductivity and electrical resistivity of these alloys have already been studied in [11, 19]. However, it was considered desirable to make a more detailed investigation of the physical properties of these alloys, and it is with this that the present article is concerned.

The specimens were prepared by sintering a mixture of tantalum carbide, tungsten, and carbon black in the course of hot compaction at a temperature of 2500-2600°C. It was found by x-ray phase analysis of the specimens so produced that up to the composition $(TaC)_{0.3} \cdot (WC)_{0.7}$ the specimens consisted of a solid solution of WC in TaC and had a crystal lattice of the NaCl type.

The physical properties were investigated in the TaC−WC solid-solution range, since this enabled the effect of the type of crystal lattice to be excluded and the composition dependence of the physical properties to be determined.

The results of chemical analysis showed that the combined carbon content in TaC−WC alloys diminishes as the tungsten content increases. This is in good agreement with the results in [22], in which the phase equilibria in the system Ta−W−C were studied, and it was shown that the single-phase TaC−WC solid-solution field contracts with rise in the tungsten content in the solid solution and the latter becomes impoverished in carbon. The specimens whose physical properties have been investigated in the present research had the following compositions:

Alloy number	Composition
1	$Ta_{0.9} \ W_{0.1} \ C_{0.998}$
2	$Ta_{0.8} \ W_{0.2} \ C_{0.997}$
3	$Ta_{0.7} \ W_{0.3} \ C_{0.976}$
4	$Ta_{0.6} \ W_{0.4} \ C_{0.970}$
5	$Ta_{0.5} \ W_{0.5} \ C_{0.962}$
6	$Ta_{0.4} \ W_{0.6} \ C_{0.948}$
7	$Ta_{0.3} \ W_{0.7} \ C_{0.903}$
8	$TaC_{0.993}$
9	$WC_{0.991}$

The above specimens were used to investigate the concentration dependence of the specific electrical resistivity, the Hall coefficient, the absolute thermo-emf, the magnetic susceptibility,

the microhardness, and the lattice constant at room temperature.
The specific electrical resistivity was measured by a compensation method; the coefficient of the thermo-emf was determined
with reference to copper and then converted to the absolute value;
the magnetic susceptibility was determined by the Gouy method at
increasing values of the magnetic field strength, which made it
possible to eliminate the effect of ferromagnetic impurities; and
the Hall coefficient was measured with dc in a constant magnetic
field of strength 18 kOe. The microhardness was determined on
a PMT-3 apparatus. The method of preparing the specimens for
the measurements has been described in [5-7]. All measurements
were made on the same specimens.

The concentration dependence of the physical properties is
depicted graphically in Fig. 1. Attention is drawn to the fact that
the increase in electrical resistivity with rise in tungsten content

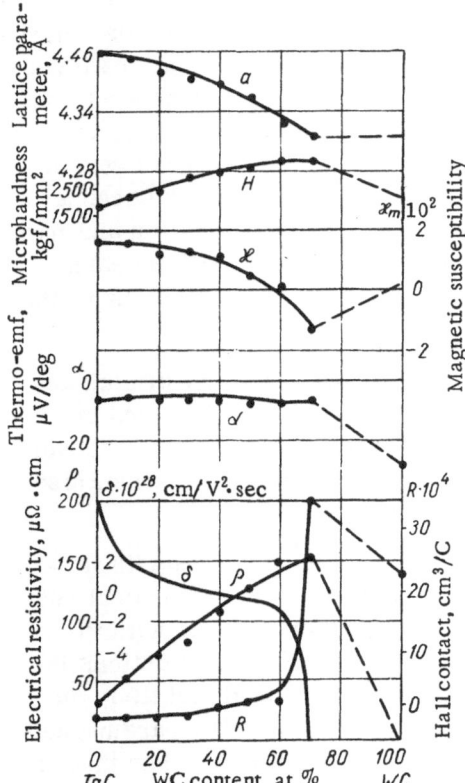

Fig. 1. Concentration dependence
of the physical properties of alloys
in the TaC—WC system.

of the solid solution is accompanied by an increase in the Hall
coefficient, which as a result changes sign from negative to positive. This indicates that over the whole range of variation of
tungsten concentration the conductivity of the alloys investigated
has a mixed electron—hole character, with the result that it changes
over from being predominantly n-type to being predominantly p-
type as the tungsten content increases. Figure 1 shows graphically,
in addition to the curves for ρ and R, the concentration dependence
of the quantity [7]

$$\delta = n_- u^2_- - n_+ u^2_+ = \frac{R}{\rho^2 e},$$

which represents the numerator in the expression for the Hall
coefficient in the two-band approximation. This quantity is easily
calculated from the experimental results for R and ρ. From its
sign and numerical value we can assess qualitatively which type
of carrier predominates in the conduction process. As can be seen
from the graph, the nature of the conductivity changes most sharply at the boundaries of the range of tungsten concentrations investigated. On the other hand the coefficient of the thermo-emf does
not change sign over the whole of the concentration range, and it
is for this reason that a discrepancy exists between the signs of
R and α. Generally speaking, this discrepancy occurs because
the thermo-emf depends on the ratio of the concentration and mobility of the carriers in a different way from the Hall constant,
and in addition it depends also on the degree of overlapping of the
bands [22].

The absolute value of the magnetic susceptibility diminishes
as the tungsten content of the solid solution increases, and specimens with maximum tungsten content are diamagnetic. It should
be noted that the microhardness increases at the same time, also
reaching its maximum value at this composition.

Characteristically, the solid solution decomposes with further increase in the tungsten content. There may be several reasons for the observed reduction in the paramagnetism with increase in the tungsten content of the TaC—WC solid solutions investigated. Since the paramagnetism of transition metals is due
mainly to the paramagnetism of the ions [18], and the paramagnetism of tungsten is approximately three times weaker than that of
tantalum [2, 17], the replacement of tantalum atoms by tungsten

atoms should naturally lead to a reduction in the paramagnetism of the solution. Moreover, bearing in mind that the microhardness increases with rise in the tungsten content of the solution, it may be assumed that the proportion of covalent bonding in the complex carbide also increases; this too should lead to a reduction in the paramagnetism and even to the appearance of diamagnetism. It follows from the results obtained that as the solution becomes increasingly saturated with tungsten atoms, the electrical conductivity diminishes and the positive value of the Hall constant increases, owing to a decrease in the concentration of conduction electrons. And since their paramagnetism is approximately three times greater than their diamagnetism [18], this should in general lead to a reduction in the paramagnetism of the solution. It is also possible that some part is played in this process by a reduction in the Van Vleck paramagnetism due to an increase in the strength of the covalent bonding [3].

In addition to measuring the parameters mentioned above at room temperature, we also investigated the temperature dependence of the specific electrical resistivity and the coefficient of the thermo-emf of the same alloys in the temperature range from 77 to 300°K. The specific electrical resistivity was measured by a compensation method, while the thermo-emf was determined with reference to copper. The results are shown in Figs. 2 and 3.

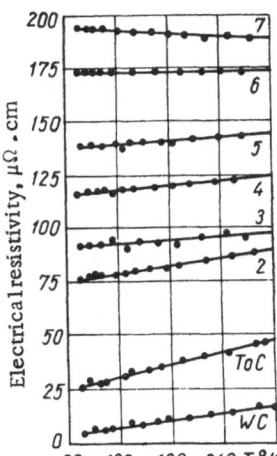

Fig. 2. Temperature dependence of the electrical resistivity of alloys of the TaC−WC system in the range 77-300°K.

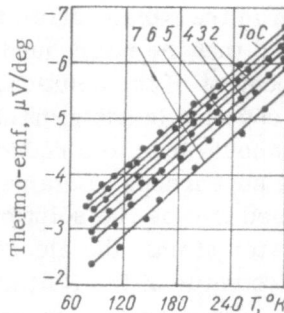

Fig. 3. Temperature dependence of the coefficient of the thermo-emf of alloys of the system TaC—WC in the range 77-300°K.

In complex carbides having a relatively small tungsten content the electridal resistivity increases with rise in temperature, the slope of the ρ vs T curves diminishing with increase in the tungsten content of the alloys. Carbides of composition $Ta_{0.4}W_{0.6}C_{0.948}$ and $Ta_{0.3}W_{0.7}C_{0.903}$ behave anomalously. In the first case the electrical resistivity remains practically unchanged over the temperature range investigated, while in the second case it decreases with rise in temperature.

The change in the temperature dependence of the specific electrical resistivity with increase in the tungsten content of the solution-is evidently due to the complex nature of the scattering of current carriers. In solid solutions at comparatively low temperatures, in addition to the scattering of current carriers at the thermal lattice vibrations, no small part may also be played by scattering at impurity ions such as tungsten ions and randomly substituting tantalum ions in the lattice. In this type of scattering the mobility is proportional to $T^{3/2}$ [1]. At small concentrations of tungsten atoms in the lattice, scattering at the thermal vibrations clearly predominates, and the electrical resistivity increases with rise in temperature. With increase in the content of these atoms the part played by scattering at impurities becomes greater, and the angle of slope of the ρ vs T curves becomes less. When the content of tungsten atoms reaches a maximum and the lattice is on the point of breakdown, scattering at impurities and defects plays the major part, and the angle of slope of the curves even becomes negative. It is to be expected that at fairly high temperatures the electrical resistivity will continue to increase with rise in temperature in all the alloys investigated.

The results obtained and their qualitative elucidation are in agreement with the general concepts currently accepted in regard to the electron structure of carbides [8, 12].

According to these concepts it is possible to distinguish two bands in the energy spectrum of carbides: (1) a band of electron states producing the Me−Me bond; and (2) a band of electron states producing the Me−C bond. In tantalum carbide close to the stoichiometric composition, there is, according to [4], one electron per metal atom in the Me−Me band responsible for the quasi-metallic properties of the carbides. On replacing some of the tantalum atoms by tungsten atoms in the TaC−WC solid solution, there is an increase in the number of electrons used by the metal atoms in the Me−Me bond; this results in an increase in the proportion of covalent bonds in the Me−Me band with rise in tungsten concentration [9]. This concept is in good agreement with the assumption about the statistical weight of atoms having stable d^5 configurations being greater in tungsten than in tantalum [8].

Such an increase in the proportion of covalent bonding in the Me−Me band of a complex carbide is accompanied by a reduction in the conduction-electron concentration, as is confirmed by the change in sign of the Hall coefficient and the increase in the specific electrical resistivity. Simultaneously with the greater localization of the electrons in the covalent Me−Me bond and the resulting reduction in the concentration of conduction electrons, electrons are removed from the Me−C band; this is evidently the reason for the regular reduction in the combined carbon content of the solid solution with rise in the tungsten content (see the table). As was indicated above, the results for the microhardness and magnetic properties of the alloy system investigated also confirm the conclusion about the strengthening of the covalent bonding in the Me−Me band of the solid solution.

Literature Cited

1. A. I. Ansel'm, Introduction to the Theory of Semiconductors [Russian translation], Fizmatgiz, Moscow and Leningrad (1962).
2. Ya. G. Dorfman, Magnetic Properties and the Structure of Matter [in Russian], Gostekhizdat, Moscow (1955).
3. Ya. G. Dorfman, Diamagnetic and Chemical Bonding [in Russian], Fizmatgiz, Moscow (1961).

4.   L. B. Dubrovskaya and I. I. Matvienko, Fiz. Metal. i Metalloved., 19:199
      (1965).
5.   S. N. L'vov and V. F. Nemchenko, Informpis'mo IMSS Akad. Nauk UkrSSR,
      No. 272 (1960).
6.   S. N. L'vov, V. F. Nemchenko, and V. I. Marchenko, Pribory i Tekh. Eksperim.,
      No. 2, 159 (1961).
7.   S. N. L'vov, V. F. Nemchenko, and G. V. Samsonov, Poroshkovaya Met., No. 4,
      3 (1962).
8.   G. V. Samsonov, Poroshkovaya Met., No. 12, 49 (1966).
9.   V. I. Trefilov, in: The Physical Nature of the Brittle State [in Russian],
      Naukova Dumka, Kiev (1965), p. 76.
10.   A. S. Umanskii and V. I. Fadeeva, Fiz. Metal. i Metalloved., 20:719 (1965).
11.   V. F. Funke and V. S. Popov, Izv. Akad. Nauk SSSR, Metally, No. 6 (1966).
12.   H. Biltz, Z. Physik, 153:338 (1958).
13.   R. Conte, Rapp. CEAR, No. 2854 (1965).
14.   D. Z. Deddneore, J. Am. Ceram. Soc., Ceram. Abs., 48:354 (1965).
15.   I. Fumitake, T. Takashi, and T. Hideo, J. Phys. Soc. Japan, 19:136 (1964).
16.   H. Goretzki, Doctoral Dissertation, Vienna (1963).
17.   H. Kojima, R. S. Tebble, and D. E. Williams, Proc. Roy. Soc., A260:237 (1961).
18.   C. T. Kriessman, Rev. Mod. Phys., 25:122 (1953).
19.   B. T. Matthias, V. D. Compton, and Corenzwit, J. Phys. Chem. Solids, 1:188
      (1956).
20.   E. Rudy and F. Benesovsky, Planseeber. Pulvermet., 8:12 (1960).
21.   E. Rudy, El. Rudy, and F. Benesovsky, Monatsh. Chem., 93:1176 (1962).
22.   H. E. Smidt, Z. Metallk., 49:113 (1958).

Thermal Expansion of Solid Solutions in the Systems ZrC–NbC and HfC–TaC

I. G. Barantseva and V. N. Paderno

A study has been made of the thermal expansion of pseudobinary carbide alloys in the systems ZrC–NbC and HfC–TaC, which form continuous series of solid solutions. The temperature dependence of the relative elongation is linear or almost so and it displays no unusual features. In both systems there is a reduction in the coefficient of thermal expansion of the alloys as compared with the individual carbides. The alloy $Zr_{0.6}Nb_{0.4}C$ in the one system and the alloy $Hf_{0.3}Ta_{0.7}C$ in the other have the lowest coefficients. The decrease in the coefficient of thermal expansion of the alloys is in line with current concepts of the electron structure of the compounds concerned.

The present article deals with an investigation into the thermal expansion of pseudobinary carbide alloys of the systems ZrC–NbC and HfC–TaC which form continuous series of solid solutions.

Specimens for the investigation were prepared by the reduction of mixtures of the oxides of the appropriate metals with carbon black, the reduction products being simultaneously sintered in a hot-compaction press. The specimens so obtained were given a homogenizing anneal. The homogeneity and single-phase nature of the specimens were checked by x-ray and metallographic methods. The specimens were cylindrical in form, with a diameter of 8 mm and a height of 150-20 mm. The expansion was measured with a quartz dilatometer in an argon atmosphere. The accuracy of the measurements was 1.3%.

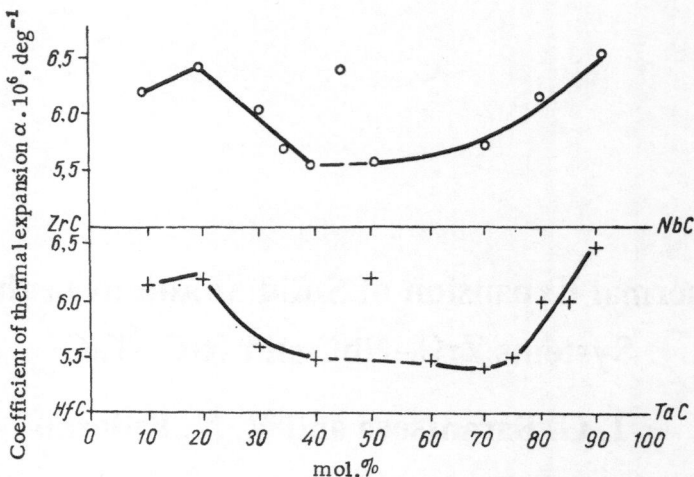

Fig. 1. Concentration dependence of the coefficient of thermal expansion of solid solutions in the systems ZrC−NbC and HfC−TaC.

The temperature dependence of the relative elongation is linear or almost so and it displays no unusual features. Figure 1 shows curves of the concentration dependence of the coefficient of thermal expansion of solid solutions in the systems ZrC−NbC and HfC−TaC.

The similarity between the curves is noteworthy. In both systems there is a decrease in the coefficient of thermal expansion of the solid solutions as compared with the individual carbides. The alloys $Zr_{0.6}Nb_{0.4}C$ and $Hf_{0.3}Ta_{0.7}C$ have the lowest coefficients of expansion in their respective systems. Alloys containing equiatomic (or almost equiatomic) amounts of the metallic components ($Zr_{0.55}Nb_{0.45}C$ and $Hf_{0.50}Ta_{0.50}C$) have coefficients close to those of the individual carbides.

We have previously communicated [1] the results of an investigation into the electrical resistivity of the same solid solutions, and these results revealed maximum values of the specific electrical resistivity in the alloys $Zr_{0.6}Nb_{0.4}C$ and $Hf_{0.3}Ta_{0.7}C$.

It may be assumed that the reduction in the coefficient of thermal expansion, like the increase in the specific electrical resistivity, of the solid solutions as compared with the individual carbides is due to special features of the materials investigated.

They are in accordance with Biltz's observation [2] that a reduction occurs in the number of free electrons in the compounds concerned at about 4.25 valence electrons per atom in them. Actually, the number of valence electrons per atom is 4.20 for the alloy $Zr_{0.6}Nb_{0.4}C$ and 4.35 for the alloy $Hf_{0.3}Ta_{0.7}C$. The reduction in the number of free electrons must be accompanied by an increase in the number of electrons localized in the atom nuclei and participating in the covalent bonds. This concept is in line with the decrease in the coefficient of thermal expansion and the increase in the specific electrical resistivity that occur in the solid solutions investigated.

The increase in the coefficient of expansion in alloys with equiatomic contents of the metallic components requires further study and analysis.

Literature Cited

1.   Yu. B. Paderno, I. G. Barantseva, and V. N. Paderno, Poroshkovaya Met., No. 2 (1967).
2.   H. Biltz, Z. Physik, 153:338 (1958).

Study of the Galvanomagnetic and Electrical
Properties of TiC–TiO, TiN–TiO, TiC–TiN,
and ZrC–ZrN Alloys

V. S. Neshpor, G. M. Klimashin,
and V. P. Nikitin

A study has been made of the specific electrical resistivity, thermo-
emf, thermal conductivity, and Hall coefficient of the monocarbides
of titanium and zirconium and of their alloys with the monoxides and
mononitrides. Relationships between the changes in physical properties of the alloys and their composition have been established.

It has been shown in [1, 2, 7, 12, 13] that the monocarbides of
transition metals of group IV which are close to the stoichiometric
composition and which have been produced by synthesis from high-
purity elements in a high vacuum [3] or by chemical gas-phase decomposition [13], differ considerably in their electrophysical properties from commercially pure monocarbides of the same metals
obtained by reduction of the oxides, by hot compaction without the
use of a specially purified protective atmosphere, or by synthesis
from commercially pure components [14, 15, 23]. Specimens of
pure monocarbides of transition metals of group IV close to the
stoichiometric composition have a comparatively low (for metals)
concentration of conduction electrons (approximately 10^{21} cm^{-3})
and a high specific electrical resistivity (of the order of $2-4 \times 10^{-4}$
$\Omega \cdot$ cm), in good agreement with the very low density of states at
the Fermi level in these compounds [32, 37], and this is confirmed

by experimental results reported in [34, 39, 40] for single crystals of these compounds produced by the electric arc plasma method (a variation of the Verneuil method). Specimens of commercially pure carbides of group IV metals have a considerably lower specific resistivity (of the order of 4-6 × 10^{-5} Ω · cm) and a higher concentration of conduction electrons (about 10^{22} cm^{-3}); this is evidently due to the presence of metallic and metalloid impurities having a higher effective valence which causes a displacement of the Fermi level in the direction of a higher density of states.

It has been found in particular that admixtures of group V transition metals in pure TiC and ZrC [8], and also the presence of nitrogen [40] and oxygen [4] in these compounds, lead to an increase in the concentration of conduction electrons and to a reduction in the specific electrical resistivity. Deviation of the composition of monocarbides of group IV metals from the stoichiometric produces the same result [1, 2, 7, 12, 13, 28]. An investigation of the electrophysical properties of alloys of group IV metal carbides having a specific electrical resistivity of the order of 40-100 × 10^{-6} Ω · cm (regarded by us as "commercially pure") with carbides of group V metals [41] or nitrides [16] has shown that the electrical resistivity of the alloys is higher than that of the original group IV metal carbides.

In the present research we have investigated the specific electrical resistivity, thermo-emf, thermal conductivity, and Hall coefficient of alloys of TiC with TiN and TiO, ZrC with ZrN, and, for comparison, alloys of TiN with TiO.

The initial monocarbides, close to the stoichiometric composition, were prepared, as already described in [4], by direct synthesis from iodide titanium and zirconium and acetylene soot in a vacuum of 5 × 10^{-6} mm Hg. The results of chemical and quantitative spectrographic analyses showed that the initial TiC had the following composition: Ti 79.70, C_{total} 19.51, C_{free} 0.10, C_{comb} 19.43, N 0.12, O 0.10, Si 0.02, Al 0.01, Mg 0.0085, Cu 0.0027, Fe 0.057, Mn 0.0021, Co 0.001, Ni 0.02, Cr 0.003, V 0.0067, Sn 0.001, Sb 0.037, Pb 0.001, Bi 0.0001, Cd 0.0001, Ag 0.0001, Zn 0.004, As 0.03, Li 0.0004, and B 0.004 wt.%.

It is possible that a certain amount of the impurities was introduced into the analytical sample during grinding in an agate mortar. ZrC contained Zr 88.90, C_{total} 11.50, C_{free} 0.12, N 0.06,

O 0.08%, and, according to the results of qualitative spectrographic analysis, small amounts of Ti, Sb, and Si and traces of Fe, Nb,
Ni, Mg, Ca, and Hf (in all not more than 0.159%). TiO, containing
74.2% Ti and 0.25% N, was prepared by reaction between analyti-
cally pure TiO and iodide titanium in a vacuum of 5×10^{-6} mm Hg.
The oxygen content (25.6%) was determined by difference. To produce the carbonitride alloys, we used TiN and ZrN powders prepared at the Donets Chemical Reagent Plant by nitriding powdered
metals. Chemical analysis showed that ZrN contained 89.35% Zr,
10.55% N, and 0.05% C and TiN contained 82.65% Ti, 17.35% N, and
0.03% C.

The alloys were prepared by sintering rectangular briquets
compacted under a pressure of 1500 kg/cm^2 from mixtures of the
powdered raw materials (monocarbide, mononitride, or monoxide)
at 2200-2250°C in a vacuum furnace (under a pressure of 10^{-5}-5 $\times$
10^{-6} mm Hg) having a pyrolytic graphite heater, and then analyzed
by chemical (for the main components), x-ray, and densitometric
methods (Tables 1-4).

All of the specimens produced in the manner described above
consisted of homogeneous solid solutions having a cubic structure
of the NaCl type. Good resolution of the $K_{\alpha_1 \alpha_2}$ doublets in the x-
ray diffraction patterns of the specimens showed that they were in
equilibrium. The composition of the alloys (as given by chemical
analysis) was reduced to the arbitrary sections $TiC_{0.96}-TiN_{0.72}$,
$ZrC_{0.98}-ZrN_{0.77}$, $TiC_{0.96}-TiO_{1.03}$, and $TiN_{0.72}-TiO_{1.03}$ (corresponding to the composition of the original materials) and expressed
in molar percentages of the binary compounds. The change in lattice parameter in the carbonitride systems reveals a small positive deviation from Vegard's straight-line law. This deviation
is greater in the $TiC_{0.96}-TiO_{1.03}$ system and is particularly marked
in the $TiN_{0.72}-TiO_{1.03}$ system, for which the relationship between
the lattice parameter and composition is represented by a curve
with a maximum. The considerable positive deviation of the lattice parameter in the systems containing TiO is possibly due to
the relaxing effect on the interatomic bond strength of the metallic
vacancies which, as will be shown below, are present in these
systems.

From the data on the pycnometric density, chemical composition, and lattice parameter, we have calculated the number of
atoms of the components in the unit cell [27] and the vacancy con-

TABLE 1. Chemical Composition and Number of Atoms per Unit Cell of Alloys of the System $TiC_{0.96}$–$TiN_{0.72}$

Alloy No.	Chemical composition wt.%				Formula composition	$TiC_{0.96}$, mol.%	$TiN_{0.72}$, mol.%	Lattice parameter, Å	Pycnometric density, g/cm³	Number of atoms per unit cell				Number of metalloid vacancies n_x†	Concentration of current carriers, cm⁻³	Concentration of valence electrons
	Ti	C	N	Σ*						Σ	n_{Ti}	n_e	n_N			
1-P	80.3	19.55	0.39	100.24	$TiC_{0.96}N_{0.02}$	98.0	2.0	4.322	4.881	7.88	3.99	3.84	0.06	0.10	$5.6 \cdot 10^{21}$	7.94
2-P	81.1	15.45	3.65	100.20	$TiC_{0.79}N_{0.15}$	80.4	19.6	4.307	4.869	7.65	3.97	3.11	0.60	0.29	$1.4 \cdot 10^{22}$	7.93
3-P	80.5	12.10	7.25	99.85	$TiC_{0.59}N_{0.31}$	62.3	37.7	4.289	4.913	7.49	3.93	2.34	1.22	0.44	$1.8 \cdot 10^{22}$	7.93
4-P	81.0	8.49	10.97	100.46	$TiC_{0.42}N_{0.46}$	41.9	58.1	4.269	4.900	7.27	3.88	1.61	1.78	0.60	$1.8 \cdot 10^{22}$	7.94
5-P	84.1	4.32	11.43	99.85	$TiC_{0.22}N_{0.48}$	23.0	77.0	4.249	–	–	–	–	–	–	$1.9 \cdot 10^{22}$	7.28
6-P	82.6	0.02	17.33	99.95	$TiN_{0.72}$	0	100	4.242	5.235	7.08	4.10	0	2.98	1.02	$2.4 \cdot 10^{22}$	7.59

* Total from analysis.

† $n_x = 4 - (n_e + n_N)$.

TABLE 2. Chemical Composition and Number of Atoms per Unit Cell of Alloys of the System $ZrC_{0.98} - ZrN_{0.77}$ *

Alloy No.	Chemical composition, wt.%				Formula composition	$ZrC_{0.98}$, mol.%	$ZrN_{0.77}$, mol.%	Lattice parameter, Å	Pycnometric density, g/cm³	Number of atoms per unit cell				Number of metalloid vacancies, n_x	Concentration of current carriers, cm⁻³	Concentration of valence electrons
	Zr	C	N	Σ						n_Σ	n_{Zr}	n_e	n_N			
1-IV	88.3	11.48	0.32	100.10	$ZrC_{0.98}N_{0.02}$	96.9	3.1	4.696	6.519	7.93	3.97	3.88	0.09	0.03	$3.1 \cdot 10^{21}$	8.04
2-IV	88.4	9.40	2.34	100.14	$ZrC_{0.80}N_{0.17}$	78.1	21.9	94.675	5.635	7.80	3.97	3.17	0.68	0.15	$1.0 \cdot 10^{22}$	8.06
3-IV	88.7	6.70	4.55	99.95	$ZrC_{0.57}N_{0.33}$	56.8	43.2	24.656	6.775	7.65	4.00	2.30	1.34	0.36	$2.2 \cdot 10^{22}$	7.96
4-IV	88.5	4.63	6.83	99.96	$ZrC_{0.40}N_{0.50}$	36.2	63.8	4.626	6.902	7.60	3.99	1.59	2.01	0.40	$3.6 \cdot 10^{22}$	8.12
5-IV	89.1	1.88	8.90	99.88	$ZrC_{0.16}N_{0.65}$	15.4	84.6	4.596	7.007	7.27	4.00	0.65	2.62	0.73	$4.7 \cdot 10^{22}$	7.91
6-IV	89.4	0.03	0.52	99.95	$ZrN_{0.77}$	0	100	4.572	7.127	7.11	4.03	0	3.08	0.92	$5.5 \cdot 10^{22}$	7.83

*See footnote to Table 1.

TABLE 3. Chemical Composition and Number of Atoms per Unit Cell of Alloys of the System $TiC_{0.96}-TiO_{1.03}$*

Alloy No.	Chemical composition, wt.%					Formula composition	$TiC_{0.96}$, mol.%	$TiO_{1.03}$, mol.%	Lattice parameter, Å	Pycnometric density, g/cm³	Number of atoms per unit cell				Number of metalloid vacancies n_x†	Number of metallic vacancies	Concentration of current carriers, cm⁻³	Concentration of valence electrons
	Ti	C	N	Σ	O†						n_Σ	n_{Ti}	n_e	n_o				
1-1	80.3	19.55	0.39	100.24	~0	$TiC_{0.96}N_{0.02}$	~100	~0	4.322	4.881	7.88	3.98	3.84	0	0.10	0.02	$5.6 \cdot 10^{21}$	7.94
2-1	80.0	16.48	0.34	96.84	3.2	$TiC_{0.82}O_{0.12}$	85.9	14.1	4.313	4.903	7.74	3.96	3.25	0.48	0.21	0.04	$6.0 \cdot 10^{22}$	8.07
3-1	79.1	12.37	0.39	91.86	8.1	$TiC_{0.62}O_{0.31}$	65.9	34.1	4.303	4.989	7.70	3.95	2.45	1.23	0.25	0.05	$1.4 \cdot 10^{23}$	8.42
4-1	77.2	8.97	0.32	86.43	13.6	$TiC_{0.46}O_{0.52}$	47.9	52.1	4.273	4.980	7.59	3.77	1.75	1.96	0.29	0.23	$1.8 \cdot 10^{23}$	9.12
5-1	74.5	2.90	0.22	77.62	22.4	$TiC_{0.16}O_{0.90}$	15.7	84.3	4.203	4.937	7.09	3.43	0.53	3.09	0.35	0.57	$1.6 \cdot 10^{23}$	10.07
6-1	74.2	0.02	0.25	74.47	25.6	$TiC_{1.03}N_{0.01}$	~0	~100	4.148	4.919	6.69	3.28	0	3.37	0.59	0.72	$6.2 \cdot 10^{22}$	10.24

* See footnote to Table 1.

† Oxygen content estimated by difference, $(100 - \Sigma)\%$.

‡ Corrected for nitrogen impurity.

TABLE 4. Chemical Composition and Number of Atoms per Unit Cell of Alloys of the System $TiN_{0.72}$–$TiO_{1.03}$*

Alloy No.	Chemical composition, wt.%					Formula composition	$TiN_{0.72}$, mol.%	$TiO_{1.03}$, mol.%	Lattice parameter, Å	Pycnometric density, g/cm³	Number of atoms per unit cell				Number of metalloid vacancies n_x†	Number of metallic vacancies	Concentration of current carriers, cm⁻³	Concentration of valence electrons
	Ti	N	C	Σ	O						n_Σ	n_{Ti}	n_N	n_O				
1-III	82.6	17.33	0.02	99.95	~0	$TiN_{0.72}$	100	0	4.242	5.235	7.08	4.10	2.96	0	1.02	0	$2.4 \cdot 10^{22}$	7.59
2-III	77.8	15.45	0.03	93.28	6.7	$TiN_{0.68}O_{0.26}$	86.2	13.8	4.251	5.269	7.68	3.96	2.69	1.03	0.28	0.04	$2.0 \cdot 10^{22}$	8.96
3-III	76.6	12.31	0.02	88.92	11.1	$TiN_{0.55}O_{0.43}$	70.0	30.0	4.256	5.236	7.70	3.89	2.14	1.67	0.19	0.11	$2.3 \cdot 10^{22}$	9.33
4-III	75.4	8.82	0.04	84.26	15.7	$TiN_{0.40}O_{0.68}$	51.5	48.5	4.226	5.216	7.44	3.66	1.47	2.31	0.22	0.34	$3.5 \cdot 10^{22}$	9.78
5-III	75.1	4.10	0.03	79.23	20.8	$TiN_{0.18}O_{0.83}$	25.0	75.0	4.168	5.010	6.90	3.43	0.63	2.84	0.26	0.57	$4.4 \cdot 10^{22}$	9.89
6-III	74.2	0.25	0.02	74.47	25.6	$TiO_{0.03}N_{0.01}$	0	100	4.148	4.919	6.69	3.28	0.04	3.37	0.59	0.72	$6.2 \cdot 10^{22}$	10.24

* See footnotes to Tables 1 and 3.

† Corrected for carbon impurity.

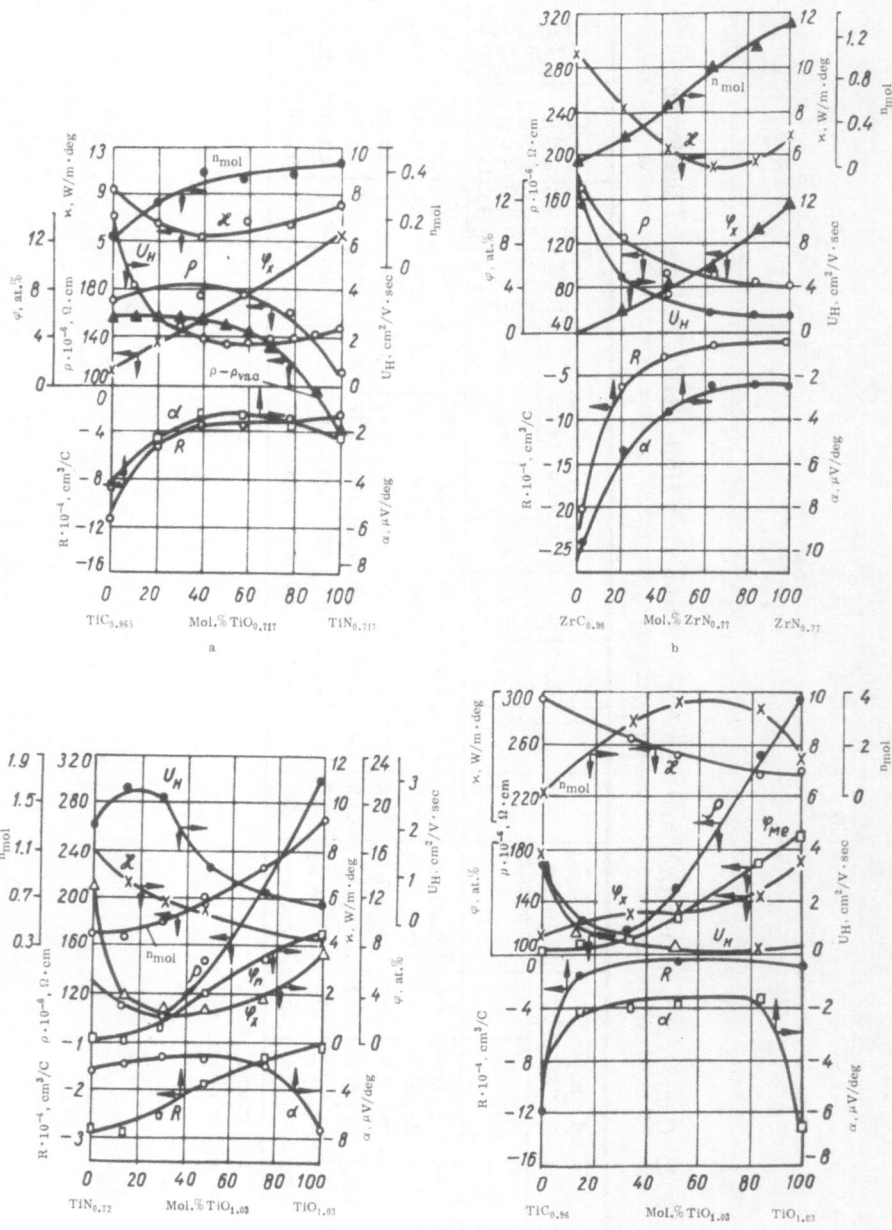

Fig. 1. Composition dependence of the specific electrical resistivity (ρ), thermal conductivity ($\varkappa$), molar concentration of free electrons (n_{mol}), Hall mobility (u_H), and concentration of metalloid vacancies (φ_x) in the alloys: a) $TiC_{0.96}-TiN_{0.72}$; b) $ZrC_{0.98}-ZrN_{0.77}$; c) $TiC_{0.96}-TiO_{1.03}$; d) $TiN_{0.72}-TiO_{1.03}$.

centrations in the metallic and metalloid sublattices of the solid solutions.

Carbonitride systems in which transition metals are present are characterized by unary defectiveness (Fig. 1), i.e., by the presence of vacancies in only one metalloid sublattice, whereas the metal sublattice remains complete within the limits of accuracy of density measurement, just as in the case of the nonstoichiometric carbides of metals of groups IV and V (except for VC) [17]. The concentration of metalloid vacancies in the systems $TiC_{0.96} - TiN_{0.72}$ and $ZrC_{0.98} - ZrN_{0.77}$ increases monotonically as we pass from the monocarbides, whose composition is close to the stoichiometric, to the less stoichiometric nitrides.

The oxycarbide and oxynitride alloys of titanium ($TiC_{0.96} - TiO_{1.03}$ and $TiN_{0.72} - TiO_{1.03}$) exhibit binary defectiveness, i.e., vacancies are present in both metalloid and metallic sublattices. Binary defectiveness is characteristic of pure TiO over the whole of its homogeneity range, and in TiO which is close to the stoichiometric composition there is approximately the same number of metallic and metalloid vacancies [31, 33]. A monotonic increase occurs in the concentration of metallic vacancies with rise in the relative content in the alloy of the component with high valence, i.e., oxygen; according to [31], this is attributable to a tendency in the system for the total number of valence electrons occupying the antibonding orbitals to decrease. The concentration of metalloid vacancies in the system $TiC_{0.96} - TiO_{1.03}$ increases comparatively slowly with rise in the relative oxygen content in the range of compositions closer to the complete monocarbide, but it increases more rapidly in the composition range close to the more defective oxide, in which the higher valence of oxygen begins to play the major part. The change in the concentration of metalloid vacancies is particularly noteworthy in the system $TiN_{0.72} - TiO_{1.03}$, in which both the extreme binary components have defective metalloid sublattices. In alloys close to TiN, an increase in the relative oxygen content leads to a rapid reduction in the concentration of metalloid vacancies resulting from the occupation by oxygen atoms of vacant sites in the metalloid sublattice based on the nitride; after this the concentration of metalloid vacancies varies very little in the middle range of compositions, but it again increases markedly in the composition range approaching the monoxide. The latter is due to the same causes as those operating in the $TiC_{0.96} - TiO_{1.03}$ system discussed above.

The specific electrical resistivity of the alloys was measured by a dc compensation method. A correction for the porosity of the specimens was made by means of the equation for the conductivity of a multiphase substance proposed by Odelevskii [21] and improved upon by Skorokhod [43] for substances with a high porosity. The thermal conductivity of the alloys was measured by a relative method under stationary-heat-flow conditions [14, 18]. The effect of porosity was taken into account by means of the simple relationship [36]

$$\chi_0 = \chi_p(1 - P), \tag{1}$$

where χ_0 and χ_p are the thermal conductivities of porosity-free specimens and specimens with a porosity P (in fractions of unity); this relationship is well satisfied in the case of isotropic and uniformly distributed porosity. The Hall coefficient was measured at a constant current in a constant stabilized magnetic field with a strength of 8–9 kOe by means of a three-probe compensation method, using a R33 resistance box to compensate for the transfer voltage. The Hall potential difference was measured with a high-sensitivity dc millimicrovoltmeter of type F-118 (having a sensitivity of 0.05 μV/div), which was used as a null instrument in compensating for the transfer voltage. The accuracy of the measurements was ±2%. A correction for the effect of porosity was introduced [35] through the relationship

$$R_0 = R_{meas} / (1 - P), \tag{2}$$

where R_0 and R_{meas} are the values of the Hall coefficient for porosity-free and porous specimens.

To judge from the sign of the Hall coefficient and the thermo-emf, all the alloys investigated have electron conduction, and the nature of the change in the Hall coefficient and the thermo-emf with composition of the alloys is almost identical in the two cases. Assuming unary conduction, we have calculated the volume concentration of conduction electrons

$$n = 1/R_0 e,$$

where e is the electron charge (see Tables 1–4) and their Hall mobility

$$U_p = R/\rho,$$

where ρ is the specific electrical resistivity (Fig. 1), and also the concentration of free electrons per formula unit of the alloy

$$n_{mol} = M \,/\, R_{meas} \; eN \gamma_{meas}$$

where M is the molecular weight of the alloy, N is Avogadro's number, and γ_{meas} is the specific weight of the specimen. The dependence of n_{mol} on the composition of the alloy is shown in Fig. 1.

It follows from a consideration of Tables 1–4 and Fig. 1 that the concentration of free electrons in the alloys investigated increases with rise in the concentration of the component having the higher valence, i.e., nitrogen in the carbonitride systems and oxygen in the oxycarbide and oxynitride systems. A certain reduction in the free-electron concentration in oxygen-rich alloys of the system $TiC_{0.96} - TiO_{1.03}$ will be discussed below. Note should be made of the very sharp reduction in the Hall coefficient and the corresponding increase in free-electron concentration in alloys of monocarbides of group IV metals with nitride and monoxide, with the nitrogen and oxygen content, and of the less-marked change in these characteristics in oxynitride alloys. On comparing the change in free-electron concentration with the change in molar concentration of the valence electrons given in Tables 1–4, we note the following. There is a similarity in slope between the change in valence-electron concentration* and in n (or n_{mol}) in the system $TiN_{0.72} - TiO_{1.03}$, although the two are not in direct proportion. In the system $TiC_{0.96} - TiO_{1.03}$ the free-electron concentration increases substantially with rise in the valence-electron concentration to a maximum in the oxygen-rich alloys, and then diminishes somewhat. In both carbonitride systems the combined effect of the relative content of the component with the higher valence (nitrogen) and of that with the lower valence (carbon) and of the metalloid vacancies leads to the valence-electron concentration being almost constant over the whole concentration range of the alloys (it even decreases somewhat with increase in the molar proportion of nitride in the alloy), while the values of n or n_{mol} increase considerably. This indicates the difficulty of considering the electrical properties of the group of alloys investigated in terms of

---

*In calculating the valence-electron concentration we have taken account of all the valence electrons of the components of the alloy which are outside the inert-gas shell.

the valence of the components and the stable-band model. The in-
applicability of the stable-band model and of the interpretation of
the transport properties in defective binary monocarbides of metals of groups IV and V has been demonstrated in [5, 7, 1, 2]. In
order to determine the possibility of using the stable-band model,
we have estimated the change in effective mass m^*/m_0 (where m_0
is the mass of a free electron) in alloys of the system $ZrC_{0.98}$ –
$ZrN_{0.77}$ by an approximate method similar to that proposed in [7].
The absolute differential thermo-emf (α) of alloys with a high residual resistance satisfies the relationship [33]

$$\alpha = \frac{\pi^2 k^2 T}{3e} \left(\frac{\partial \ln \rho}{\partial \varepsilon} \right) \varepsilon_F, \qquad (3)$$

whence the logarithmic derivative of the specific electrical resistivity with respect to the energy at the Fermi level ε_F can be
calculated directly from the experimentally measured values of α.
To determine ε_F we used the formula for an expressed electron
gas:

$$E_F = \frac{h^2}{2m_0\varphi} \left(\frac{3}{8\pi} n \right)^{2/3}, \qquad (4)$$

where $\varphi = m^*/m_0$, h is Planck's constant, and n is the volume concentration of free electrons, taking into account the fact that the
current-carrier concentration in the alloys concerned ($n = 10^{21}$–
10^{22} cm^{-3}) considerably exceeds the degeneracy limit. Values of
$(\partial \ln \rho/\partial \varepsilon)$, m_0 were calculated by graphical differentiation on the
basis of the dependence of the logarithm of the specific electrical
resistivity of the carbonitride alloys of zirconium on the hypotheti-
cal value of ε_F calculated from the experimental values of n at
$m^* = m_0$ (i.e., $\varphi = 1$).

By differentiating Eq. (4) it can be shown that $(\partial \ln \rho/\partial \varepsilon) \varepsilon_F$,
$m^* = \varphi (\partial \ln \rho/\partial \varepsilon) \varepsilon_F$, m_0. On the assumption that $(\partial \ln \rho/\partial \varepsilon) \varepsilon_F$, m^*
should be equal to the experimental value of $(\partial \ln \rho/\partial \varepsilon)\varepsilon_F$ calculated
from Eq. (2) for an alloy of given composition (with given values of
α and n), values of φ were determined from the relationship

$$\varphi = m^*/m_0 = (\partial \ln \rho/\partial \varepsilon) \varepsilon_F, \ m^*_{exp} / (\partial \ln \rho/\partial \varepsilon) \varepsilon_F, \ m_0 \text{ calc.} \qquad (5)$$

The values obtained for m^*/m in relation to the composition of the
alloys in the system $ZrC_{0.98}$ – $ZrN_{0.77}$ provide evidence that the ef-

fective mass of the conduction electrons in alloys of this system varies considerably with change in composition and that the stable-band model cannot be used in dealing with the transport properties of alloys of this type. Taking account of the values obtained for $\varphi = m^*/m_0$, the density of states at the Fermi level in zirconium carbonitrides was determined from the formula [11]

$$N(\varepsilon) = \frac{3}{2} \cdot \frac{n_{mol}}{\varepsilon_F} \text{ states/eV} \cdot \text{mole,} \tag{6}$$

where ε_F was calculated from Eq. (4). It may be noted that in the case of the monocarbide $ZrC_{0.98}$ the value obtained for $N(\varepsilon)$ is not in bad agreement with the theoretical value as given by the density-of-states curve [37] for TiC with a composition close to the stoichiometric. It can be seen that the density of states at the Fermi level in zirconium carbonitrides increases considerably with rise in the molar fraction of nitride in the alloys, in agreement with the nature of the concentration dependence of the magnetic susceptibility of ZrC − ZrN alloys [30].

Because of the difficulty of using the band model to interpret the properties of the alloys investigated, it is more convenient to utilize the concepts developed by Samsonov [18, 25] regarding the stable configurations of bonding electrons in metals and alloys and previously used [19] in discussing the electron specific heat of quasimetallic compounds of the same type as the transition-metal carbides.

It is known from [6] that the compounds with which we are concerned crystallize in an NaCl-type structure, in which the metalloid atoms occupy octahedral positions in the closest cubic packing of the metal atoms, so that each metalloid atom (in this case carbon, nitrogen, or oxygen) is surrounded by six metal atoms. According to [42], suitable orbitals of the metalloid in the given positions for bond formation with metal atoms consist of a combination of three mutually orthogonal p orbitals, each of which is occupied by a single electron, and of one s orbital, so that the four valence electrons of carbon in the carbide form a bonding configuration sp^3 (which is by no means dissimilar to the tetrahedral sp^3 hybridized orbital). The metal atoms in the structure in question form bonds with the metalloid through the hybridized d^2sp^3 orbitals, which have octahedral symmetry. However, the four valence electrons of titanium or zirconium are not enough to fill all of the possible states in the combination of these orbitals. It may be assumed that the part of the hybridized d^2sp^3 orbital which is genetically bonded to the d state of the metal remains substantially free in the carbides, while the valence electrons of the metal occupy mainly that part of the combination of hybridized orbitals which is genetically bonded to the s and p states of the metal, i.e., the actual electron configuration of the metal atoms in monocarbides of group IV metals may be represented as d^0sp^3, while the stable chemical bond between the metal and carbon is due to the interaction of the metal and carbon electrons in the sp^3 states, which, according to [25], are very stable. The neutral d^0 configurations are also stable [22, 25, 24]. This assumption is in line with quantum-mechanical calculations of the electron distribution among the states in titanium carbides [29, 32], according to which the d states in the energy spectrum of titanium monocarbide are not filled, while the Fermi level lies in the region of a deep minimum on the density-of-states curve close to the filled sp states. The stable electron configurations responsible for the formation of valence bonds [24, 25] make no contribution to the density of states of the collectivized electrons to which are due the low electron specific heat and diamagnetism of carbides of group IV metals [19]. When carbon atoms are replaced in carbonitride alloys by nitrogen, which disposes of five valence electrons, the "superfluous" (as compared with carbon) electron in nitrogen upsets the combination of stable electron configurations that is characteristic of carbides of group IV metals. It cannot enter the stable bonding sp^3 configuration of the metalloid and is redistributed among the system of Me−X bonds (where X = C or N) by occupying the antibonding superincumbent states (as is indicated in particular by the fall in melting point [10] and Debye temperature [16]); this leads to destabilization of the d^0 electron configurations of the metal atoms and to an increase in the number of collectivized electrons with rise in the relative content of nitrogen in the carbonitride alloys, in spite of the fact that the total number of valence electrons in carbonitride alloys remains almost unchanged (see Tables 1-4). This is also responsible for the increase in the density of states of the collectivized electrons. The electron conduction of the alloys over the whole composition range investigated indicates that the Fermi level lies close to the lower part of the filled d state bands. It is clear that on this effect is superimposed the simultaneous reduction of the screening of the metal —metal bonds by the valence electrons of the metalloid in consequence of the increase in the number of metalloid vacancies, which should lead, according to [1, 2, 7, 12], to a further rise in the number of collectivized electrons. However, in carbonitride systems this latter phenomenon evidently has less significance in comparison with the replacement of carbon by a metalloid with a large number of valence electrons (nitrogen), since a similar dependence of the free-electron concentration is found in oxycarbide alloys of titanium, in which metalloid vacancies are formed simultaneously with the formation of an equivalent number of metallic vacancies.

The oxycarbide alloys of transition metals of group IV should behave in a similar way; here the replacement of the tetravalent carbon by the hexavalent oxygen also upsets the stability of the sp^3 bonding configurations and the d^0 configurations of the metal, resulting, as was shown above, in an increase in the number of delocalized (free) electrons. The concentration of free electrons might be expected to rise more sharply in TiC —TiO alloys with increase in the molar fraction of TiO than it would in TiC —TiN alloys with increase in the molar fraction of TiN, the valence of oxygen being higher than that of nitrogen. However, the experimental results show that this is true only at a relatively small oxygen content (up to about 40 mole % TiO), after which the rise in free-electron concentration becomes slower; after passing through a maximum, the concentration falls again in the oxygen-rich alloys. This may be due to the fact that the oxygen atoms, having a high electronegativity, may exhibit, in addition to a donor effect, an acceptor effect with the formation of stable s^2p^6 configurations [25], which should be accompanied by a reduction in the number of collectivized electrons and an increase in the number of localized electrons in the alloy. Clearly, the effect in question should become more marked with increase in the relative oxygen content in the alloys, since oxygen is evidently responsible for the extreme character of the dependence of the concentration of the free electrons in the system $TiC_{0.96}$ —$TiO_{1.03}$ on the composition of the alloys.

An increase in the concentration of free electrons also takes place in the system $TiN_{0.72}$ —$TiO_{1.03}$ with increase in the relative content of the higher-valence component (oxygen). However, in contrast to the preceding system, this relationship is monotonic and less marked; this may be due to the fact that both the nitride

and monoxide of titanium are compounds in which the staility of
the electron configurations is to a greater or less extent destroyed.
In each of the compounds concerned the free-electron concentration is high, while only the quantitative change in the number of
valence electrons plays any part in the formation of the oxynitride
alloys [and it is this change which is responsible for the similarity
of slope of the change in free-electron concentration and valence-
electron concentration in the alloys (see Table 4 and Fig. 1)]. The
absence of an extreme relationship between the free-electron concentration and the alloy composition in the titanium oxynitride system (in contrast to the oxycarbide system) may be due to the smaller difference in electronegativity (X) of nitrogen and oxygen (ΔX =
0.5) as compared with the difference in the case of carbon and
oxygen (ΔX = 1.0) [44]; as a result of this the acceptor properties
and the tendency for the formation of stable s^2p^6 configurations
in the oxygen atoms are considerably less marked in oxynitride
than in oxycarbide alloys.

Attention may be drawn to the change in specific electrical
resistivity with composition of the alloys investigated, as it is
somewhat different from the usual change in continuous series of
solid solutions. In the zirconium carbonitride system, the specific electrical resistivity (see Fig. 1) falls continuously with increase in molar fraction of the nitride, in spite of the increase
in the concentration of metalloid vacancies, i.e., of the additional
contribution due to scattering of conduction electrons. In the
titanium carbonitride system, the specific electrical resistivity
varies little with composition in the region rich in carbon, but it
begins to fall with further increase in the molar fraction of nitride. The contribution made by metalloid vacancies to the specific electrical resistivity of titanium carbonitride alloys has
been estimated by the method already described in [16], using
experimental values of the volume concentration of conduction
electrons. The dependence of the specific electrical resistivity
of titanium carbonitrides, after deduction of the contribution made
by scattering at metalloid vacancies, on the alloy composition is
shown in Fig. 1. This value falls continuously with increase in
the molar fraction of nitride, i.e., alloys of the system $TiC_{0.96}$ −
$TiN_{0.72}$ behave similarly to those of the system $ZrC_{0.98}$ − $ZrN_{0.77}$,
with the sole difference that the metalloid vacancies in the latter
system have a smaller scattering section owing to the sharper in-

crease in the volume concentration of free electrons which screen the disturbing potential of the scattering centers, with increase in the relative nitrogen content. It is possible that this difference in the change in free-electron concentration with composition of the alloys in the titanium and zirconium carbonitride systems is due to the smaller acceptor capacity of zirconium [26] as compared with titanium. The specific electrical resistivity is related to the concentration (n) and mobility (u) of the current carriers by the equation $1/\rho = neu$. As can be seen from Fig. 1, the current-carrier mobility falls in both carbonitride systems with increase in the molar fraction of nitride, as a result of the increasing contribution made by scattering at metalloid vacancies with rising concentration of the latter. The fall in the rate of reduction of mobility on passing to alloys having a large relative nitrogen content is due to the gradual decrease in the scattering section of the metalloid vacancies resulting from a rise in the conduction-electron concentration. Thus, the reduction in the specific electrical resistivity of carbonitride alloys with increase in the relative nitrogen content is entirely due to the rise of the free-electron concentration.

It may be noted that in the alloys of the system $ZrC_{0.92} - ZrN_{0.84}$ previously investigated, which were made from commercially pure materials [16], a maximum was found in the specific electrical resistivity both as measured and as corrected for the contribution made by the metalloid vacancies. This maximum is evidently due to the contribution made by the scattering of current carriers at the impurities present, which may also have a significant influence on the concentration of free electrons in carbides of group IV metals.

In the systems $TiC_{0.96} - TiO_{1.03}$ and $TiN_{0.72} - TiO_{1.03}$ the specific electrical resistivity passes through a minimum in the composition region low in oxygen, though the reason for the appearance of this minimum is different in the two systems. In the first case the mobility falls sharply (see Fig. 1) with increase in the relative oxygen content in the region of comparatively low oxygen contents. In this region a sharp increase takes place in the concentration of free electrons, which makes a major contribution to the change in electrical resistivity and leads to its reduction. At higher relative oxygen contents the mobility falls considerably less sharply, although the electron concentration varies only slightly

in this composition range, and it begins to decrease at approximately 60 mol.% $TiO_{1.03}$. In consequence the specific electrical resistivity of the alloys having a relatively high oxygen content increases with rise in the molar fraction of monoxide. The sharp
fall in mobility in this composition range, in which (as can be seen
from Fig. 1) the concentration of defects in the two sublattices
does not change very much, is evidently due to the fact that on
substituting the hexavalent oxygen for the tetravalent carbon, a
marked disturbance occurs in the stability of the electron configurations in the system of bonds based on titanium monocarbide,
accompanied by an increase in the density of states of the collectivized electrons.

In the system $TiN_{0.72} - TiO_{1.03}$ an increase in the relative oxygen content is accompanied by a moderate increase in the concentration of free electrons, while the mobility passes through a
maximum in the region of comparatively small oxygen contents
as a result of a sharp reduction in the concentration of metalloid
vacancies in this composition range, where the concentration of
metallic defects does not change very much. In this composition
range the specific electrical resistivity of the alloys falls with increase in the molar fraction of the monoxide. At higher oxygen
contents the mobility falls considerably as a result of an increase
in both metallic and metalloid vacancies, which produces an increase in the electrical resistivity for a comparatively small
change in the concentration of conduction electrons.

The thermal conductivity of alloys in all the systems investigated either falls monotonically with increase in the relative
content of the component with the higher valence (and at the same
time the higher number of vacancies) or it passes through a shal-
low minimum which is evidently due to the contribution of the lattice component; as already shown in [20], this accounts for approximately half (or more) of the total thermal conductivity of
compounds of the type we are dealing with. As the Lorentz number for carbides and nitrides [20] (and also evidently monoxides)
does not have a definite value, it is not yet possible to make a more
detailed analysis of the concentration dependence of the thermal
conductivity of the alloys investigated.

Literature Cited

1.   A. I. Avgustinik, in: Studies in the Field of the Chemistry of Silicates and
     Oxides [in Russian], Nauka, Moscow and Leningrad (1965), p. 241.

2. A. I. Avgustinik, Izv. Akad. Nauk SSSR, Neorg. Mat., 2:1439 (1966).
3. A. I. Avgustinik, G. M. Klimashin, and L. V. Kozovskii, Izv. Akad. Nauk SSSR, Neorg. Mat., 1:830 (1965).
4. A. I. Avgustinik, in: Studies in the Field of the Chemistry of Silicates and Oxides [in Russian], Nauka, Moscow and Leningrad (1965), p. 244.
5. A. I. Avgustinik et al., Izv. Akad. Nauk SSSR, Neorg. Mat., 3:286 (1967).
6. G. B. Bokii, Introduction to Crystal Chemistry [in Russian], Izd. MGU, Moscow (1954), p. 350.
7. O. A. Golikova et al., Fiz. Tverd. Tela, 7:12 (1965).
8. O. A. Golikova et al., Fiz. Tverd. Tela, 9:1557 (1967).
9. I. N. Danisina et al., Zh. Priklad. Khim. (1967).
10. F. Laves, in: The Theory of Phases in Alloys [Russian translation].
11. J. Simon, Electrons and Photons [Russian translation], IL, Moscow (1962), p. 106.
12. V. S. Neshpor and G. M. Klimashin, Izv. Akad. Nauk SSSR, Neorg. Mat., 1:1545 (1965).
13. V. S. Neshpor, V. S. Daviydov, and B. G. Ermakov, Poroshkovaya Met., No. 12, 65 (1967).
14. V. S. Neshpor, Zh. Priklad. Khim., 37:2375 (1964).
15. V. S. Neshpor, Ukrainsk. Fiz. Zh., No. 5, 839 (1960).
16. V. S. Neshpor, Yu. N. Vil'k, and I. N. Danisina, Poroshkovaya Met., No. 1, 89 (1967).
17. V. S. Neshpor and S. S. Ordon'yan, Izv. Akad. Nauk SSSR, Neorg. Mat., 1:173 (1965).
18. V. S. Neshpor, V. P. Nikitin, and V. V. Rabotnov, Poroshkovaya Met. (1967).
19. V. S. Neshpor and G. V. Samsonov, Fiz. Metal. i. Metalloved. (1967).
20. V. S. Neshpor, Izv. Akad. Nauk SSSR, Neorg. Mat., Vol. 3 (1967).
21. V. I. Odelevskii, Zh. Tekhn. Fiz., 21:682 (1951).
22. I. F. Pryadko, Poroshkovaya Met., No. 12, 61 (1966).
23. G. V. Samsonov and Ya. S. Umanskii, Hard Compounds of Refractory Metals [in Russian], Metallurgizdat, Moscow (1957).
24. G. V. Samsonov, Poroshkovaya Met., No. 12, 49 (1966).
25. G. V. Samsonov, Izv. Akad. Nauk SSSR, Neorg. Mat., 3:17 (1967).
26. G. V. Samsonov, V. S. Neshpor, and G. A. Kudintseva, Radiotekhnika i Elektronika, No. 2, p. 631 (1957).
27. Ya. S. Umanskii, X-Ray Metallography [in Russian], Metallurgizdat, Moscow (1960), p. 342.
28. A. Auskern, S. Aronson, J. Sadovsky, and F. Salzano, J. Phys. Chem. Solids, 27:613 (1966).
29. H. Biltz, Z. Physik, 153:338 (1958).
30. H. Bittner, H. Goretzki, F. Benesovsky, and H. Nowotny, Monatsh. Chem., 94:518 (1963).
31. S. Denker, J. Phys. Chem. Solids, 25:1397 (1964).
32. B. Ern and A. Switendik, Phys. Rev., 137:1927 (1965).
33. W. Cool, J. Mat. Sci., 1:261 (1966).
34. L. Hollander, J. Appl. Phys., 32:996 (1961).
35. H. Juretschke and R. Steinitz, J. Phys. Chem. Solids, 4:118 (1958).
36. A. Lorb, J. Am. Ceram. Soc., 37(II):96 (1954).
37. R. Lye and E. Logothetis, Phys. Rev., 137:A1927 (1965).

38. N. Mott and H. Jones, The Theory of the Properties of Metals and Alloys, Oxford (1936), p. 310.
39. J. Piper, J. Appl. Phys., 33:2394 (1962).
40. J. Piper, Compounds of Interest in Nuclear Reactor Technology (edited by Waber and Chiotti), The Met. Soc., AIME, Michigan (1964), p. 29.
41. E. Rudy and F. Benesovsky, Planseeber. Pulvermet., 8:72 (1960).
42. R. Rundle, Acta Cryst., 1:180 (1948).
43. V. V. Skorokhod, Powder Metallurgy, 12:188 (1932).

The Influence of Production Conditions on Some Physical Properties of Tungsten Carbide and of a Complex Carbide Based on It

A. N. Kruchinskii

The microhardness and abrasive power of tungsten carbide and of a complex TiC—WC carbide have been determined for various conditions of reduction and carbidization. It was found that both increase with rise in the reduction and carbidization temperatures. This increase in microhardness and abrasive power of high-temperature WC can be attributed to the greater mobility of the tungsten and carbon atoms, which leads to healing of the macro- and microdefects in the grains during reduction and carbidization, particularly in the case of WC produced at temperatures close to the recrystallization temperatures.

In the production of tungsten carbide by the reduction of its oxides, evaporation of the volatile component (oxygen) from the chemical compound (WO_3) must be regarded as a source of submicroscopic cracks and voids. When carbide is made from the metallic powder, the same amount of porosity is retained.

It has been found in [1, 2] that the physicomechanical and utilization properties of VK8V alloys based on high-temperature WC (1800-1850°C), as manifested in the drilling of rocks of medium hardness, substantially exceed those of VK15 alloys based on carbide produced at moderate temperatures; this is due to the pres-

TABLE 1. The Effect of Production Conditions
on the Microhardness and Abrasive Power
of WC and a Complex TiC—WC Carbide

| Carbide | Reduction tempera-ture of WO_3, °C | | Carbidiza-tion tem-perature, °C | Abrasive power, g/g | Micro-hardness, kg/mm² |
	first	second			
WC	700—750	870—900	1380—1400	1.1	1980
			1500—1550	1.16	2240
			1600—1650	1.2	2400
			1800—1850	2.68	2450
			2200	2.75	2510
WC	1200	—	1380—1400	1.5	1900 *
			1800—1850	5.1	2340 *
				5.1	2700
TiC — WC	700—750	800—850	2000	4.1	2400
			2200	4.14	2450

*Microhardness determined on WC powder wetted in AKR.

ence of fewer defects in the high-temperature WC resulting from its formation at a high temperature close to the recrystallization temperature. This in turn enables higher physicomechanical properties to be obtained at a lower cobalt content. Hence the strength characteristics of VK8V are higher than those of the ordinary VK8 alloy, which cannot replace VK15 alloy in these conditions owing to its lower strength in bending and under impact.

Since the defectiveness of WC must be accompanied by a corresponding hardness characteristic, which falls with increase in the number of defects, the microhardness and abrasive power of WC have been determined in relation to reduction and carbidization conditions. The hardness measurements were made either directly on grains of carbide powder (unground high-temperature WC) wetted in AKR or on specimens produced by hot compaction followed by annealing to remove the internal stresses set up as a result of the compaction process.

The abrasive power of the materials was measured on an ASZ-4 apparatus. In this method the carbide grains being investigated are held between rotating steel and glass disks and a certain amount of glass is ground from the latter in a predetermined time. The abrasive power was expressed as the ratio of the volume of glass ground to the volume of the sample ground, i.e., the specific abrasive power was determined. Boron carbide powder was used as the control abrasive.

The results of measuring the microhardness (load on the indentor 50 and 100 g, time of holding under load 10 sec) and the abrasive power are given in Table 1.

It emerges from the data we have obtained that the microhardness of WC is governed by the conditions under which the tungsten anhydride is reduced and by the carbidization temperature of the tungsten. It increases with rise in both these temperatures and reaches high values (up to 2700 kg/mm^2).

The change in abrasive power, like that of the microhardness is determined by the reduction and carbidization conditions, reaching maximum values when the tungsten and WC are produced at temperatures close to the recrystallization temperatures and when WC particles with a more perfect crystal lattice are obtained.

The reduction in the number of defects in WC with rise in temperature and the corresponding increase in its hardness and abrasive power can be ascribed to the greater mobility of the tungsten and carbon atoms, which results in an increase in the pycnometric weight of WC and in the healing of the micro- and macrodefects in the grains (particularly in carbides produced at temperatures close to the recrystallization temperature, which is responsible for the greater mobility of the atoms). The lower microhardness of high-temperature WC can be explained by the carbide grains having less strength than those that have not compacted.

Literature Cited

1.   G. S. Kreimer, O. S. Safonova, and E. M. Bogino, Trudy VNIITS, Tverdye Splavy, 41, 1 (1959).
2.   G. S. Kreimer, S. A. Khudosovtsev, O. S. Sofonova, and E. M. Bogino, Trudy VNIITS, Tverdye Splavy, 2, 3 (1960).

A Study of Recrystallization Processes
in Refractory Compounds
I. P. Kushtalova

A study of the effect of volume deformation and surface hardening on
the properties of the carbides of titanium, niobium, zirconium, molybdenum, and tungsten has revealed the presence of slip lines. The temperature at which recrystallization begins has been determined from
the appearance of the first point discontinuities in the continuous lines
of the x-ray diffraction diagrams. The relative temperature at which
recrystallization begins in these carbides (T_r/T_{mp}) was found to be
0.5. This is due to distinctive features of the structural changes that
occur in refractory compounds on deformation and also to their high
bond energy as compared with ductile metals and alloys.

Interest in the study of softening and recrystallization processes in refractory compounds is determined by their use as the
basis of metal-cutting tools and also in high-temperature alloys
[3-5]. However, the deformation processes and the mechanism
and kinetics of recrystallization have not yet been adequately investigated in refractory compounds.

In the present research a study has been made of the effect
of volume deformation and surface hardening on the properties of
the carbides of titanium, niobium, zirconium, molybdenum, and
tungsten, using cylindrical specimens prepared by hot compaction
of powders of the refractory compounds. Specimens with the least
porosity (2-5%) were chosen. Volume deformation was effected
by compressing the specimens on all sides under a quasi-hydro-

TABLE 1. Some
Characteristics of the
Recrystallization
Processes in the
Carbides
Investigated after
Surface Deformation

Carbide	T_{mp}, °K	T_r, °K	T_r/T_{mp}
TiC	3420	1820	0.53
ZrC	3803	1950	0.51
NbC	4033	2150	0.53
Mo_2C	2838	1450	0.51
WC	3143	1650	0.52

static pressure of 50 kbar at the Institute of Superhard Materials.
The degree of deformation produced in this way was 12-15%. Surface hardening was carried out by grinding the specimens on a plane-
grinding machine of the ZG-71 type, the degree of hardening in this
case being 7-10%.

Electron-microscopic examination of the deformed refractory
compounds revealed the presence of slip lines. Propagation of the
slip bands is restricted by the existence of grain boundaries, at
which the regular structure is destroyed by the slip plane. This
is found even when the slip bands develop in two adjacent grains
which differ only slightly in orientation.

X-ray diffraction diagrams were taken in 57.3-mm-diameter
RKD camera to determine the temperature at which recrystallization begins. The specimen was placed at an angle of 20-25° to the
incident x-ray beam. In all cases the photographs were taken in
CuK_α radiation. To reduce the fogging due to the secondary characteristic radiation, two films were fixed in the camera, and the
second of these was used in the research. Soft radiation was
chosen so as to prevent penetration of the rays to a depth greater
than the thickness of the hardneed layer in the case of surface deformation. The temperature at which recrystallization begins
was determined from the appearance of the first point discontinuities on the continuous lines of the x-ray patterns (see Table 1).

The compounds being investigated were annealed in a vacuum
of 10^{-3} mm Hg. The results given in Table 1 show that the relative

temperature at which recrystallization begins in the carbides is approximately the same, being about 0.5, whereas for ductile metals and alloys it is 0.3–0.4 [2]. This difference evidently arises from features peculiar to the structural changes taking place in the refractory compounds on deformation and also from their greater bonding energy as compared with ductile metals and alloys.

The increase in the relative recrystallization temperature of the refractory compounds in comparison with metals may also be compared with the reduction in compounds of metals with nonmetals of the weight of the unlocalized part of the valence electrons of the metals [6]. Thus, in the case of titanium carbide, part of the unlocalized electrons of the titanium atoms pass into the localized state and participate in the formation of sp^3 hybrid functions of the bond of the carbon atoms (such a transfer has been demonstrated in [1]); the same occurs in the carbides of molybdenum and tungsten.

The reduction in the weight of the unlocalized part of the valence electrons on the formation of compounds of transition metals with nontransition metals restricts the possibility of the $s \rightleftharpoons d$ exchange and increases the rise in energy necessary for exciting the stable configurations formed by the localized part of the valence electrons; this leads to an increase in the ratio T_r/T_{mp}.

Literature Cited

1.   M. P. Arbuzov and B. V. Khaenko, Poroshkovaya Met., No. 4, 74 (1966).
2.   A. A. Bochvar, Metal Science [in Russian], Oborongiz, Moscow (1956).
3.   S. S. Gorelik et al., Izv. VUZ Tsvetnaya Met., No. 4 (1962).
4.   T. M. Golovinskaya and V. M. Pelepelin, Dopovidi Akad. Nauk UkrSSR, No. 1 (1963).
5.   B. D. Grozin, V. N. Bakul, and V. N. Pelepelin, Priklad. Mekhanika, Vol. 9, No. 1 (1963).
6.   G. V. Samsonov, Ukrainsk. Khim. Zh., 31:1233 (1965).

VI. THERMODYNAMIC PROPERTIES
OF CARBIDES

The Thermodynamic Properties of Tantalum Carbide in the Homogeneity Range

E. A. Guseva, A. S. Bolgar, V. A. Gorbatyuk, and V. V. Fesenko

The heat content of the carbides $TaC_{0.70}$, $TaC_{0.73}$, $TaC_{0.85}$, and $TaC_{0.99}$ has been measured in the range 1200-2200°K by the method of mixing, using a massive calorimeter. From the results obtained and from data available in the literature, values of the heat content, specific heat, and increase in entropy have been tabulated for $TaC_{0.70}$, $TaC_{0.73}$, and $TaC_{0.85}$ for the range 1200-2200°K; while for $TaC_{0.99}$ values of the enthalpy, specific heat, entropy, and reduced thermodynamic potential have been tabulated for the range 298-2500°K.

Measurements of the specific heat of tantalum carbide containing 6.26 wt.% carbon in the temperature range 54-295°K have been reported in [3]. Standard values of the entropy and specific heat calculated from the data in [3] are $S^0_{298} = 10.11$ and $C_{p,298} = 8.79$ cal/mole · deg.

Levinson [5] has obtained the following equation for tantalum carbide containing 92.14% Ta, 6.21% C, 0.2% Fe, 0.5% Nb, and 0.8% W in the range 1296-2843°K:

$$H^0_T - H^0_{310° K} = -24.07 + 5.942 \cdot 10^{-2}T + 3.442 \cdot 10^{-6}T^2 \text{ cal/g.} \qquad (1)$$

The following equation is recommended [4] for the specific heat of tantalum carbide close to the composition $TaC_{1.0}$ in the

317

range 298-2000°K:

$$C_p = 7.28 + 1.65 \cdot 10^{-3}T \text{ [cal/mole} \cdot \text{deg]}. \qquad (2)$$

At temperatures above 1200°K the data given in [5, 4] for the specific heat of tantalum carbide differ by 30-40%.

In the present research we have investigated the temperature dependence of the heat content and specific heat of specimens of tantalum carbide of four different compositions.

The initial carbide powders were produced by synthesis from the elements. Chemical analysis showed that the specimens had compositions corresponding to the formulas $TaC_{0.70}$, $TaC_{0.78}$, $TaC_{0.85}$, and $TaC_{0.99}$. The phase compositions of the specimens were checked before and after the measurements by x-ray analysis, using the standard method.

The investigations of the temperature dependence of the enthalpy were carried out in a high-temperature vacuum apparatus [1] by the method of mixing, using a massive calorimeter. The apparatus consisted of a high-temperature furnace, the calorimetric system, and the measuring arrangement. A system of screens and blinds protected the calorimeter from radiation from the furnace. The temperature of the specimen was measured with an OPM-019 optical pyrometer. The error in temperature measurement in the range 1200-2200°K did not exceed 0.8%. The quantity of heat introduced into the calorimeter was determined by a copper resistance thermometer. The true value of the change in resistance was found by a semigraphical method [2]. The calorimeter system was calibrated by the usual method from the amount of electric power consumed. In determining the amount of heat introduced into the calorimeter, account was taken of the losses of heat by the specimens after the time of their fall into the calorimeter. It may be noted that this correction did not exceed 3% of the total amount of heat even at the maximum temperature of 2200°K. The total error in determining the enthalpy did not exceed 1.5%.

The possibility of using the apparatus to investigate the specific heat and heat content of refractory compounds was checked by tests on specimens of high-purity tantalum and tungsten. The results of the measurements of the heat content of these two metals in the range 1200-2200°K agreed within 1.5-2% with the data in [6].

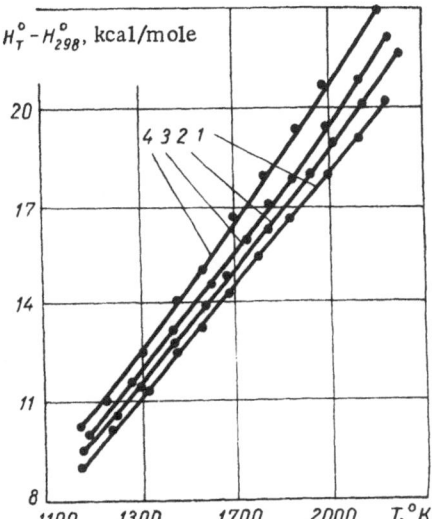

$H_T^o - H_{298}^o$, kcal/mole

Fig. 1. Experimental results of the determination of the heat content of the tantalum carbides: 1) $TaC_{0.70}$, 2) $TaC_{0.78}$, 3) $TaC_{0.85}$, and 4) $TaC_{0.99}$.

At equal temperatures the heat content of cubic tantalum carbide decreases with reduction in carbon content (Fig. 1), thus indicating a fall in the total strength of the chemical bonds Me−C and Me−Me in this compound with increasing carbon deficiency. The reduction in strength of the metal−carbon chemical bond due to the decrease in the number of individual Me−C bonds evidently cannot be offset by the increase in the total strength of the metal−metal bond in the carbide resulting from the removal of the screening of the individual Me−Me bonds and the increase in the number of electrons producing these bonds.

The results obtained for the heat content of the tantalum carbides in the range 1200-2200°K can be satisfactorily represented by the equations:

$$TaC_{0.70}: H_T^0 - H_{298}^0 = 8.954T + 0.681 \cdot 10^{-3}T^2 - 2730 \text{ cal/mole,} \qquad (3)$$

$$TaC_{0.78}: H_T^0 - H_{298}^0 = 9.098T + 0.809 \cdot 10^{-3}T^2 - 2780 \text{ cal/mole,} \qquad (4)$$

$$TaC_{0.85}: H_T^0 - H_{298}^0 = 9.180T + 0.976 \cdot 10^{-3}T^2 - 2880 \text{ cal/mole,} \qquad (5)$$

$$TaC_{0.99}: H_T^0 - H_{298}^0 = 9.384T + 1.211 \cdot 10^{-3}T^2 - 2900 \text{ cal/mole.} \qquad (6)$$

The results of measuring the enthalpy of the carbide $TaC_{0.99}$ agree within 1.5% with the data given in [5].

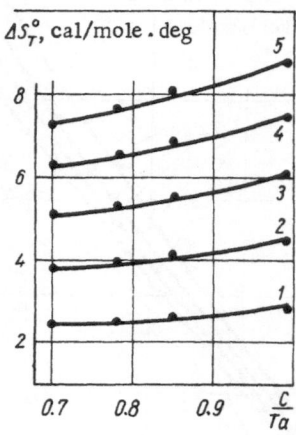

Fig. 2. Isotherms of the increase in the entropy of tantalum carbide in the homogeneity range: 1) 1400; 2) 1600; 3) 1800; 4) 2000; 5) 2200°K.

By differentiating Eqs. (3)-(6), we have calculated the specific heat of the carbides $TaC_{0.70}$, $TaC_{0.78}$, $TaC_{0.85}$, and $TaC_{0.99}$ in the temperature range investigated, and by mathematical treatment of the results so obtained, the following equation, which describes within an accuracy of 3% the specific heat of cubic tantalum carbide in the homogeneity range for temperatures from 1200 to 2200°K, was derived:

$$C_p = 7.3 - 1.5 \cdot 10^{-3}T + (0.773 + 1.476 \cdot 10^{-3}T)\, e^x \text{ cal/mole} \cdot \text{deg}, \qquad (7)$$

where x is the atomic ratio of carbon and tantalum.

TABLE 1. Thermodynamic Properties of Tantalum Carbide in the Homogeneity Range

T, °K	$TaC_{0.70}$			$TaC_{0.78}$			$TaC_{0.85}$		
	$H_T^0 - H_{298}^0$, cal/mole	C_p cal/mole ·deg	ΔS_T^0	$H_T^0 - H_{298}^0$, cal/mole	C_p cal/mole ·deg	ΔS_T^0	$H_T^0 - H_{298}^0$, cal/mole	C_p cal/mole ·deg	ΔS_T^0
1200	9000	10.6	0.0	9300	11.0	0.0	9540	11.5	0.0
1300	10 060	10.7	0.9	10 410	11.2	0.9	10 700	11.7	0.9
1400	11 140	10.9	0.7	11 540	11.4	1.7	11 880	11.9	1.8
1500	12 230	11.0	2.4	12 680	11.5	2.5	13 080	12.1	2.6
1600	13 340	11.1	3.1	13 850	11.7	3.3	14 310	12.3	3.4
1700	14 460	11.3	3.8	15 020	11.8	4.0	15 550	12.5	4.2
1800	15 590	11.4	4.5	16 220	12.0	4.7	16 810	12.7	4.9
1900	16 740	11.5	5.1	17 420	12.2	5.3	18 080	12.9	5.6
2000	17 900	11.7	5.7	18 650	12.3	6.0	19 380	13.1	6.3
2100	19 080	11.8	6.2	19 890	12.5	6.6	20 700	13.3	6.9
2200	20 270	12.0	6.8	21 150	12.7	7.1	22 040	13.5	7.5
2300	21 470	12.1	7.3	22 420	12.8	7.7	23 400	13.7	8.1

TABLE 2. Thermodynamic Properties of
Tantalum Carbide $TaC_{0.99}$

T, °K	$H_T^0 - H_{298}^0$, cal/mole	c_p, cal/mole · deg	S_T^0, cal/mole · deg	Φ_T'', cal/mole · deg
298	0	8.8	10.1	10.1
300	16	8.8	10.2	10.1
400	1000	10.6	13.0	10.5
500	2100	11.4	15.4	11.2
600	3260	11.8	17.6	12.1
700	4450	12.0	19.4	13.0
800	5660	12.1	21.0	13.9
900	6780	12.2	22.4	14.9
1000	7820	12.3	23.7	15.9
1100	8980	12.4	24.9	16.7
1200	10 100	12.5	26.0	17.6
1300	11 340	12.6	27.0	18.3
1400	12 610	12.8	27.9	18.9
1500	13 900	13.0	28.8	19.6
1600	15 210	13.3	30.0	20.2
1700	16 550	13.5	30.5	20.7
1800	17 910	13.7	31.3	21.3
1900	19 300	14.0	32.0	21.9
2000	20 710	14.2	32.7	22.4
2100	22 140	14.5	33.4	22.9
2200	23 600	14.7	34.1	23.4
2300	25 080	15.0	34.8	23.9
2400	26 590	15.2	35.4	24.3
2500	28 120	15.4	36.0	24.8

From the results obtained for the specific heats of the four
carbides, we have determined the entropy increase in these compounds in the range 1200-2300°K, and several isotherms of these
increases are shown in Fig. 2. The isotherms are curved, thus
indicating the nonlinear character of the variation in strength of
the metal—carbon chemical bond with decrease in combined-car-
bon content in the compound.

By using the data in the literature quoted above [3, 5] and the
results of our own heat-content measurements on the carbides,
we have calculated from the well-known thermodynamic relationships the principal thermodynamic properties of $TaC_{0.70}$, $TaC_{0.78}$,
and $TaC_{0.85}$ in the range 1200-2300°K and those of $TaC_{0.99}$ in the
range 298-2500°K. The results of the calculations are given in
Tables 1 and 2.

Conclusions

1. The heat content of the carbides $TaC_{0.70}$, $TaC_{0.78}$, $TaC_{0.85}$,
and $TaC_{0.99}$ has been measured over the range 1200-2200°K by the
method of mixing, using a massive calorimeter.

2. The temperature dependence of the specific heat of cubic tantalum carbide TaC_x in the homogeneity range can be described in the temperature interval 1200-2200°K by the equation: $C_p = 7.3-1.5 \times 10^{-3} T + (0.773 + 1.476 \times 10^{-3} T)e^x$ cal/mole · deg.

3. From the present results and data given in the literature, values of the heat content, specific heat, and increase in entropy have been tabulated for the carbides $TaC_{0.70}$, $TaC_{0.78}$, and $TaC_{0.85}$ over the temperature range 1200-2300°K, while for $TaC_{0.99}$ the enthalpy, specific heat, entropy, and reduced thermodynamic potential values have been tabulated for the range 298-2500°K.

Literature Cited

1. E. A. Guseva et al., Teplofiz. Vys. Temp., 4:649 (1966).
2. M. M. Popov, Thermometry and Calorimetry [in Russian], Izd. MGU (1954).
3. K. K. Kelley, J. Am. Chem. Soc., 62:818 (1940).
4. O. Kubaschewski and E. Evans, Metallurgical Thermochemistry, Pergamon Press, London and New York (1958).
5. L. S. Levinson, J. Chem. Phys., 39:1550 (1963).
6. D. R. Stull and G. C. Sinke, Thermodynamic Properties of the Elements (Advances in Chemistry Series, Vol. 18), American Chemical Society, Washington, D. C. (1956).

A Study of the Dependence
of Some Thermodynamic Properties
# of Niobium and Zirconium
Carbides on Temperature and Composition

V. V. Fesenko, A. G. Turchanin,
and S. S. Ordan'yan

Values of the enthalpy of niobium carbides of the compositions $NbC_{0.99}$, $NbC_{0.91}$, $NbC_{0.85}$, $NbC_{0.75}$ and zirconium carbides of the compositions $ZrC_{0.99}$, $ZrC_{0.75}$, $ZrC_{0.69}$ have been determined with an accuracy of 1.1% by the mixing method. Equations for the temperature dependence of the enthalpy and specific heat have been calculated for the whole temperature range investigated from the experimental results.

The carbides of niobium and zirconium are widely used in various new fields of technology. This accounts for the great interest that is being shown in their thermodynamic properties.

The most reliable data on the enthalpy and specific heat of NbC having almost stoichiometric composition are given in [11] for the temperature range 1200-2700°K. Similar results at lower temperatures are reported in [13, 12]. The authors of [14] have investigated the specific heat of three samples of niobium carbide in the homogeneity range at temperatures from 10 to 320°K. Gel'd and Kusenko [2] studied the change in the heat content of NbC in the homogeneity range from 300 to 1800°K. These properties of the NbC samples of various computations have not been studied at higher temperatures.

The temperature dependence of the enthalpy and specific heat of ZrC of almost stoichiometric composition has been determined in [1, 4, 10], but no information is available in the literature on these properties in carbides of other compositions.

In the present research we have investigated the change in heat content of four specimens of NbC with compositions in the homogeneity range and three specimens of ZrC at temperatures from 1300 to 2500°K.

The niobium carbide specimens were produced by compacting and sintering powder mixtures consisting of NbC of limiting composition and metallic niobium. Sintering was carried out in a vacuum furnace at 2200-2600°K in a vacuum of 10^{-5} mm Hg [8].

The zirconium carbide specimens investigated were obtained by synthesis from the elements, followed by compaction and sintering of the powders so produced in a vacuum furnace at 2200-2300°K and a vacuum of 5×10^{-5} mm Hg. The compositions of the specimens are given in Table 1.

The results of chemical analyses were checked by x-ray methods. The specimens produced were cylindrical in shape and weighed 4-5 g. Their porosity did not exceed 20%.

The temperature dependence of the change in enthalpy was determined by means of a high-temperature vacuum calorimetric apparatus, the design and principle of operation of which have been described in detail in [3]. The total relative error in determining the enthalpy did not exceed 1.1%. None of the specimens changed

TABLE 1. Composition of
Specimens of Niobium
and Zirconium Carbides

Formula	Nb, %	Zr, %	C_{tot}, %	C_{free}, %
$Nb_{0.75}$	91.20	—	8.80	—
$Nb_{0.86}$	90.05	—	10.01	—
$Nb_{0.91}$	89.5	—	10.47	—
$Nb_{0.99}$	88.27	—	11.70	0.3
$Zr_{0.69}$	—	91.68	8.30	—
$Zr_{0.76}$	—	90.96	9.0	—
$Zr_{0.99}$	—	88.10	11.40	0.5

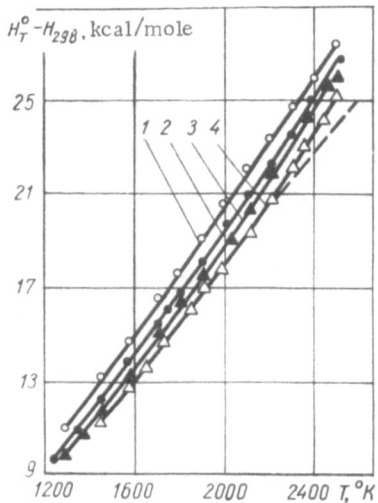

Fig. 1. Temperature dependence of the enthalpy of niobium carbides: 1) $NbC_{0.99}$; 2) $NbC_{0.91}$; 3) $NbC_{0.86}$; 4) $NbC_{0.75}$.

very much during the investigation (< 0.005%), and, as the results of x-ray and chemical analyses showed, the composition of the specimens did not alter within the limits of errors of the analyses. The results of the heat-content measurements made on niobium carbides are presented in Fig. 1.

The results we obtained for $NbC_{0.99}$ agree within 1% with those reported in [11] and are 2-3% below those given in [13, 2].

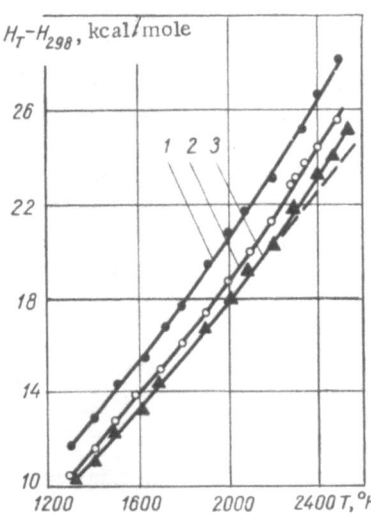

Fig. 2. Temperature dependence of the enthalpy of zirconium carbides: 1) $ZrC_{0.99}$; 2) $ZrC_{0.75}$; 3) $ZrC_{0.69}$.

The experimental results for the enthalpy of zirconium carbides are shown in Fig. 2. A comparison of these results for zirconium carbide of almost stoichiometric composition with the results of other authors shows that our enthalpy values agree within 1.2% with those given in [10, 4].

A more rapid increase in the enthalpy of nonstoichiometric carbides of both niobium and zirconium occurs at temperatures above 2200°K than at 1300-2200°K. Although the experimental values of the enthalpy of the stoichiometric carbides can be represented over the whole of the temperature range investigated by the power series $A + BT + CT^2$, for nonstoichiometric carbides this three-term series can be used only in the range 1300-2200°K. At higher temperatures the experimental values of the enthalpy of nonstoichiometric carbides differ from the results obtained by extrapolating the above three-term series.* This difference increases with rise in temperature. The effect is more marked, the poorer the carbides are in carbon.

The experimental results show that the temperature dependence of the enthalpy (cal/mole · deg) and of the specific heat (cal/mole) can be represented by the following equations:

$$NbC_{0.99}: H^0_T - H_{298} = 11.583T + 0.522 \cdot 10^{-3}T^2 - 4773; \tag{1}$$

$$c_p = 11.58 + 1.044 \cdot 10^{-3}T; \tag{2}$$

$$NbC_{0.91}: H^0_T - H^0_{298} =$$
$$= 11.505T + 0.460 \cdot 10^{-3}T^2 + 1.072 \cdot 10^5 \cdot \exp\left(-\frac{14.261 \cdot 10^3}{T}\right) - 5458; \tag{3}$$

$$c_p = 11.51 + 0.920 \cdot 10^{-3}T + 1529 \cdot 10^6 T^{-2} \cdot \exp\left(-\frac{14.261 \cdot 10^3}{T}\right); \tag{4}$$

$$NbC_{0.86}: H^0_T - H^0_{298} =$$
$$= 11.301T + 0.416 \cdot 10^{-3}T^2 + 4.680 \cdot 10^5 \cdot \exp\left(-\frac{16.300 \cdot 10^3}{T}\right) - 5381; \tag{5}$$

$$c_p = 11.30 + 0.832 \cdot 10^{-3}T + 7628 \cdot 10^6 T^{-2} \cdot \exp\left(-\frac{16,300 \cdot 10^3}{T}\right); \tag{6}$$

---

*In Figs. 1 and 2 this extrapolation was carried out for the carbides with minimum carbon content.

$$NbC_{0.75}: H_T^0 - H_{298}^0 =$$

$$= 10.09T + 0.569 \cdot 10^{-3}T^2 + 8.685 \cdot 10^6 \cdot \exp\left(-\frac{22.907 \cdot 10^3}{T}\right) - 4439; \qquad (7)$$

$$c_p = 10.10 + 1.138 \cdot 10^{-3}T + 1989 \cdot 10^8 T^{-2} \cdot \exp\left(-\frac{22.907 \cdot 10^3}{T}\right); \qquad (8)$$

$$ZrC_{0.99}: H_T^0 - H_{298}^0 = 7.329T + 1.663 \cdot 10^{-3}T^2 - 7749; \qquad (9)$$

$$c_p = 7.329 + 3.326 \cdot 10^{-3}T; \qquad (10)$$

$$ZrC_{0.76}: H_T^0 - H_{298}^0 =$$

$$= 7.762T + 1.174 \cdot 10^{-3}T^2 + 1.659 \cdot 10^6 \cdot \exp\left(-\frac{19.957 \cdot 10^3}{T}\right) - 1595; \qquad (11)$$

$$c_p = 7.762 + 2.348 \cdot 10^{-3}T + 3311 \cdot 10^7 T^{-2} \cdot \exp\left(-\frac{19.957 \cdot 10^3}{T}\right); \qquad (12)$$

$$ZrC_{0.69}: H_T^0 - H_{298}^0 =$$

$$= 8.057T + 0.964 \cdot 10^{-3}T^2 + 1.619 \cdot 10^8 \cdot \exp\left(-\frac{31.397 \cdot 10^3}{T}\right) - 2031; \qquad (13)$$

$$c_p = 8.057 + 1.928 \cdot 10^{-3}T + 7623 \cdot 10^9 T^{-2} \cdot \exp\left(-\frac{31.397 \cdot 10^3}{T}\right). \qquad (14)$$

The third term in Eqs. (3), (5), (7), (11), and (13), which takes account of the more rapid increase in the enthalpy of nonstoichiometric carbides above 2300°K, was calculated by the method proposed in [7, 9]. It should be noted that a similar increase in the rate of rise of the enthalpy and specific heat also occurs in pure metals [7, 5] and borides [6]. The phenomenon is usually attributed to vacancy formation at high temperatures.

Conclusions

1. The enthalpy values of specimens of niobium carbide of four different compositions and of zirconium carbide of three different compositions have been determined in the homogeneity range at temperatures of 1300–2500°K.

2. Equations for the temperature dependence of the enthalpy and specific heat have been derived from the experimental results for the whole temperature range investigated.

Literature Cited

1. A. S. Bolgar, E. A. Guseva, and V. V. Fesenko, Poroshkovaya Met., No. 4, 40-43 (1967).
2. P. V. Gel'd and F. G. Kusenko, Izv. Akad. Nauk SSSR, Otd. Tekh. Nauk, Met. i Toplivo, No. 2, 79-86 (1960).
3. E. A. Guseva et al., Teplofiz. Vys. Temp., 4:649-652 (1966).
4. P. B. Kantor and E. N. Fomichev, Teplofiz. Vys. Temp., 5:48-51 (1967).
5. V. A. Kirillin et al., Teplofiz. Vys. Temp., 3:395 (1965).
6. V. A. Kirillin, Teplofiz. Vys. Temp., 2:710-715 (1964).
7. Ya. A. Kraftmakher, in: Investigations at High Temperatures [in Russian], Nauka, Novosibirsk (1966).
8. V. S. Neshpor, Zh. Neorg. Khim., 37:2375-2382 (1964).
9. V. Ya. Chekhovskoi, Zh. Fiz. Khim., 39:2947 (1966).
10. L. Levinson, J. Chem. Phys., 42:2891-2892 (1965).
11. L. Levinson, J. Chem. Phys., 39:1550 (1963).
12. R. McDonald and F. Oetting, Proc. Meeting Interagency Chem. Rocket Propulsion Group Thermochem., New York (1963), p. 213.
13. L. Pankratz, W. Weller, and K. Kelley, US Bur. Mines Rep. Invest., No. 6446 (1964), p. 9.
14. T. Sandenav and E. Storms, J. Phys. Chem. Solids, 27:217-218 (1966).

Temperature Dependence of the Congruently Evaporated Compositions of the Zirconium Monocarbide Phase and Their Evaporation Rate

T. A. Nikol'skaya, R. G. Avarbe, and Yu. N. Vil'k

Zirconium monocarbide has been found to be capable of undergoing congruent evaporation. The temperature dependence of the congruently evaporated compositions of the ZrC_x phase can be described by the simple equation

$$\log (1 - x) = -0.1696 - \frac{1633}{T},$$

and the evaporation rate by the equation

$$\log V_{congr} = 8.555 - \frac{38134}{T}.$$

Zirconium carbide is one of the refractory compounds most widely used in present-day high-temperature technology. In establishing the optimum conditions for the use of zirconium carbide components, it is particularly important to know the rate of loss of weight per unit surface area for a fixed chemical composition and in the absence of diffusion in the component, i.e., the rate of congruent evaporation of the material being investigated.

Zirconium monocarbide is known to be capable of congruent evaporation [1, 3-5]. Thermodynamic analysis of the evaporation of the zirconium monocarbide phase in vacuum confirms this. However, the papers cited give no indication of the nature of the

functional temperature dependence of the congruently evaporated
compositions of the zirconium monocarbide phase, as it is generally assumed that congruent evaporation takes place close to the
stoichiometric composition. The evaporation rate has therefore
been investigated only at compositions close to the stoichiometric
one and over fairly narrow temperature ranges. There is no information about the evaporation of the ZrC_x phase at compositions
that differ considerably from the stoichiometric. Assumptions
about the evaporation of the zirconium monocarbide phase in the
vicinity of $ZrC_{1.00}$ at all temperatures are inconclusive.

The present investigation was undertaken with the object of
determining the temperature dependence of the congruently evaporated compositions of the zirconium monocarbide phase and of
measuring the rate of congruent evaporation in the absence of the
influence of diffusion factors on the process.

To determine the character of the evaporation of zirconium
monocarbide, we prepared specimens with seven different compositions having X = C/Zr ratios between 0.96 and 0.74 (Table 1).
The nitrogen content was less than 0.5 wt.%, and the amount of
oxygen and other impurities, estimated by difference, did not exceed 0.3-0.4 wt.%. The specimens, prepared by pressing through
a mouthpiece die the initial mass of powders and subsequently sintering at 2100°C in a TVV-4 furnace, were in the form of cylindrical rods 2.4-2.5 mm in diameter and 30-50 mm long. The specimens were placed in a water-cooled evaporating chamber equipped

TABLE 1. Composition of Specimens Undergoing
Evaporation

Specimen No.	Results of chemical analysis, wt.%		X = C/Zr	a, kX	a, kX
	C_{total}	N			
1 *	11.68	0.65	1.013	—	4.6905
2	11.20	0.53	0.963	4.6850	4.6883
3	10.30	0.50	0.877	4.6883	4.6855
4	10.12	0.55	0.861	4.686	4.685
5	9.65	0.40	0.814	4.683	4.682
6	9.05	0.59	0.760	4.676	4.676
7	8.83	0.31	0.739	4.666	4.6715

* $C_{free} = 0.09\%$.

TABLE 2. Compositions of Congruently Evaporated Phases of Zirconium Carbide

Specimen No.	T°, K	τ, h	Initial composition, wt.%		Final composition, wt.%			X_{mean}
			C	N	C	N	$X = C/Zr$	
1	2273	8	10.30	0.50	10.25	0.47	0.871	0.87
		10	10.15	0.55	10.23	0.46	0.870	
		5	9.65	0.40	10.03	0.42	0.850	
		5	11.2	0.53	10.64	0.48	0.910	
		10	10.30	0.50	10.20	0.47	0.867	
			9.65	0.40				
2	2388	2	10.30	0.50	10.10	0.45	0.857	0.86
		0.3	10.30	0.50	10.30	0.51	0.876	
		3	10.12	0.55	10.15	0.47	0.862	
		5	10.12	0.55	10.08	0.44	0.855	
		6	9.65	0.40	10.07	0.40	0.854	
3	2498	10	10.30	0.50	10.05	0.43	0.853	0.85
		2	9.65	0.40	9.87	0.37	0.835	
		5	9.65	0.40	9.98	0.35	0.850	
4	2613	3	10.30	0.50	9.93	0.43	0.841	0.84
		5	10.30	0.50	9.94	0.45	0.843	
		4	10.12	0.55	9.92	0.43	0.840	
		2	9.65	0.40	9.96	0.39	0.843	
		8	11.20	0.53	9.98	0.42	0.845	
5	2728	2	10.30	0.50	9.86	0.43	0.835	0.83
		2	10.15	0.55	9.79	0.47	0.828	
6	2843	1,5	10.30	0.50	9.90	0.47	0.838	0.82
		2	10.30	0.50	9.72	0.43	0.821	
		1	10.15	0.55	9.65	0.48	0.815	
		2	9.65	0.40	9.70	0.34	0.818	
		3	9.65	0.40	9.71	0.39	0.820	
7	2958	0.15	10.30	0,50	9.95	0.43	0.843	0.81
		0.30	10.30	0,50	9.73	0.43	0.823	
		0.50	10.12	0.55	9.62	0.45	0.812	
		0.80	10.12	0.55	9.59	0.43	0.809	
		1	9.65	0.40	9.57	0.38	0.808	
8	3073	10	10.30	0.50	9.49	0.42	0.800	0.80
		25	10.15	0.55	9.52	0.43	0.803	

with a tube with optical glass for observing the specimen and connected to a vacuum system (vacuum of $1-2 \times 10^{-5}$ mm Hg). The specimens were heated by the direct passage of a current through them. The temperature of the specimen was measured by means of an OMP-043 optical micropyrometer, a correction being made for the absorption in the optical glass. The specimens were held at a given temperature for a predetermined time. X-ray, metallographic, and chemical analyses showed that the zirconium monocarbide phase is in fact capable of congruent evaporation, though at temperatures above 2000°C the congruently evaporated com-

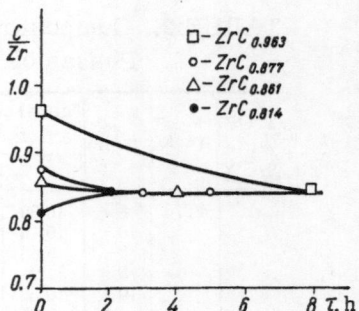

Fig. 1. Establishment of the congruently evaporated composition $ZrC_{0.84}$ with time at 2613°K.

TABLE 3. Experimental Results in the Investigation of the Evaporation Rate of ZrC_x

$T°$, K	Loss in weight ΔP, g	Mean area S, cm²	τ, sec	$\frac{\Delta P}{S}$, g/cm²	v, g/cm² · sec
2613	0.0061 0.0083 0.0179 0.0189 0.0272	2.735 2.510 2.730 2.828 2.770	1800 3600 7200 7200 10 800	0.00223 0.00331 0.00656 0.00668 0.00983	$9.1 \cdot 10^{-7}$
2843	0.0807 0.1341 0.2712 0.2692 0.2562	2.990 2.660 2.740 1.690 2.550	1800 3600 7200 7200 7200	0.0270 0.0504 0.0990 0.1001 0.1005	$1.4 \cdot 10^{-5}$
3073	0.2458 0.4824 0.5208 0.5872 0.6562	2.560 2.680 2.480 2.330 2.600	600 1200 1500 1800 1800	0.0960 0.1800 0.2100 0.2520 0.2524	$1.4 \cdot 10^{-4}$

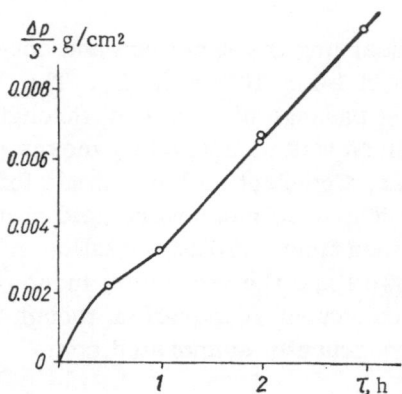

Fig. 2. Loss in weight from unit surface area of the specimen as a function of time at 2613°K.

position departs markedly from the stoichiometric. At each of eight temperatures investigated in the range 2270–3070°K, the congruently evaporated composition was reached at a constant time for the various carbon compounds in the original phases (Table 2). Figure 1 illustrates the attainment of the congruently evaporated composition at 2613°K.

It was found that the congruently evaporated compositions of the zirconium monocarbide phase are temperature dependent, and the following equation, which represents the dependence very well, was derived by the method of least squares:

$$\log (1 - X) = - 0.1696 - 1633/T, \tag{1}$$

where X = C/Zr.

The temperature range investigated was limited on the high-temperature side by the solidus line on the Zr − C constitutional diagram [2].

The evaporation rate measured for three congruently evaporated compositions in the temperature range, mentioned at corresponding temperatures in accordance with Eq. (1), were determined by Langmuir's method from the loss in weight of the specimen after it had been held in vacuum for a definite period of time. It was found that as a result of evaporation the area of the specimen changes owing to the alteration in its diameter. The reduction, which is practically imperceptible at low temperatures, becomes measurable at 2700–3000°K and represents 20–30% of the original area. Hence it is assumed that the effective evaporation area is equal to the arithmetical mean of the initial and final values. Moreover, the loss in weight per unit surface area is somewhat greater

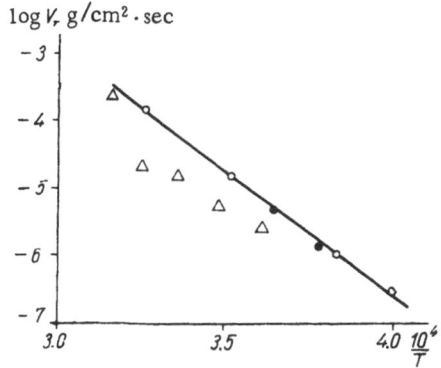

Fig. 3. Logarithm of the rate of congruent evaporation of the zirconium monocarbide phase as a function of the inverse temperature. Data from: O) present authors; Δ) Bolgar and Fesenko [1]; ●) Pollock [4].

in the initial stage than it is subsequently, as Fig. 2 shows. This is evidently due to oxidation of the specimen surface resulting from liberation of adsorbed gases and to removal of the oxidation products. At the end of gas evolution, corresponding to the restoration of the original vacuum, the loss in weight per unit surface area becomes proportional to the evaporation time. To eliminate the effect of the initial stage of evaporation, which is accompanied by gas evolution, on the result, the origin of coordinates is transferred to the point at which proportionality of the loss in weight per unit surface area is first observed, and the evaporation rate is determined from the slope of the line so obtained. The experimental results (Table 3), when treated by the method of least squares, give the following expression for the rate of congruent evaporation of the ZrC_x phase:

$$\log V_{cong} = 8.555 - 38,134/T. \tag{2}$$

As can be seen from Fig. 3, our results for the rate of congruent evaporation of the zirconium monocarbide phase are in good agreement with those of Pollock [4], but are somewhat higher than Bolgar's [1].

Conclusions

1. It has been found by experiment that the zirconium monocarbide phase is capable of congruent evaporation; this takes place at 2200–3100°K in the case of compositions other than the stoichiometric.

2. The temperature dependence of the congruently evaporated compositions of the ZrC_x phase can be represented by the simple equation (1).

3. The rate of congruent evaporation of ZrC_x can be expressed by Eq. (2).

Literature Cited

1.   A. S. Bolgar and V. V. Fesenko, Poroshkovaya Met., No. 1, 145 (1963).
2.   Yu. N. Vil'k, R. G. Avarbe, and A. I. Avgustinik, Zh. Priklad. Khim., 38:2 (1965).
3.   L. A. Goffman, G. M. Kibler, T. F. Lynon, and B. D. Accione, WADD, TR-60-646 (1963), p. 11.
4.   B. D. Pollock, J. Phys. Chem., 65:731 (1961).
5.   G. L. Vidale, AD-259 (1961), p. 469.

Mass-Spectrometric Study of the Evaporation
of Zirconium Carbide at High Temperatures

V. V. Torshina, G. N. Smolina,
S. L. Dobychin, R. G. Avarbe, and Yu. N. Vil'k

The evaporation of zirconium carbide containing 9.4-11.5 wt.% C has been studied in the temperature range 2000-2900°C. ZrC dissociates into its elements. The vapor phase of ZrC comprises Zr and C_1 atoms and C_2 and C_3 molecules. The pressure of the components of the vapor phase is a complex function of composition and temperature.

Carbides of transition metals of subgroups IV and V—Ti, Nb, Hf, Ta, and Zr — are some of the most refractory compounds, which accounts for their use in high–temperature technology. However, the possibility of using a particular carbide is determined to a considerable extent by the rapidity and nature of its evaporation. The object of the present investigation was to study the evaporation of zirconium carbide by means of a mass-spectrometer, which enables us to determine the composition of the vapor phase and the pressure of the components in the temperature range 2000-2900°K. The research was carried out on a MKh-1308 (SSSR) mass-spectrometer designed for the study of the evaporation of refractory compounds. The source of the molecular vapors was a tungsten Knudsen cell. The ratio of the evaporation area to the area of the effusion hole (0.5 mm in diameter) was 1000-500. The effusion cell was heated by electron bombardment. The maximum power dissipated by the cell was 2 kW. The temperature of the cell was determined by an OPK-57 optical pyrometer trained on a hole simulating an ideal black body. The reproducibility of the temperature reading was 2%, and the accuracy of measurement ±2% (according to the certificate). The correction for absorption at the pyrometer window was determined from the standard heating body SI-10-300.

Both the relative and the absolute methods were used in determining the vapor pressure of the substances investigated. In the first case the Langmuir—Knudsen formula

$$P = 17.14 \frac{Q}{K, t, S_{ef}} \sqrt{\frac{T}{M}}, \quad \text{if } \alpha = 1, \tag{1}$$

was used to calculate the pressure, where P is the pressure (mm Hg), Q is the weight of substance evaporated (g), K is the Clausing coefficient for the effusion hole, S_{ef} is the area of the effusion hole (cm^2), t is the time to evaporate the sample (sec), T is the evaporation temperature (°K), and M is the molecular weight of the compound being evaporated.

In the second case the vapor pressure was determined by use of Eq. (2). Silver was used as the standard substance, its vapor pressure having been fairly well investigated in [3]:

$$P_i = \frac{I_i^+ T_i}{I_{Ag} T_{Ag}} \cdot \frac{\sigma_{Ag}}{\sigma_i} \cdot \frac{\eta_{Ag}}{\eta_i} P_{Ag}, \tag{2}$$

where P_i is the pressure of the compound under investigation, P_{Ag} is the pressure of silver at T_{Ag}, I_i is the ionic current of i-mass of the compound being studied, taking account of the coefficient of expulsion of the secondary electrons when using the electron multiplier for recording the ionic currents, I_{Ag}^+ is the same for silver, η_i is the relative abundance of the i-mass isotope, η_{Ag} is the same for silver, σ_i is the ionization cross section of the i-mass atom or molecule [6], and σ_{Ag} is the same for silver [6].

In order to check the MKh-1308 mass-spectrometer, we determined the vapor pressures of silver, platinum, and nickel (see Table 1). The evaporation of zirconium carbide was investigated on specimens with a total carbon content of 11.5-9.8 wt.%. The specimens were in the form of powders or compacts sintered from powder, using a 10% solution of polyvinyl alcohol as bonding agent.

TABLE 1. Verification of the Mass-
Spectrometric Method of Determining
Vapor Pressures

Substance	$T°$, K	Vapor pressure, mm, Hg		Method of investigation by the Knudsen procedure	Reference
		Exper-imental	Calcu-lated		
Ag	1640	1.540	1.537	Complete combustion	[3]
	1660	2.029	2.150		
Ni	2230	2.090	2.014	Relative method	[8]
Pt	2450	0.040	0.030	Ditto	[1]

The evaporation of the powders was studied in the temperature
range in which reaction between the zirconium carbide and the
tungsten cell is negligible. Before evaporation began the specimens were annealed at 2100°K. Chemical and x-ray structural
analyses were carried out on the specimens before and after the
experiments.

In order to determine the composition of the vapor phase
and the pressure of the components, mass-spectra of the vapors
over zirconium carbide were recorded both at a single temperature and for various temperatures; the dependence of the ion intensity on the electron energy was also determined. The mass-
spectra of the vapor phase of all the ZrC specimens consisted,
over the whole temperature range investigated, of the ions Zr^+,
C_1^+, C_2^+, and C_3^+; the ZrC^+ ion was not observed. The dependence
of the intensity of the ions present on the electron energy revealed
that the molecular composition of the vapor phase of zirconium
carbide comprises Zr, C_1, C_2, and C_3 and that in addition to the
ions C_1^+, C_2^+, and C_3^+, which have appearance potentials equal to the
ionization potentials of C_1, C_2, and C_3, there are also ions with appearance potentials of 20, 22, and 24, the origin of which is not yet
clear (see Fig. 1).

The pressure of the components of ZrC vapor was determined
by using the relative method of determining vapor pressures by
ionizing molecules and atoms with electrons having an energy
$U_e = 17$ eV.

Each experiment was carried out on a new area of the given
sample having an initial total carbon content of 11.4%. In addition,

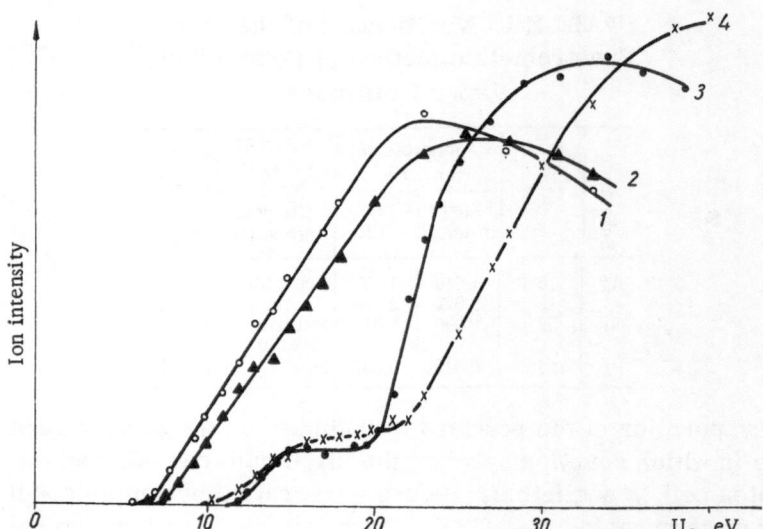

Fig. 1. Dependence of the intensity of the ions Zr^+ (1), Ag^+ (2), C_1^+ (3), and C_2^+ (4) on the electron energy.

the temperature dependence of the vapor pressure of zirconium over ZrC was determined under conditions in which the total carbon content of the carbide remained practically constant.

The experimental results indicate that the constant vapor pressure of zirconium over ZrC at a given temperature is attained practically instantaneously, whereas the vapor pressure of carbon becomes constant only after 3 h. The vapor pressure of C_3 cannot be measured with reasonable accuracy owing to the high background of the apparatus of the mass concerned.

Some reduction in the vapor pressure of C_1, C_2, and C_3 takes place at constant temperature at the initial time (approximately 3 h) with constant vapor pressure of zirconium; this is evidently due to combustion of carbon which is not bound in the carbide lattice. The results obtained are the same whether the sample is in the form of powder or a compact. For specimens that differ little in carbon content, the relationship $\ln P = f(1/T)$ has the same slope over the temperature range 2100–2900°K; calculations show that the heat of sublimation of zirconium from ZrC coincides, within the limits of experimental error of ±5 kcal/mole, with that of pure zirconium [4].

The results of the present investigation show that in the range 2000-2900°K ZrC dissociates into its constituent elements, which then evaporate. The pressure of the components is a para-function of temperature and composition. There is apparently a constant evaporated composition of the ZrC for each temperature. The pressures of the constituents of the vapor phase have not been used to calculate the thermodynamic functions, since it is proposed later to determine these quantities more accurately by obtaining more accurate values of the ratio σ_i / σ_{Ag}.

Conclusions

1. The evaporation of zirconium carbide being a total carbon content of 11.5-9.4 wt.% has been investigated in the temperature range 2000-2900°K.

2. The carbide evaporates by dissociation into its constituent elements.

3. The vapor phase of ZrC comprises Zr and C_1 atoms and C_2 and C_3 molecules.

4. The pressure of the components of the vapor phase is a function of composition and temperature.

5. Each temperature apparently has its own constant evaporated composition of the solid phase of ZrC.

Literature Cited

1. S. Dushman, The Scientific Foundations of Vacuum Technique [Russian translation], IL, Moscow (1950), p. 588.
2. N. M. Karnaukhova, Zh. Tekhn. Fiz., 34:1906 (1964).
3. Thermodynamic Properties of Inorganic Substances [in Russian], Atomizdat, Moscow (1965).
4. Thermodynamic Properties of Individual Substances [in Russian], Izd. Akad. Nauk SSSR, Moscow (1962).
5. V. V. Fesenko and A. S. Bolgar, The Evaporation of Refractory Compounds [in Russian], Metallurgiya, Moscow (1966).
6. V. D. Ovas and P. D. Robertson, J. Am. Chem. Soc., 78:546 (1956).
7. B. D. Pollock, J. Phys. Chem., 65:731 (1961).
8. H. L. Johnston, J. Am. Chem. Soc., 62:1382 (1940).

Thermodynamic Properties of Titanium Carbide of Variable Composition at High Temperatures

V. V. Fesenko and A. G. Turchanin

Values of the enthalpy of titanium carbide specimens of four different compositions in the homogeneity range have been determined experimentally in the temperature range 1300-2500°K. Equations have been derived for the temperature dependence of the enthalpy, specific heat, and change in entropy of the four carbides. It is concluded that the more rapid increase in enthalpy that occurs in some nonstoichiometric carbides is due to the formation of thermal vacancies.

The excess enthalpy of titanium carbide the composition of which is close to the stoichiometric one has been determined in [5, 4]. The authors are not aware of any data on the thermodynamic functions of this carbide with various concentrations of carbon in the homogeneity range.

In the present communication we report the results of an experimental study of the enthalpy and the calculation of the thermodynamic functions of titanium carbides of four different compositions in the homogeneity range at temperatures of 1300-2500°K. The specimens for the investigation were prepared by synthesis from metallic titanium and lamp black powders in vacuo at 1900°K. The carbide powders obtained in this way were compacted in a cylindrical die and sintered at 2200-2300°K in a vacuum of 5×10^{-6} mm Hg. The chemical composition of the specimens, determined

TABLE 1. Composition of
Titanium Carbide
Specimens

Car-bide	Ti, %	C_{total}, %	C_{free}, %
$TiC_{0.64}$	86.1	13.78	Not observed
$TiC_{0.71}$	84.7	15.08	The same
$TiC_{0.82}$	82.6	17.07	» »
$TiC_{0.99}$	80.0	19.9	0.54

after sintering, is given in Table 1. That the specimens were single phase was checked by an x-ray method. The specimens weighed 4-5 g, and their porosity did not exceed 18%.

The enthalpy of the carbides was determined by the method of mixing, using the vacuum high-temperature calorimeter apparatus described in [2, 3]. The total relative error in determining the enthalpy was less than 1.1%.

The negligible change in weight of the specimens during the investigation indicated that their composition had remained constant, and this was confirmed by chemical analysis of the specimens after the tests.

TABLE 2. Results of Determining the Enthalpy of Titanium Carbide in the Homogeneity Range (cal/g-formula)

$TiC_{0.64}$		$TiC_{0.71}$		$TiC_{0.82}$		$TiC_{0.99}$	
T, °K	$H_T^0 - H_{298}^0$	T, °K	$H_T^0 - H_{298}^0$	T, °K	$H_T^0 - H_{298}^0$	T, °K	$H_T^0 - H_{298}^0$
1298	9820	1416	11 570	1378	11 400	1295	11 190
1405	11 000	1515	12 730	1500	13 080	1396	12 550
1499	12 060	1611	13 840	1612	14 430	1505	13 930
1623	13 470	1700	14 970	1723	15 880	1596	14 870
1702	14 450	1805	16 030	1815	17 070	1689	16 440
1801	15 570	1916	17 650	1905	18 380	1805	17 960
1912	16 870	1986	18 540	2001	19 410	1910	19 340
1991	17 890	2106	20 120	2121	21 100	1999	20 670
2088	19 290	2197	21 380	2212	22 390	2114	22 460
2211	20 810	2298	22 940	2323	24 010	2203	23 550
2258	21 550	2400	24 430	2432	25 770	2300	25 100
2325	22 660	2455	25 300	2515	27 010	2402	26 730
2394	23 640	2501	26 060	—	—	2488	28 210
2470	24 790	—	—	—	—	—	—

The results of the enthalpy study on the titanium carbides are presented in Table 2. It may be noted that the values we obtained for $TiC_{0.99}$ agree within 0.5% at all temperatures with those reported in [5, 4].

It is assumed that the more rapid increase in the enthalpy of the carbides $TiC_{0.71}$ and $TiC_{0.64}$ in the high-temperature region is due to the formation of thermal vacancies. The additional heat content associated with their formation can be expressed by the equation

$$\Delta(H_T^0 - H_{298}^0) = UA \exp\left(-\frac{U}{RT}\right), \tag{1}$$

where U is the energy of vacancy formation, and A is the entropy factor which determines the concentration of defects.

It is to be noted that all the points of the relationship

$$\ln \Delta (H_T^0 - H_{298}^0) = f(-1/T) \tag{2}$$

for the carbides $TiC_{0.71}$ and $TiC_{0.64}$ lay close to a straight line; this enabled us to calculate U/R in Eq. (1) and to describe the observed effect by a separate term in the equations for the temperature dependence of the enthalpy. Values of $\Delta(H_T^0 - H_{298}^0)$ were determined as the difference between the experimental values of the enthalpy of the carbides at each given temperature and the enthalpy values obtained by linear extrapolation of the three-term expression (of the form $A + BT + CT^2$) which is valid at lower temperatures.

Equations for the temperature dependence of the enthalpy, obtained by treating the experimental data in a "Minsk-22" electronic computer, are as follows:

$$TiC_{0.64}$$
$$H_T^0 - H_{298}^0 = 8.076T + 1.091 \cdot 10^{-3}T^2 + 1.203 \cdot 10^8 \cdot \tag{3}$$
$$\exp(-29\,860/T) - 2506 \quad \text{cal/g-formula};$$

$$TiC_{0.71}$$
$$H_T^0 - H_{298}^0 = 8.110T + 1.192 \cdot 10^{-3}T^2 + 1.028 \cdot 10^7 \cdot \exp(-24\,513/T) -$$
$$- 2254 \quad \text{cal/g-formula};$$

$$TiC_{0.82} \tag{4}$$
$$H_T^0 - H_{298}^0 = 8.144T + 1.392 \cdot 10^{-3}T^2 - 2344 \quad \text{cal/g-formula}; \tag{5}$$

$$TiC_{0.99}$$
$$H_T^0 - H_{298}^0 = 8.178T + 1.581 \cdot 10^{-3}T^2 - 2009 \quad \text{cal/g-formula}. \tag{6}$$

TABLE 3. Thermodynamic Functions of Titanium Carbide*

$T°$, K	TiC$_{0,64}$			TiC$_{0,71}$			TiC$_{0,82}$			TiC$_{0,99}$		
	$H_T^0 - H_{298}^0$	c_p	$S_T^0 - S_{1200}^0$	$H_T^0 - H_{298}^0$	c_p	$S_T^0 - S_{1200}^0$	$H_T^0 - H_{298}^0$	c_p	$S_T^0 - S_{1200}^0$	$H_T^0 - H_{298}^0$	c_p	$S_T^0 - S_{1200}^0$
1200	8756	10.69	0	9194	10.97	0	9433	11.48	0	10 080	11.97	0
1300	9836	10.91	0.86	10 300	11.21	0.88	10 590	11.76	0.93	11 290	12.29	0.97
1400	10 940	11.13	1.68	11 440	11.45	1.72	11 780	12.04	1.81	12 540	12.60	1.89
1500	12 060	11.35	2.45	12 590	11.69	2.52	13 000	12.32	2.65	13 810	12.92	2.77
1600	13 210	11.57	3.19	13 370	11.92	3.28	14 250	12.60	3.46	15 220	13.24	3.62
1700	14 380	11.81	3.90	14 980	12.21	4.01	15 520	12.88	4.23	16 460	13.55	4.43
1800	15 560	12.07	4.58	16 200	12.49	4.72	16 830	13.16	4.97	17 830	13.87	5.21
1900	16 780	13.37	5.24	17 460	12.81	5.41	18 150	13.43	5.69	19 240	14.18	5.97
2000	18 050	12.73	5.89	18 790	13.17	6.07	19 510	13.71	6.39	20 670	14.50	6.71
2100	19 340	13.19	6.52	20 120	13.58	6.72	20 900	13.99	7.06	22 240	14.82	7.42
2200	20 690	13.81	7.14	21 500	14.09	7.37	22 310	14.27	7.72	23 630	15.13	8.12
2300	22 110	14.63	7.77	22 940	14.69	8.00	23 750	14.55	8.36	25 260	15.45	8.80
2400	23 630	15.74	8.42	24 450	15.41	8.65	25 220	14.83	9.99	26 720	15.77	9.46
2500	25 720	17.22	9.09	26 030	16.27	9.29	26 720	15.10	9.60	28 320	16.08	10.11

*$H_T^0 - H_{298}^0$ in cal/g-formula; c_p and $S_T^0 - S_{1200}^0$ in cal/g-formula · deg.

The mean deviation of the experimental data from the calculated curves did not exceed 0.15%.

By differentiating Eqs. (3)-(6) with respect to temperature, we obtained the following equations for the specific heat of the carbides investigated (in cal/g-formula · deg):

$$TiC_{0.64}: c_p = 8.076 + 2.182 \cdot 10^{-3}T + 3591 \cdot 10^9 T^{-2} \cdot \exp(-29\,860/T); \tag{7}$$

$$TiC_{0.71}: c_p = 8.110 + 2.334 \cdot 10^{-3}T + 2519 \cdot 10^8 T^{-2} \cdot \exp \cdot (-24513/T); \tag{8}$$

$$TiC_{0.82}: c_p = 8.144 + 2.784 \cdot 10^{-3}T; \tag{9}$$

$$TiC_{0.99}: c_p = 8.178 + 3.162 \cdot 10^{-3}T. \tag{10}$$

By using the known thermodynamic relationship [1], we also calculated equations for the temperature dependence of the change in entropy of the compounds investigated (in cal/g-formula · deg):

$$TiC_{0.64}: S_T^0 - S_{1200}^0 = 8.076 \ln T - 2.182 \cdot 10^{-3}T +$$
$$+ 4029 \exp(-29\,860/T) \cdot (1 + 29\,860/T) - 59.814; \tag{11}$$

$$TiC_{0.71}: S_T^0 - S_{1200}^0 = 8.110 \ln T + 2.384 \cdot 10^{-3}T +$$
$$+ \exp(-24\,510/T) \cdot (1 + 24\,510/T) - 60.296; \tag{12}$$

$$TiC_{0.82}: S_T^0 - S_{1200}^0 = 8.144 \ln T + 2.784 \cdot 10^{-3}T - 61.018; \tag{13}$$

$$TiC_{0.99}: S_T^0 - S_{1200}^0 = 8.178 \ln T + 3.162 \cdot 10^{-3}T - 61.710. \tag{14}$$

Table 3 contains values of the thermodynamic properties of the carbides investigated calculated at temperature intervals of 100°K in the range 1300-2500°K from Eqs. (3)-(14).

The accuracy of the enthalpy values given in Table 3, is as already stated, 1.1%. The error in determining the specific heat and the change in entropy is 2-3% over the whole temperature range for the carbides $TiC_{0.82}$ and $TiC_{0.99}$ and over the range 1200-1700°K for the carbides $TiC_{0.71}$ and $TiC_{0.64}$. Above 1700°K the accuracy of the values of the specific heat and the change in entropy is 3-5%.

Conclusions

1. Values of the enthalpy of specimens of titanium carbide of four different compositions in the homogeneity range have been determined at temperatures from 1300 to 2500°K.

2. Equations for the temperature dependence of the enthalpy, specific heat, and change in entropy have been calculated.

3. The more rapid rise in the enthalpy that is observed in certain nonstoichiometric carbides is ascribed to the formation of thermal vacancies.

Literature Cited

1. L. V. Gurvich et al., Thermodynamic Properties of Individual Substances [in Russian], Izd. Akad. Nauk SSSR, Moscow (1962).
2. E. A. Guseva et al., Teplofiz. Vys. Temp., 4:649 (1966).
3. A. G. Turchanin, S. S. Ordan'yan, and V. V. Fesenko, Poroshkovaya Met., No. 9, 23 (1967).
4. L. S. Levinson, J. Chem. Phys., 42:2891 (1965).
5. B. F. Naylor, J. Am. Chem. Soc., 68:370 (1946).

Thermodynamic Properties of the Higher Manganese Carbide

V. N. Eremenko, G. M. Lukashenko, and R. I. Polotskaya

The thermodynamic properties of the higher manganese carbide in Mn_7C_3 have been determined by measuring the emf of the high-temperature galvanic cells

$$(-)Mn_{solid} \left| \begin{matrix} KCl = NaCl \\ MnCl_2 \end{matrix} \right| [Mn_7C_3 + C_{graphite}](+)$$

In the range 1000-1150°K the isobar-isothermal potential of the formation of the carbide is represented by the equation

$$\Delta Z° = -22.470 - 0.492\ T \quad \text{kcal/mole.}$$

The standard heats and entropies of formation are, respectively, -19.8 ± 1.2 kcal/mole and 1.1 ± 0.4 cal/deg · mole.

The constitutional diagram of the manganese – carbon system has not been completely determined [5], and it has not been finally established how many manganese carbides exist, their composition, and the temperature limits of their stability. A short review of recent information in this question has been given by Alekseev and Shvartsman [1].

However, there is no doubt about the existence of the carbide Mn_7C_3 which is isostructural with Cr_7C_3 and is stable in equilibrium with graphite up to 1100°C. The present communication

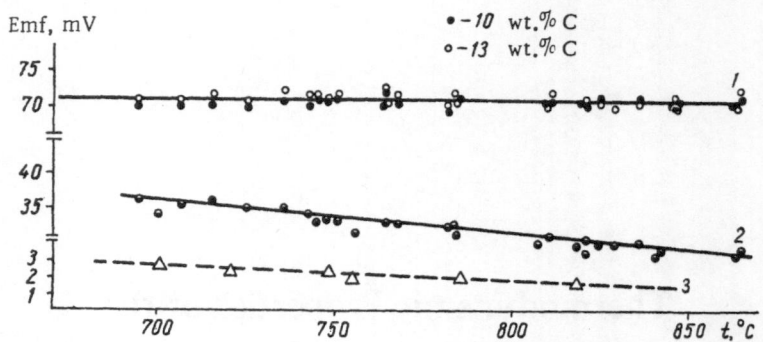

Fig. 1. Temperature dependence of the emf of galvanic cells for Mn—C alloys containing: 1) 10-13 wt.% C, 2) 6 wt.% C, 3) 2 wt.%.

deals with the determination of its thermodynamic properties by the emf method.

Since there are no carbides richer in carbon and the solubility of manganese in graphite is probably negligible, the thermodynamic functions of the formation of Mn_7C_3 by the reaction

$$7Mn_{\text{ solid}} + 3C_{\text{graphite}} \rightarrow Mn_7C_3$$

can be obtained directly by measuring the electromotive force of the concentration galvanic cells of the amalgam type

$$(A) \quad (-) \, Mn_{\text{solid}} \left| \begin{array}{c} KCl - NaCl \\ + MnCl_2 \end{array} \right| [Mn_7C_3 + C_{\text{graphite}}] \, (+).$$

The method of measurement has already been described in detail in [4, 3].

An undoubted merit of the emf method is that there is no need to prepare stoichiometric carbides, it being sufficient to produce two-phase alloys of arbitrary gross composition in which the compound being studied (carbide) is in equilibrium with the pure less-noble component (carbon). We obtained such alloys from 99.8% Mn, previously remelted in an arc furnace, and atomic graphite produced by the powder-metallurgy method. Sintering was carried out in closed graphite crucibles for 20 h at 1050°C in vacuo. According to McCabe and Hudson [8], this procedure ensures that the reaction between manganese and graphite proceeds to completion with the formation of Mn_7C_3.

During the process the specimens lose part of the manganese, so becoming enriched in carbon, but their phase composition should not alter.

The results of the measurements show that (Fig. 1, curve 1) the emf values are well reproducible with cyclic variation of the temperature, that they are in close agreement for alloys of different composition within the boundaries of the one particular homogeneous field, and that they do not vary from experiment to experiment, i.e., the conditions of reversible operation of the galvanic cell are practically fulfilled. For comparison we also give (Fig. 1, curves 2 and 3) emf values for alloys of other compositions corresponding to other phase fields.

By treating the results by the method of least squares, it was possible to express the temperature dependence of the emf of the element (A) in the form of the linear equation

$$E = 69.58 + 0.001525T \; [\text{mV}]$$

with a mean square deviation of the experimental values of not more than ± 0.7 mV.

Using the ordinary formulas of the emf method, we calculated the isobaric−isothermal potential, entropy, and heat of formation of the carbide Mn_7C_3 from β-Mn and graphite in the range 1000−1150°K:

$$\Delta Z^0 = (-22.470 - 0.492T) \pm 425 \; \text{cal/mole},$$
$$\Delta S^0 = 0.49 \pm 0.36 \; \text{cal/deg} \cdot \text{mole},$$
$$\Delta H^0 = -22.47 \pm 1.2 \; \text{kcal/mole}.$$

The errors in the values obtained were taken as $\pm$ twice the mean square deviations, calculated by the formulas of the method of least squares, which gives a 95% confidence interval.

The thermodynamic properties of Mn_7C_3 have been studied by the method of vapor-pressure measurement (see Table 1). The values of ΔZ^0 which we have calculated for 1100°C are in complete agreement with the results of McCabe and Hudson [8], although according to Fujishiro and Goksen [7] the isobar potential of formation is more negative. As the accuracy of determination is not given in the literature, it is hard to assess the degree of agreement between the results, particularly when we bear in mind that

TABLE 1. Isobaric – Isothermal
Potential of Formation of
Mn_7C_3

$T, °K$	$\Delta Z^0 = f\ (T)$	$\Delta Z^0_{1100° K}$, cal/g-C atom
1075—1235	5130—11,64 T	—7700 [8]
1075—1250	—1000—6,6 T	—8250 [6]
1025—1145	—	—9380 [7]
1000—1150	—7490— —0,164 T	—7670± ±150 *

* Authors' data.

the temperature dependence of ΔZ^0, and consequently the values of the heat and entropy of formation of the carbide, vary markedly from author to author. Actually these values have not been reliably determined in vapor-pressure experiments. By using the estimates of the high-temperature entropy and enthalpy of Mn_7C_3 made by Fujishiro and Goksen [7], it is possible to calculate the standard values of the thermodynamic properties of this carbide for its formation from pure d-manganese and graphite:

$$\Delta H^0_{298} = -19.8 \pm 1.2 \ \text{kcal/mole},$$

$$\Delta S^0_{298} = 1.1 \pm 0.4 \ \text{cal/deg} \cdot \text{mole},$$

$$\Delta Z^0_{298} = -20.13 \pm 0.45 \ \text{kcal/mole} = -6.75 \ \text{kcal/g-C atom}.$$

The calculated value of the standard heat of formation does not conflict with Fujishiro and Goksen's estimate [7]:

$$\Delta H^0_{298} = -26.8 \pm 10 \ \text{kcal/mole}.$$

It should be noted that inaccurate values of the heat and isobaric potential of formation of manganese carbide are given in the handbook [2].

Literature Cited

1. V. I. Alekseev and L. A. Shvartsman, in: Problems of Metals Science and Physics of Metals [in Russian], Vol. 8, Metallurgizdat, Moscow (1964), p. 281.
2. U. D. Veryatin et al., Thermodynamic Properties of Inorganic Substances [in Russian], Atomizdat, Moscow (1965).
3. V. N. Eremenko, G. M. Lukashenko, and V. P. Sidorko, Izv. Akad. Nauk SSSR, Met. i Gornoe Delo, No. 6, 151 (1964).

4.   V. N. Eremenko, G. M. Lukashenko, and V. P. Sidorko, Poroshkovaya Met., No. 5, 49 (1964).

5.   M. Hansen and K. Anderko, The Structure of Binary Alloys [Russian translation], Vol. 1, Metallurgizdat, Moscow (1962).

6.   J. E. Butler, C. L. McCabe, and H. W. Paxton, Trans. Met. Soc., AIME, 221:479 (1961) (cited in [1]).

7.   N. A. Goksen and Fujishiro, Trans. Met. Soc. AIME, 227:542 (1963).

8.   C. L. McCabe and R. G. Hudson, J. Metals, 9:17 (1957).

Enthalpy and Specific Heat
of Silicon Carbide at High Temperatures

V. A. Kirillin, A. E. Sheindlin,
and V. Ya. Chekhovskoi

Enthalpy values of α-SiC have been determined over a wide range of
temperatures by the method of mixing, using a massive calorimeter
with an isometric casing. Equations have been derived for the temperature dependence of the enthalpy and specific heat of SiC over the
range 273.15-2900°K.

Experimental studies of the enthalpy and specific heat of silicon
carbide have been made at temperatures not exceeding 1500-1800°K
[3, 5, 6, 8-12]. However, silicon carbide is known to be a promising constructional material for use in many new fields of technology, in which 1800°K is not the maximum temperature encountered. The authors of the present article have attempted to broaden the limits of the data on the enthalpy and specific heat of SiC
to the maximum temperature of its existence, 2800-2900°K [1], at
which rapid decomposition of this material sets in. We have calculated the enthalpy and specific heat of SiC free from impurities
by using the results of measurements of the enthalpy and specific
heat of α-SiC containing 12 mass% of free carbon [2] in the range
1100-2850°K. It should be pointed out that we have already [7]
made a similar preliminary calculation based on the single assumption that the additivity law applies to the enthalpy and specific heat of SiC and free carbon. Now, having at our disposal

TABLE 1. Results of Chemical
Analysis of Specimens after
the Experiments

Content, mass %	Speci-men 1	Speci-men 2	Speci-men 3
Silicon carbide	96.30	87.07	63.69
Free carbon	0.33	12.00	33.04
Free silicon	0.27	None	None
Iron	1.97	0.73	0.78

evidence of the existence of additivity in these two substances, we have made a more accurate calculation of the enthalpy and specific heat of SiC by using recent very accurate experimental data on the enthalpy of carbon [13] measured over the wide range 1200-2600°K and have found agreement of the results obtained with those of other authors in the range 300-1000°K.

In order to verify that the additivity law is applicable to the specific heat of SiC and free carbon, additional experiments were made to determine the enthalpy and specific heat on specimens No. 1 and 3 (Table 1), in which the free carbon content differs considerably from that of specimen No. 2, which was used in [2]. In these additional experiments use was made of the method of mixing (a massive calorimeter with isomeric casing being employed). The experiments were conducted in an argon atmosphere.

The use of compact specimens, produced by hot compaction, made it possible to carry out the investigation without the use of capsules. The specimens were in the form of a truncated zone (30 mm high, 18-20 mm in diameter) with an axial hole for mea-

TABLE 2. Experimental Enthalpy Results

Speci-men No.	T, °K	$H_T - H_{273,15}$, kcal/kg	T_n°, K	$H_T - H_{T_n}$, kcal/kg	q, kcal/kg
1	1560.9	342.40	299.48	338.44	0.97
	1607.2	355.05	299.16	351.14	1.09
2	1354.2	293.24	299.26	289.32	0.41
	1373.2	300.44	299.30	296.51	0.46
3	1949.2	536.69	302.17	532.30	2.32
	2025.2	567.37	301.23	563.13	2.64
	2126.2	606.65	302.92	602.12	3.25

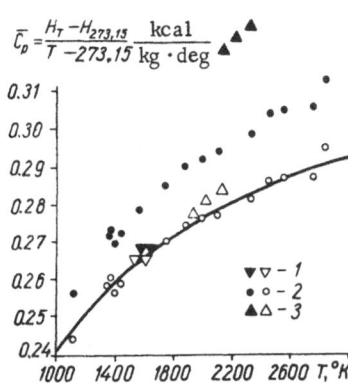

Fig. 1. The effect of free carbon on the mean specific heat of SiC: 1, 2, 3) results obtained on specimens No. 1, 2, and 3, respectively; solid symbols) experimental points obtained on specimens containing free carbon as shown in Table 1; open symbols) points obtained by calculation, using the additivity law, of the specific heat (in the absence of free carbon).

suring the temperature. The method of measurement and the apparatus employed were the same as those described in [2]. The results of the enthalpy measurements are given in Table 2.

Using the data in Table 2 and our earlier results [2] for specimen No. 2, we calculated the enthalpy of SiC containing no free carbon, by means of the additivity rule. The enthalpy of carbon was taken from [13]. The newly obtained data for SiC containing no free carbon are plotted in Fig. 1, open symbols being used for the mean specific heat. These points agree satisfactorily among themselves, thus indicating the existence of additivity of the specific heat of SiC and free carbon to the extent that the latter was present as impurity (as much as 33 mass%). For comparison, Fig. 1 also shows values of the mean specific heat (solid symbols) of SiC containing various amounts of free carbon. These lie above the mean curve for the specific heat of SiC containing no free carbon by 17 and 4-7%, respectively, for specimens No. 3 and 2.

The results obtained for the enthalpy of SiC are in agreement with those of the most reliable measurements of other authors [3, 5, 8, 11] at lower temperatures. A critical analysis of investigations of the enthalpy and specific heat of SiC available in the literature revealed that some of them [9, 10, 12] are less reliable. Apart from systematic errors, unintentional contamination of the test specimens by impurities probably played a part in these cases. The most satisfactory agreement (within about 2%) was found between the results of a study of the enthalpy of SiC obtained in the range 400-1200°K [3, 5, 8, 11]. In this connection it should be noted that the experimental results Maxsimenko and Polubelova

[3] obtained on green SiC are discrepant, differing from the results obtained by the same investigations on black SiC by approximately 10%, although the SiC content of the specimens was the same in both cases (98.4%).

From the results of calculating the enthalpy of SiC containing no free carbon, the most reliable results of other investigators [4, 5, 8, 11], and the true specific heat of SiC at 273.15°K (c_p = 5.820 cal/mole · deg) according to low-temperature measurements [6], we found the following empirical equations for the enthalpy and specific heat of SiC in the temperature range 273.15-2900°K (molar mass = 40.101; 1 cal = 4.184 J):

$$H_T - H_{273.15} = 13.250\,(T - 273.15) - 4687 \log T\ 273.15 + {} \\ + 5070 \exp\,(-5680/T)\quad \text{cal/mole,} \tag{1}$$

$$c_p = 13.250 - 2035 T^{-1} + 288 \cdot 10^5 T^{-2} \exp\,(-5680/T)\quad \text{cal/mole · deg.} \tag{2}$$

The empirical equations (1) and (2) were found by using the equalization method proposed in [4] for substances whose Debye temperature exceeds 500°K (the Debye temperature for SiC is 1192°K [8]). Figure 2 shows the results of this treatment. The mean specific-heat curve calculated by means of Eq. (1) is shown as a continuous line in Figs. 1 and 2. The deviation of Magnus's experimental points [8] for the enthalpy from the mean values calculated from Eq. (1) is not more than 1%. Magnus's investigation was carried out on a specimen consisting of gray crystals whose chemical composition was not given. The experimental points

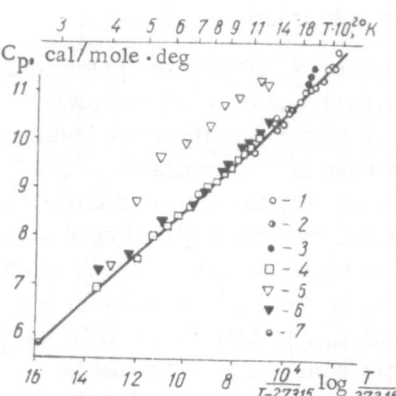

Fig. 2. Method of equalizing the mean specific-heat data for SiC: 1, 2, 3) experimental points, corrected for the free carbon content, for specimens 1, 2, and 3, respectively; 4) experimental points from [2]; 5, 6) experimental points for black and green SiC [3]; 7) true specific heat of SiC at 273.15°K according to data in [1].

from [3], obtained in a study of black SiC, have a considerable scatter, the deviation reaching 2%. The smoothed values for the enthalpy of α-SiC obtained in [5] in the range 300-1800°K, on a high-purity specimen (99.73 mass% SiC) lie systematically above the results calculated from Eq. (1) by an average of 2%, the amount increasing with rise in temperature.

The maximum deviation of the enthalpy values of SiC obtained on specimens No. 1 and 2 from those calculated from Eq. (1) does not exceed 1% except for the last point at 2843°K, at which temperature rapid decomposition of the SiC was observed. The experimental points obtained from specimen No. 3 lay somewhat higher (0.86-1.7%). The reason for this must be assumed to lie in the insufficiently accurate chemical-analysis data in the case of this specimen. By taking into account the error in measuring the enthalpy by the method of mixing (the maximum random errors in an individual measurement are 0.9 and 1.3% in the ranges 1300-2300 and 2300-2800°K, respectively) [12], an error in the enthalpy of carbon [13] of about 0.4%, the error in chemical analysis, and the simplified version of the effect on the enthalpy of regarding free carbon as the only impurity, it is possible to estimate approximately the error in calculating the enthalpy of SiC by means of Eq. (1) as ±(1-2)% in the range 1200-2900°K and ±(0.5-1)% in the range 300-1200°K.

Literature Cited

1. I. S. Kainarskii and E. V. Degtyareva, Carborundum Refractories [in Russian], GNTI, Khar'kov (1963).
2. V. A. Kirillin, A. E. Sheindlin, and V. Ya. Chekhovskoi, Teplofiz. Vys. Temp., 2:1 (1964).
3. M. S. Maksimenko and A. S. Polubelova, The Specific Heat and Thermodynamic Functions of Silicon Carbide and Boron Carbide [in Russian], 38, Goskhimizdat (1955), p. 30.
4. V. Ya. Chekhovskoi, Zh. Fiz. Khim., 39:12 (1965).
5. G. L. Humphrey, S. S. Todd, I. P. Coughlin, and E. G. King, US Bur. Mines Rep. Invest., No. 4888 (1952), p. 23.
6. K. K. Kelly, J. Am. Chem. Soc., 63:4 (1941).
7. V. A. Kirillin, A. E. Sheindlin, and V. Ya. Chekhovskoi, "Enthalpy and heat capacity of some solid materials at extremely high temperatures," in: High-Temperature Technology, Proc. Internat. Symposium, Washington, Butterworth (1964), pp. 471-484.
8. A. Magnus, Ann. Physik, 70:303 (1923).

9.   W. Miehr, H. Immke, and I. Kratzert, Tonind. Z., 50:1671 (1926).
10. A. S. Russell, Physikal. Z., 13:59 (1912).
11. E. B. Walker, C. T. Ewing, and R. R. Miller, J. Chem. Eng. Data, 7:4 (1962).
12. O. Weigel, Nachr. K. Ges. Wiss. Göttingen, Math.-Phys. Kl., 3:299 (1915).
13. E. D. West and S. Tshihara, "A calorimetric determination of the enthalpy of
     graphite from 1200 to 2600°K," in: Advances in Thermophysical Properties
     at Extreme Temperatures and Pressures, Third Symposium on Thermophysical
     Properties, New York (1965), pp. 146-151.

VII. MECHANICAL AND ABRASIVE PROPERTIES OF CARBIDES

Investigation of the Hot Hardness
of Cast Zirconium Carbide

E. M. Savitskii, A. A. Kul'bakh,
and N. A. Evstyukhin

Measurements of the hot hardness of cast specimens of zirconium carbide with a porosity approaching zero and a density close to the theoretical value are reported. The hardness was determined by a static
method in the temperature range 900-1650°C. With rise in temperature both cast and hot-compacted zirconium carbide undergo softening.

Investigations of the hot hardness of zirconium carbide specimens prepared by hot compaction have been reported in [2].

Zirconium carbide specimens for measuring the hardness by
a dynamic method had a porosity of 1-1.5% and a density at room
temperature of 6.47-6.54 g/cm^3. Specimens for measuring the
hardness by a static method had a somewhat greater porosity,
2.5%, and a mean density of 6.45 g/cm^3. In both cases the porosity was determined by pycnometric and metallographic methods. In the present article we communicate the results of measurements of the hot hardness of cast zirconium carbide specimens, the porosity of which was practically nil and the density ap-
proached the theoretical value of 6.66 g/cm^3. Cast zirconium carbide alloys and specimens of them having the required shape and
size were melted under an atmosphere of purified argon in an
MIFI-9-3 arc furnace, the materials consisting of spectrographi-

cally pure carbon and iodide zirconium of the following composition (wt.%):

Zr	Hf	O	Al	Fe	Ni	Mo	Cu	C
99.85	0.04	0.03	0.02	0.002	0.002	0.005	0.001	0.03

To obtain homogeneity the alloys were subjected to prolonged remelting (1.5-2 h). A copper hearth with holes in it was used to produce specimens of a predetermined shape and size, and in this way we were able to obtain cast specimens of cylindrical shape which required no additional treatment. The ends of specimens 10 mm in height and 14 mm in diameter were made parallel by means of a grinding machine followed by treatment with fine emery paper and polishing on a cloth with a suspension of chromic oxide in water.

Hot Hardness of Cast ZrC Specimens of Almost Stoichiometric Composition. Measurement of the hot hardness by the static method was carried out with a sapphire pyramid on the UST-2 apparatus described in [1]. The hardness was determined by the static method in the temperature range 900-1650°C, since "splitting" of the pyramid occurred at low temperatures and softening of the sapphire set in at higher ones.

As in the case of hot-compacted specimens, cast specimens of zirconium carbide begin to soften considerably at 900°C. The static hardness of the carbide (containing 10.34 wt.% C) falls sharply from 220 kg/mm^2 at 950°C to 110 kg/mm^2 at 1200°C and 85 kg/mm^2 at 1450°C. The relationship between the hardness of ZrC and carbon content that was established for hot-compacted specimens was fully confirmed by the tests on cast specimens (Fig. 1). The static hardness of ZrC decreases with reduction in the carbon content that was established for hot-compacted specimens was fully confirmed by the tests on cast specimens (Fig. 1). The static hardness of ZrC decreases with reduction in the carbon content in the homogeneity range: the less carbon there is in the ZrC, the more markedly it softens and the greater the fall in its hot hardness. Thus, for example, the static hardness of zirconium carbide containing 9.01 wt.% C falls from 180 kg /mm^2 at 950°C to 20 kg /mm^2 at 1300°C. The hot static hardness of zirconium carbide containing 9.78 wt.% C falls from 190 kg/mm^2 at 950°C to 100 kg/mm^2 at 1200°C and to 40 kg/mm^2 at 1600°C. Measurements of the dynamic hardness of cast ZrC specimens made on the UDT-2

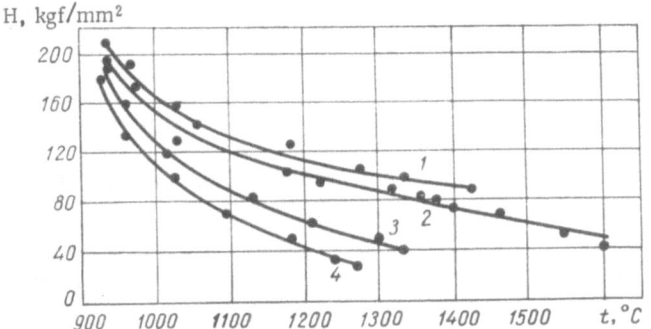

Fig. 1. Hardness of zirconium carbide as a function of temperature: 1) 10.34; 2) 9.78; 3) 9.3; 4) 9.01 wt.% C.

apparatus by the method described in [2] in the range from room temperature to 2000°C provide evidence that the cast carbide, like the hot-compacted carbide, softens considerably in the range 1400-2000°C.

The Hot Hardness of Cast Zr − ZrC Alloys. A knowledge of the high-temperature strength properties of alloys of zirconium with carbon up to the carbide composition (the so-called cermets) is essential to the study of the heat-resistant carbide coatings and of the makeup of carbide nuclear fuel. The following ten specimens

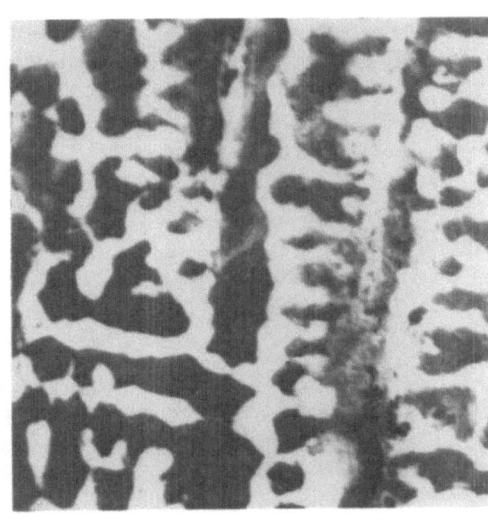

Fig. 2. Microstructure of a cast zirconium alloy containing 5.480 wt.% C. ×450.

of cast alloys were prepared for investigation by the methods described above:

Zr, wt.%	99.731	99.674	99.506	99.495	99.270	99.016	98.440	99.700	96.250	94.520
C, wt.%	0.269	0.326	0.494	0.505	0.730	0.934	1.560	2.300	3.750	5.480

The cermet alloys were subjected to chemical, x-ray, and metallographic analyses by the methods described in [2]. The microstructure of the alloy of Zr with 0.269 wt.% C contained large grains of the solid solution of carbon in zirconium, within which finely dispersed particles of precipitated carbide were visible. The amount of ZrC increases with rise in the carbon content, and in the alloy containing 5.48 wt.% C a second, dark phase (ZrC) occupies approximately half of the field of the microsection (Fig. 2). The light phase is a solid solution of carbon in zirconium.

The hot hardness of the cermet alloys of zirconium was determined by the static method. The results of some of the measurements are presented in Fig. 3, plotted on semilogarithmic coordinates and it will be seen that two or three points of inflection occur on the temperature-dependence curves of the hardness of all the alloys investigated (except that containing 5.48 wt.% C). To account for these points it is necessary to bear in mind those factors that determine the resistance of a material to deformation, viz., the type of interatomic bond, the possibility of impurity atoms being present, and the mechanisms of plastic deformation. The first, low-temperature point of inflection, which appears on the hardness curves at 100-150°C, is evidently due to the presence of interstitial impurities, mainly oxygen and nitrogen atoms.

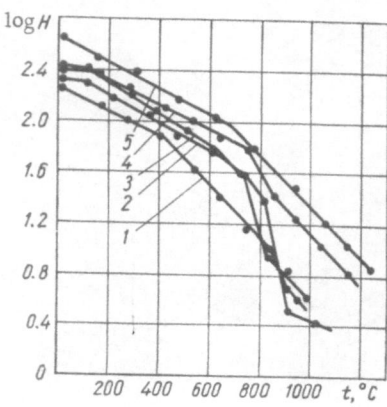

Fig. 3. Temperature dependence of the hardness of cast zirconium—carbon alloys containing: 1) 0.934, 2) 1.56, 3) 2.30; 4) 3.75, 5) 5.48 wt.% C.

The second point on the curves is apparently associated with a change in the type of deformation. In pure iodide zirconium this change from one type of deformation to another occurs at about 425°C [3]. In the cermet alloys it is displaced in the high-temperature direction, no doubt owing to the effect of the carbon atoms.

The third point of inflection on the hardness curves is observed at 800-900°C, and is connected with the $\alpha \rightarrow \beta$ transformation in zirconium.

In the alloys containing 2.3 and 3.75 wt.% C (Fig. 3) the third point of inflection is displaced in the lower-temperature direction on the curves. This is evidently due to the fact that the tests were carried out on the specimens after compression at 1100-1200°C from high temperatures to low.

Conclusions

1. Measurements of the hot hardness of cast ZrC specimens made by static and dynamic methods are reported.

2. It is shown that the hot hardness of the cast specimens corresponds in the main to that of hot-compacted specimens of ZrC and that it depends on the carbon content in the homogeneous monocarbide.

3. Considerable softening of ZrC (cast and hot-compacted) sets in the range 1400-2000°C, when deformation due to the falling ball becomes completely inelastic and the rebound height from the hot surface is greatly reduced.

4. Measurements of the static hot hardness of the cast cermets showed that it increases with the carbon content of the alloys. Explanations are given of the points of inflection observed on the temperature-dependence curves of the hardness of the alloys investigated.

Literature Cited

1. Yu. G. Godin et al., in: The Metallurgy and Metal Science of Pure Metals [in Russian], Vol. 5, Atomizdat, Moscow (1956), pp. 189-198.
2. N. A. Evstyukhin and V. M. Shchavelin, in: The Metallurgy and Metal Science of Pure Metals [in Russian], Vol. 6, Atomizdat, Moscow (1967).
3. The Metallurgy of Zirconium [Russian translation from the English, edited by G. A. Meerson and N. V. Gagarinskii], IL, Moscow (1959).

The Microhardness of Transition-Metal Carbides

A. A. Ivan'ko

The results of investigating the microhardness of carbides of some of
the transition metals are reported, and the hardness variations are discussed in relation to the electron structure of the metal atoms. The
effect of the conditions of measurement on the microhardness number
is noted, and conditions for carrying out the measurements are recommended.

Microhardness is an important property of materials. However, it is often difficult to use microhardness data for comparing
the properties of different substances, because large variations in
the microhardness values reported by different investigators occur even for the same compounds. This is due not only to differences in the degree of purity of the specimens tested, and in their
production technique and method of preparation for testing, but
also to different conditions of measurement. On this account systematic researches into the determination of the microhardness
of simple substances and refractory compounds have been carried
out in the Powder Metallurgy Department of the Kiev Polytechnic
Institute.

The present communication is concerned with the microhardness of carbides of transition metals of groups IV to VI. The
microhardness was measured on a PMT-3 apparatus over a wide
range of loads on the indentor (20-150 g) and times of holding under load (5-60 sec). The chemical analyses of the specimens are
given in Table 1. The hot-compacted specimens were annealed so
as to make it possible to obtain microhardness data when the speci-

mens were in a state close to equilibrium. Microsections were prepared by the usual methods, the structure being revealed by chemical etching after polishing.

We recommend the conditions for carrying out the microhardness measurements. The purpose of choosing the optimum conditions for measuring the microhardness was to obtain results having the smallest possible errors. Thus, maintaining the specimen under load for a long time may lead to a reduction in the microhardness value owing to vibration; the use of very small loads may cause the results to be too high, while too high a load

TABLE 1. Characteristics of the Original Specimens

Carbide	Results of chemical analysis, %	Conditions of producing specimens				Porosity, %
		As hot compacted		As annealed		
		t, °C	τ, min	t, °C	τ, h	
TiC	Ti — 79.5 C_{total} — 18 C_{free} — 0.3 Fe — 0.3	2200	10	1800	3	10.3
ZrC	Zr — 90.35 C_{total} 9.4 C_{free} — 0.2	2300	15	1800	3	11.8
HfC	Hf — 93.8 C_{total} — 6.1 C_{free} — 0.1	2650	15	1800	3	6.7
VC	V — 83.5 C_{total} 16.3 C_{free} — 0.3	2070	10	1450	3	12.3
NbC	Nb — 89.78 C_{total} — 10.15 C_{free} — 0.09	2200	10	1800	3	18.4
TaC	Ta — 93.9 C_{comb} — 5.8 C_{free} — 0.3	2500	20	1800	3	20.4
Cr_3C_2	Cr — 86.6 C_{total} 13.3 C_{free} — 0.3	1260	5	900	3	16.7
Mo_2C	Mo — 94.21 C_{total} — 5.7 C_{free} — 0.3	1750	10	1450	3	16.1
WC	W — 93.65 C_{total} — 6 C_{free} — 0.3	1850	8	1450	3	

TABLE 2. The Microhardness of
Carbides

Metal	Configuration of valence electrons	Statistical weight of d^5 configurations, %	Carbide	Microhardness, kg/mm²
Ti	$3d^24s^2$	43	TiC	2640 ± 200
Zr	$4d^25s^2$	52	ZrC	2836 (4)
Hf	$5d^26s^2$	55	HfC	2640 ± 160
V	$3d^34s^2$	63	VC	2190 ± 170
Nb	$4d^45s^1$	76	NbC	2140 ± 225
Ta	$5d^36s^2$	81	TaC	1540 ± 70
Cr	$3d^54s^1$	73	Cr_3C_2	2060 ± 170
Mo	$4d^55s^1$	88	Mo_2C	1640 ± 200
W	$4d^46s^2$	94	WC	1840 ± 200

may give rise to brittle failure of the material. Consequently, in
the present article we recommended loads below those at which
brittle failure of the carbide concerned begins.

The results reported here for the microhardness of carbides
confirm the existence of a relationship between microhardness
and the load on the indentor and the time for which the load is
maintained, as was found previously for refractory compounds
and simple substances.

The results lead us to recommend for carbides a load on the
indentor of 80–100 g and a time of maintaining the load of 10 sec
(Table 2).

The carbides of group IVa metals have the highest microhardness values, whereas the carbides of group VIa metals have
comparatively low values. The microhardness values obtained
were compared with features of the electron structure, chemical
bond, and crystal structure.

It may be concluded that the hardness of carbides is determined by the probability of the very stable sp^3 configurations of
the valence electrons being formed in the carbon atoms, and this
depends on the donor-acceptor properties of the metal. On the
basis of the electron structures of TiC and TiO [1], it may be assumed that all metals of group IVa are electron donors. In the
formation of the carbides of these metals, the unlocalized electrons of the metal are transferred to the carbon atoms and are
maintained there by exchange reaction, stabilizing the sp^3 con-

figurations of carbon. This accounts for the great hardness of the carbides of group IVa metals.

Metals of group Va are weak acceptors, and this leads to some breakdown of the sp^3 configurations in carbon and to a corresponding reduction in hardness.

Metals of group VIa have a considerable statistical weight of atoms with d^5 configurations, as a result of which the sp^3 configurations of carbon break down. In consequence the strength of the Me−C bond is reduced, and at the same time the proportion of the covalent component of the Me−Me bond is increased [5].

The relatively great hardness of Cr_3C_2 as compared with the carbides of other group VIa metals must be due to a large extent to the strengthening of the C−C bond in comparison with carbides possessing the NaCl-type structure. The carbide Cr_3C_2 crystallizes in a FeB-type structure, which is characterized by the formation of zigzag chains with covalently bonded carbon atoms; whereas in carbides having the NaCl-type structure the carbon atoms occupy the octahedral sites and are practically unconnected. The relatively low microhardness of tungsten carbide may be attributed to the high statistical weight and energy stability of its d^5 configurations, which lead to a breakdown of the sp^3 configurations of carbon and to marked delocalization of the electrons.

Since, in the case of all transition-metal carbides, the microhardness is determined by the statistical weight of carbon atoms having sp^3 configurations, it is clear that its value should be measured in relation to the carbon content of the carbide, and hence further work will be directed to determining the microhardness of carbides in the homogeneity range.

Literature Cited

1. M. P. Arbuzov and B. V. Khaenko, Poroshkovaya Met., No. 4, 74 (1966).
2. G. V. Samsonov, V. S. Neshpor, and L. N. Khrenova, Fiz. Metal. i Metalloved., 8:622 (1959).
3. G. V. Samsonov, A. A. Ivan'ko, and E. N. Chupakhina, Fiz. Khim. Mekhan. Met., 2:152 (1966).
4. G. V. Samsonov and Ya. S. Umanskii, Hard Compounds of Refractory Metals [in Russian], Metallurgizdat, Moscow (1957).
5. V. I. Trefilov, in: The Physical Nature of Brittle Fracture in Metals [in Russian], Naukova Dumka, Kiev (1965).

Some Aspects of the Abrasive Wear
of Transition-Metal Carbides

A. Ya. Artamonov and G. A. Bovkun

The results of abrasive-wear tests on some carbides are presented.
The specimens were tested on a Kh4-B machine, using M. M. Khrushchov's standard procedure. A relationship has been established between the relative wear resistance of the carbides and their porosity,
hardness, degree of plasticity, and nature of the materials. The wear
resistance of the carbides is shown to decrease linearly with rise in
porosity. No unique connection was found between the hardness and
wear resistance of the carbides; this was due to one specific property
of the materials investigated, namely their great brittleness. The results obtained are discussed in terms of the configuration-localization
model.

It has been established in [5, 6] that the relative wear resistance of various materials is a fundamental characteristic of
their working properties, since it enables us to determine the resistance of the materials to failure in the heavily deformed state.
Physically, it is natural to associate the relative wear resistance
with the bonding forces in the crystal lattice, which are usually
characterized by the modulus of normal elasticity or by the square
of the coefficient of lattice regidity $m\Theta^2$ [3].

The extension of this relationship to other classes of material
is of great scientific and practical interest, as it enables a scien-
tifically based approach to be made to the creation of materials
having a high resistance to abrasive wear. In this respect refractory compounds represent a totally unexplored class of ma-

terials, although their great hardness, adequate strength, and chemical passivity over a broad temperature range are responsible for the wide application of these materials under conditions of wear by free or bonded abrasive particles.

The literature contains data on the abrasive wear resistance of materials based on the carbides of titanium and tungsten (the hard alloys VK and TK) and on chromium carbide (alloy GK-15) [2]. As regards pure carbides, no systematized information at all is available in the literature on their resistance to the scratching action of abrasive particles. A knowledge of the main features of the abrasive wear of carbides would enable us to solve many current problems connected both with the use of these materials under heavy working conditions and with the search for new highly wear-resistant materials based on carbides.

In the present research we have investigated the resistance to abrasive wear of carbides of the transition metals of groups IV to VI (TiC, ZrC, HfC, NbC, TaC, Mo_2C, and WC).

Billets of these materials were obtained by the hot-compaction method followed by annealing. Cylindrical specimens 2 mm in diameter for the abrasive wear tests were cut on an electric-spark machine with a GIT-1 impulse generator.

M. M. Khrushchov's method was employed in carrying out the wear tests. After it had been confirmed that the standard procedure was suitable for refractory compounds, the following test conditions were adopted: pressure on the specimen 0.955 MN/m^2; friction path 15 m; abrasive surface, boron carbide emery paper No. 5; and standard, steel U8A.

Since metalloceramic materials are characterized by great instability of their physicomechanical properties throughout the volume of the specimen, the abrasive wear tests were carried out many times, and the relative wear resistance was determined as the mean of a large number of identical tests.

It was found that the resistance of the carbides to abrasive wear is only small (see Table 1), although, according to [1], materials based on refractory-metal carbides acquire a high wear resistance.

On account of the linear relationship that has been found in this research between wear resistance and porosity, it is essen-

TABLE 1. Hardness and
Wear Resistance
of Refractory Compounds

Compound	Relative porosity, %	Microhardness, MN/m²	Relative wear resistance, %
TiC	3.45	30 000	0.61
	4.30		0.52
	7.10		0.52
	12.10		0.44
	15.00		0.30
	20.00		0.27
ZrC	0.30	29 300	0.45
	4.00		0.42
	5.00		0.41
	12.00		0.36
HfC	0	29 000	0.57
NbC	0	19 600	0.67
TaC	0	16 000	1.50
	4.50		0.90
Mo₂C	0	15 000	1.40
WC	4.20	17 800	0.79
	5.80		0.68
	12.50		0.53
SiC	0	33 500	0.74
B₄C	0	49 000	3.36

tial, in making comparative tests of transition-metal carbides, to
strive to produce specimens with minimum and — most important —
identical porosity.

It has been shown in [5] that the abrasive wear resistance of
commercially pure metals is directly proportional to hardness;
it is related to the modulus of normal elasticity by the equation

and is directly proportional to the square of the coefficient of lattice rigidity. The establishment of a correlation between the relative wear resistance of refractory compounds and their physicomechanical properties would provide a scientific basis for the formulation of new wear-resistant materials based on them. An attempt to establish such a relationship has been made in the present investigation.

Great brittleness is a distinguishing property of refractory
compounds and transition-metal carbides. In view of the brittle

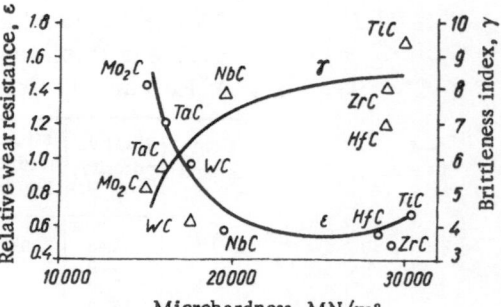

Fig. 1. Relative wear resistance, brittleness, and ductility of transition-metal carbides as a function of the microhardness of the compounds.

nature of the failure of carbides during wear on abrasive emery paper, it is to be expected that, other things being equal, the wear resistance will be higher, the less the brittleness of the compound.

In the course of abrasive wear high stresses are set up in the microcutting zone of the material, and these are responsible for the brittle failure of the friction contact by the removal of microvolumes of the material being abraded which are more or less incommensurable with the "cutting" element, depending on the brittleness of the material. It is natural that, with rise in brittleness, the individual volumes damaged by scratching the carbides with an abrasive grain should also increase. The experimental results show that the wear resistance of the carbides falls with increase in brittleness (see Fig. 1).

Brittleness, like other physicomechanical properties of refractory compounds, is determined by failures of their crystal and electron structures.

Comparison of the deformation properties of the sp elements of group IV (C, Si, Ge, Sn, and Pb) with the distribution of the valence electrons between the localized and unlocalized states, shows that their deformation properties fall sharply with increase in the stability of the sp^3 configurations and decrease in the proportion of unlocalized electrons. Thus, diamond, which has the greatest statistical weight of the most stable configurations, is extremely brittle and undergoes practically no deformation when an external load is applied. On passing to silicon and germanium, we find that the statistical weight and energy stability of the sp^3 configurations decrease, and there is a corresponding increase in the statistical weight of electrons that are easily excitable into the unlocalized state. These materials possess negligible plasticity, although it is quite appreciable in comparison with that of diamond. Further reduction in the energy stability of the sp^3 configurations in the cases of tin and lead results in the statistical weight of the sp^3 configurations of these metals becoming small (in particular, they become insufficient for the formation of the cubic diamond lattice), and so tin and lead have considerable plasticity.

Thus, it can be seen that the brittleness of these materials is determined to a large extent by the statistical weight of the stable sp^3 configurations of the electrons in the localized state, and that it falls with increase in the proportion of unlocalized electrons.

It may be concluded that for such compounds as the transition-metal carbides, whose physical properties are determined mainly by the statistical weight and energy stability of the sp^3 configurations of the carbon atoms, the relationship between the brittleness and proportion of unlocalized electrons will be the same.

It has been shown experimentally [4] that the brittleness of transition-metal carbides falls from titanium carbide to tungsten carbide. This can be explained in the following way.

When TiC is formed, the titanium does not destroy the sp^3 configurations of the carbon atoms, but, on the contrary, it transfers its own valence electrons to them. This leads to stabilization of the sp^3 configurations of the carbon atoms and to TiC having the greatest brittleness among all the carbides. On passing to zirconium and hafnium carbides, the concentration of unlocalized electrons rises on account of the great excitation of the sp^3 configurations of the carbon atoms by the d^5 configurations of the carbide-forming metal, and the brittleness of these carbides falls, although still remaining comparatively high.

When we pass to the carbides of metals of groups V and VI, the degree of excitation of the sp^3 configurations of the carbon atoms increases with rise in the statistical weight of the d^5 configurations of the metal atoms, and this leads to an increase in the proportion of unlocalized electrons and to a reduction in brittleness. Tungsten carbide, which is characterized by maximum delocalization of the valence electrons of the carbon atoms, possesses least brittleness.

On comparing the data in [4] with the experimental results of the present article, we can see that the resistance to abrasive wear of the carbides increases with increase in their brittleness. This indicates that there is a qualitative correlation between the relative wear resistance of the transition-metal carbides and the nature of the electron structure and chemical bonding of the elements in the lattices of these compounds; this correlation takes the form of a reduction in the abrasive wear resistance with increase in the localization of the valence electrons of the nonmetallic atoms in stable hybrid configurations.

Literature Cited

1.   A. Ya. Artamonov and G. A. Bovkyn, Poroshkovaya Met., No. 5, 29 (1966).
2.   V. N. Klimenko, "The wear resistance of components made of alloys based on chromium carbide," Summaries of Reports to the All-Union Seminar on "Experience in the Production and Use of Metalloceramic Antifriction and Structural Components in Mechanical Engineering," TsINTI on Automation and Mechanical Engineering of the State Committee for Mechanical Engineering of the State Planning Commission of the USSR, Moscow (1964).
3.   B. M. Rovinskii, Izv. Akad. Nauk SSSR, Otd. Tekh. Nauk, No. 9, 35 (1961).
4.   G. V. Samsonov and V. V. Stasovskaya, Poroshkovaya Met., No. 12, 95 (1966).
5.   M. M. Khrushchov and M. A. Babichev, Study of the Wear of Metals [in Russian], Izd. Akad. Nauk SSSR, Moscow (1962).
6.   M. M. Khrushchov, in: Methods of Wear Testing [in Russian], Izd. Akad. Nauk SSSR, Moscow (1962).

Hard-Facing Alloys Based on
Refractory Compounds

A. Ya. Artamonov, D. Z. Yurchenko,
and I. F. Kirilets

The results of investigations of hard-facing materials based on some transition-metal borides containing 5-15% boron carbide are reported. Iron powder was added to the hard-facing mixtures in order to reduce the brittleness of the boride and carbide phases. A relationship was established between the resistance of the hard-facing layers to abrasive wear and the structure of the layer, the grain shape of the structural constituents, their hardness, and the degree of inhomogeneity, and the quantitative ratio between the hard and soft phases in the hard-facing layer. Hard-facing materials based on the borides of titanium and tungsten displayed a high resistance to wear and can be used to protect the working surfaces of machine components operating under conditions of abrasive wear.

Researchers are at present trying to discover highly wear-resistant hard-facing materials based on compounds of some of the refractory metals, namely, borides, carbides, and silicides. Already, alloys having a high wear resistance and based on chromium carbide and boride and titanium boride (KBKh, BKh-2, TBKKh, and GK-15) have been developed.

In the present article we report the results of investigations of hard-facing materials based on the borides of the transition metals chromium, titanium, tungsten, and titanium − chromium, with addition of 5-15 wt.% boron carbide. In order to reduce the

377

brittleness of the boride and carbide phases, 10-80 wt.% of iron powder was added to the hard-facing mixtures. The powder mixture was prepared in a mixing drum in a medium of distilled alcohol for 1 h, and then it was passed through a 150-mesh sieve. The mixture was melted onto a substrate of St.3 steel by means of a UDAR-300-type machine, using graphite electrodes with a current strength of 150-200 A at a voltage of 16-20 V. The thickness of the layer produced in this way was 2-3 mm.

The quality of the hard-facing layer was assessed from the number of clearly visible defects in it (cracks, blisters, sweating, blowholes, etc.), results of metallographic analysis across the section of the layer, and from the nature of the resistance to abrasive wear as measured on the Kh4-B machine by the standard method of testing given in [6], as compared with standard specimens of steel U8A.

Investigations of hard-facing materials of type Fe + MeB + 15% B_4C showed that the mixtures melt down well, the layers produced having no clearly visible defects and their hardness being 86-90 on the Rockwell A scale. However, the resistance to abrasive wear for the diborides of Cr, Mo, Ti, and W is 1.2-1.7 and that for the binary Ti−Cr boride is 3.5.

The low resistance to abrasive wear of a number of the hard-facing materials is due to the unfavorable distribution of the soft structural constituent (solid solution) in the relatively hard eutectic base [1, 3, 7], when the individual grains or dendrites of the solid solution are enclosed in the relatively hard eutectic (an example of such a structure is shown in Fig. 1a), or when the structure contains a very large number of equiaxial crystals of the boride phase cemented together by small particles of eutectic, which leads to brittle breakdown of the layers in the course of abrasive wear (an example of such a structure is shown in Fig. 1b).

The relatively high wear resistant of hard-facing material composed of $(TiCr)B_2$ + 15% B_4C + 5% Fe (of the order of 3.5) is due to the favorable distribution − from the point of view of wear resistance − of the hard boride phase in the form of fine nonequiaxial grains having a microhardness of 2200 kg/mm^2 in a relatively hard eutectic (1200 kg/mm^2) (Fig. 1c).

Hardness measurements carried out on materials composed of Fe + MeB + 5% B_4C gave values of 86-89 Rockwell A, while those on materials based on TiB_2 gave values of 81-83 Rockwell A.

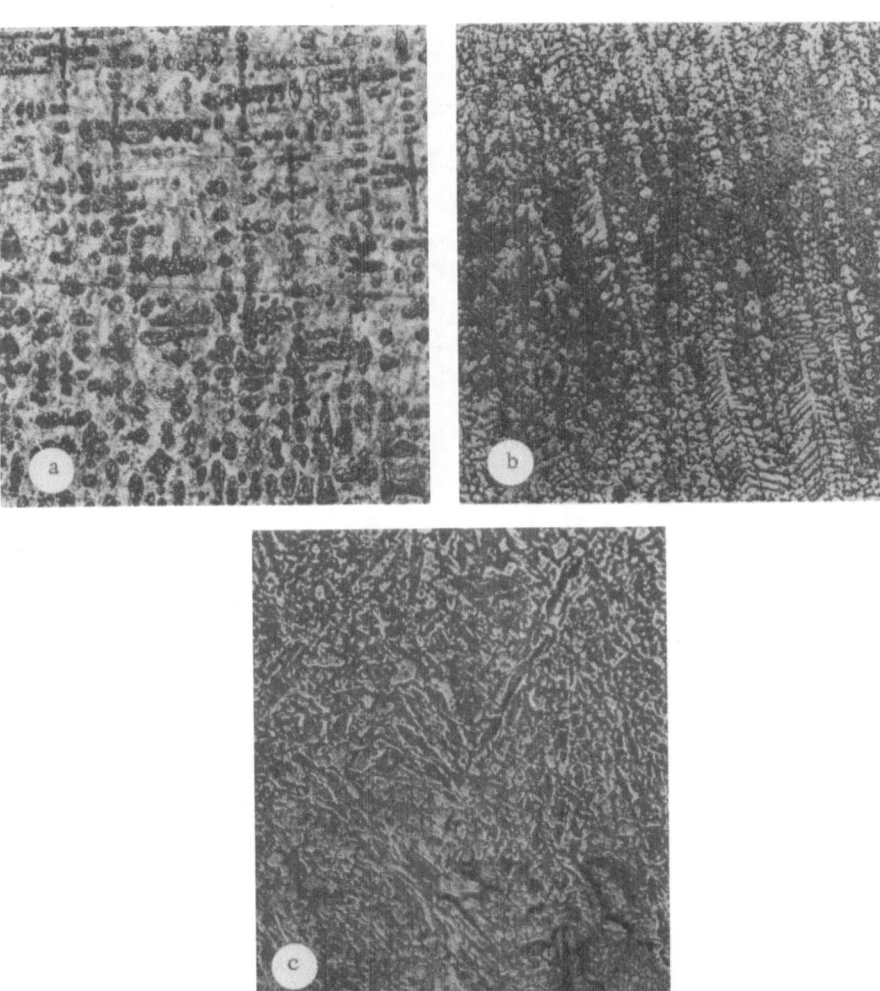

Fig. 1. Structures of hard-facing materials containing: a) W_2B_5 + 5% Fe + 15% B_4C (×200): b) Fe + 20% Mo_2B_5 + 15% B_4C (× 200); c) $(TiCr)B_2$ + 5% Fe + 15% B_4C (×200).

The most wear resistance of the systems investigated were those composed of TiB_2 + 15% Fe + 5% B_4C and Fe + 20% W_2B_5 + 5% B_4C, which had a relative wear resistance of 6-7.

The structure of highly wear-resistant hard-facing materials is shown in Figs. 2a and b, from which it can be seen that the upper zone of the material (Fig. 2a) consists of a eutectic base (H_μ = 800 kg/mm^2) with uniformly distributed acicular crystals based

Fig. 2. Structures of hard-facing materials containing: a) TiB_2 + 15% Fe + 5% B_4C
(×200);   b) Fe + 20% W_2B_5 + 5% B_4C (× 200);   c) CrB_2 + 15% Fe + 5% B_4C (×200).

on titanium boride (H_μ = 3300 kg/mm^2) and crystals of tungsten
boride (Fig. 2b) having a microhardness of 2000 kg/mm^2.

Similar indices of the resistance to abrasive wear of two
materials the hard phases in which have different microhardness
values may be attributed to a difference in diboride content in the
composition [1] of the powder mixtures and particularly in the
melted-on layer.

At the interference with the substrate the structure consists mainly of coarse grains of the solid solution surrounded by eutectic. In the middle of the hard-facing layer the number and size of the solid-solution grains decrease, whereas the eutectic content increases. The solid solution is almost absent from the upper zone of the layer.

It is interesting to note that a hard-facing layer made from a mixture containing CrB_2 + 15% Fe + 5% B_4C has a comparatively low relative wear resistance not exceeding 2.3. As can be seen from Fig. 2c, the structure of this material is characterized in the upper zone by nonuniformly distributed grains of the boride phase (H_μ = 2000 kg/mm^2) in a hard eutectic (H_μ = 1100 kg/mm^2). The nonuniform distribution of the hard phase and the isometric shape of the grains are no doubt responsible for the reduction in resistance of this hard-facing material to abrasive wear.

Conclusions

1. A study has been made of the hard-facing properties, structure, microhardness, and resistance to abrasive wear of welded-on layers having such compositions as Fe + MeB + 5% B_4C and Fe + MeB + 15% B_4C, in which the MeB content varied from 10 to 80 wt.%.

2. The principal factors determining the wear resistance of hard-facing materials are the structure of the welded-on layer, the grain shape of the structural constituents, their hardness, the degree of homogeneity, and the quantitative ratio of hard to soft phase in the layer.

3. Hard-facing mixtures consisting of TiB_2 + 15% Fe + 5% B_4C and Fe + 20% W_2B_5 + 5% B_4C have a high resistance to abrasive wear compared with existing materials and can be used to protect the working surfaces of machine components and shaping tools operating under conditions of abrasive wear.

Literature Cited

1. A. Ya. Artamonov and G. A. Bovkun, Poroshkovaya Met., No. 5, 29 (1966).
2. V. V. Grigor'eva, I. N. Sheenko, and M. K. Sheman, in: Nauchnye Trudy VNIIOChERMETa, Vol. 7, Metallurgiya (1965).
3. L. G. Zhuravlev, "Investigation of the resistance of steel to abrasive wear in relation to structure, composition, and hardness," Friction and Wear in Machines [in Russian], Vol. 8, Izd. Akad. Nauk SSSR, Moscow (1958).

4.   I. I. Iskol'dskii, Welded-on Boride Hard Alloys [in Russian], Mashinostroenie,
     Moscow (1965).
5.   Scientific and Technical Report on the Search for New Materials for Hard-Facing
     Components of the Charging Equipment of Blast Furnaces and Otner Components
     of Metallurgical Plant, Based on Chromium Carbide and Also the Borides of
     Other Metals [in Russian], VNIIOChERMET (1965).
6.   M. M. Khrushchov and M. A. Babichev, Study of the Wear of Metals [in Rus-
     sian], Izd. Akad. Nauk SSSR (1960).
7.   S. P. Shabashov, "Investigation of the wear resistance of experimental and
     standard types of hard alloys on aging," Izv. VUZ. Mashinostroenie, No. 6
     (1963).

New Polishing Materials Consisting of Quasimetallic Compounds

A. Ya. Artamonov, O. V. Tutakov, and V. V. Sychev

The behavior of micropowders of a number of quasimetallic compounds during polishing and finishing operations has been investigated. It was found that in these processes micropowders of refractory compounds and the existing abrasive materials are ground in almost the same way and have identical geometry of the grains. The polishing characteristics of the refractory compounds TiC, Cr_3C_2, etc., are no worse than those of existing abrasive materials and are not greatly inferior to synthetic diamond for polishing hardened steel.

Present-day abrasive materials do not meet the quality requirements of polished surfaces (they damage the surface by leaving deep scratches on it and reacting chemically with the material being polished). In consequence the search for new polishing materials is very pressing.

Experiments have been carried out at the Institute of Problems in Materials Science of the Academy of Sciences of the Ukrainian SSR in order to study the polishing and finishing capacity of micropowders of quasimetallic compounds. An investigation was made of the polishing capacity of micropowders with a grain size M40 of the carbides of zirconium, titanium, tungsten, chromium, and molybdenum, which were compared with such existing abrasive materials as boron and silicon carbides, electrocorundum, and synthetic diamond. The following qualitatively different materials were subjected to treatment: U8A steel (Brinell hardness 202-217), U8A steel (Rockwell hardness C 59-62), type-K8 glass (Vickers hardness 525 kg/mm²), and bronze of type Br-OTsS-58-3.

The specimens were treated by polishing them to a surface finish having mean irregularities $R_z = 5$-$6.0~\mu$. The polishing and finishing experiments were carried out on a machine of the ASMP type, using a lap made of cast iron having a pearlitic structure.

On the basis of the calculation of a single-layer distribution of grains on the lap surface (made separately for each experiment), equal portions by volume (0.03 cm³) of the micropowder of the polishing material were used. For bonding purposes, about 0.02 cm³ of a neutral lubricating medium — vaseline oil — was added to the micropowder (to eliminate the possibility of any significant physicochemical action taking place).

TABLE 1. Comparative
Characteristics of the Polishing
Capacity of Micropowders

Micropowder M40	Change in micro-irregularities R, μ		Material removed, g
	from	to	
U8A steel (Brinell hardness = 202-217); polished 10 min			
Synthetic diamond	6.1	1.6	0.0288
Boron carbide	5.9	3	0.0284
Silicon carbide	5.3	2	0.0214
Zirconium carbide	5.7	1.4	0.0306
Titanium carbide	5.2	1.7	0.0324
Chromium carbide	6.4	1.7	0.0205
Bronze Br-OTsS-58-3, polished 60 min			
Boron carbide	5.8	2.7	0.0600
Silicon carbide	5.9	2.25	0.0633
Zirconium carbide	6.3	1.60	0.0620
Titanium carbide	6.1	1.43	0.0635
U8A steel (Brinell hardness = 202-217); polished 60 min			
Synthetic diamond	0.25	0.09	0.0900
Zirconium carbide	0.27	0.08	0.0845
Titanium carbide	0.28	0.11	0.0823
Synthetic diamond*	0.25	0.06	0.0285
Zirconium carbide*	0.29	0.05	0.0347

*Polished on felt.

TABLE 2. Kinetics of Change in Grain Size of Micropowders of Refractory Compounds and Synthetic Diamond

Micropowder	3 min						5 min						7 min						10 min					
	Size, μ			Content, %			Size, μ			Content, %			Size, μ			Content, %			Size, μ			Content, %		
	a	b	c	a	b	c	a	b	c	a	b	c	a	b	c	a	b	c	a	b	c	a	b	c
Zirconium carbide	40	12	4	1	26	5	24	8	4	2	37	18	12	4	2	5	40	32	8	2	1	2	35	30
Synthetic diamond	40	12	8	1	20	15	28	8	4	3	32	10	28	4	2	3	32	12	4	4	1	3	35	7
Silicon carbide	40	16	8	5	22	4	40	16	8	2	30	5	32	12	4	2	23	8	28	8	4	4	26	12

Notes: a) maximum grain size; b) size of predominant fraction of grain in sample; c) minimum grain size.

The efficiency of the polishing powders was assessed from the change in the microirregularities on the surface of the specimens and from the amount of material removed from the specimen during the test (Table 1).

In order to increase the reliability of the results obtained, polishing tests were made on the same specimens with micropowders of zirconium and tantalum carbides (which gave the best results) having a grain size of 3 μ for comparison with synthetic diamond. The results of these tests (Table 2) indicate once again the high polishing capacity of TiC and ZrC. Specimens of bronze Br-0TsS-58-3 were polished with various micropowders of grain size M40 on aluminum, cast iron, and felt laps.

These experiments demonstrate the undesirability of using synthetic diamond for polishing unquenched steels and nonferrous materials. The possibility of using carbides of Ti, Zr, W, Cr, and Mo for these purposes, in addition to the silicon carbide and electrocorundum which are in current use, has been established. The results of finishing tests with zirconium and titanium carbides, compared with existing abrasives tested under the same conditions, are given in Fig. 1. Micropowder of grain size M40 was used in the tests.

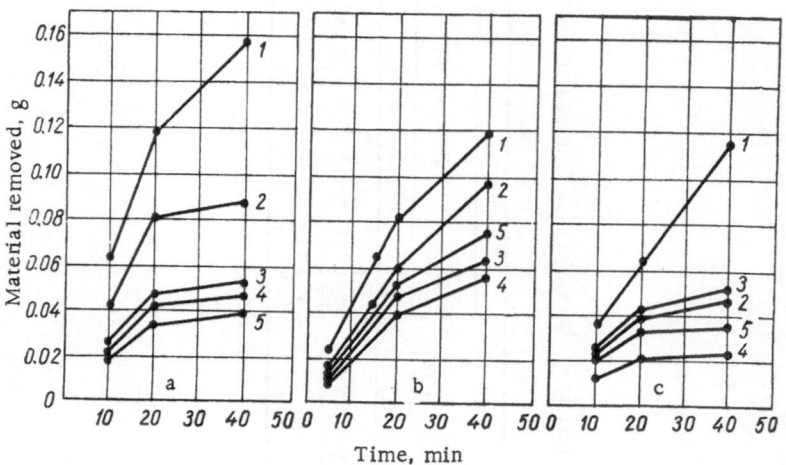

Fig. 1. Results of finishing tests with carbides of zirconium and titanium: 1) synthetic diamond; 2) green silicon carbide; 3) electrocorundum EB; 4) titanium carbide; 5) zirconium carbide; a) glass K8; b) U8 steel (Brinell hardness = 200-250 kg/mm²); c) U8 steel (Rockwell hardness c = 55-60 kg/mm²).

It is shown that, for finishing, zirconium and titanium carbides are as good as type-EB electrocorundum as regards removal of material when polishing unquenched U8 steel, quenched U8 steel (Rockwell hardness C = 55-60), and K8 glass. The roughness value falls within the limits of the ninth class of surface finish specified in the standard GOST 2789-59; when diamond, boron carbide, and silicon carbide are used, it falls into classes 9a-9b; and when zirconium carbide, titanium carbide, and electrocorundum are used, it falls into classes 9b-10a of surface finish.

The polishing capacity of abrasive grains is not determined by one single factor. It depends on the composition of the material, its strength, the surface microtopography, the combination of size and shape of the grains, the nature of the breakdown of the grain in the polishing process, and also on the quality, quantity, and degree of sharpness of the newly formed forces. Hence a study of the kinetics of grinding and the change in radius of curvature of the cutting edges and of the angles between two faces of the grain plays an important part in investigating an abrasive material.

In order to investigate the kinetics of the grinding of grains during polishing, we chose an abrasive material in which the quantitative content of percentage fractions in the micropowder having an original grain size M40 conformed with the standard GOST 9506-59.

Experiments on the grinding of the abrasive grains during polishing were carried out on an ASMP-type machine, using two laps of U8A steel quenched to a Rockwell hardness C = 65, to eliminate any uncertainty about the results of the grinding tests due to the embedding of the grains in the surface of soft and porous laps. To the polishing zone was added 0.03 cm^3 of the appropriate micropowder, moistened with 0.02 cm^3 of distilled water. The experiments lasted for 10 min, the micropowder being removed from the polishing zone after 3, 5, 7, and 10 min. In order to remove adhering particles, the steels were repeatedly heated to 60°C in aqueous solutions of hydrochloric acid with concentrations of 2:1, 10:1, and 20:1. Then the micropowder was washed in alcohol, dried, and placed on the stage of a BMD-1 biological microscope equipped with a special measuring scale in the ocular.

Measurements were made on 500-700 grains, after which the micropowder was again placed in the polishing zone and investigated a second time, and so on to the end of the experiment.

TABLE 3. Characteristics of the
Geometrical Parameters of the Grains of
Micropowders of the Original Grain
Size and after Polishing 10 Minutes

Micropowder	Angles cutting point, deg		Radius of curvature, μ	
	before test	after test	before test	after test
Zirconium carbide	124	20	2.3	0.9
Synthetic diamond	125	36	2.7	1.0
Tungsten boride	126	33	2.7	1.0
Silicon carbide	128	39	2.63	1.1

These investigations show that ZrC micropowders are ground somewhat more rapidly than micropowders of synthetic diamond and green silicon carbide (see Table 2). Consequently, micropowders of quasimetallic compounds are more suitable for general use than diamond and carbide ceramic powders.

The experimental results indicate a more favorable transformation of the grains during polishing with quasimetallic compounds than that which occurs when existing abrasive materials are used.

Besides determining the grain dimensions, we also measured the angles between the faces forming the cutting edges and the radius of curvature of these edges of grains of the micropowders of zirconium carbide, silicon carbide, and synthetic diamond, both before and after the experiment (Table 3). It is shown that during polishing ZrC is ground somewhat more rapidly than are the existing abrasives, while the geometry of its grains is similar to that of existing abrasive materials. The principal features of the cutting properties of micropowders of quasimetallic compounds are practically the same as those of existing abrasives.

Conclusions

1. It has been found that the refractory compounds ZrC, TiC, and Cr_3C_2 can be successfully used as polishing and finishing materials, since their polishing characteristics are no worse than

those of existing abrasives (SiC, B_4C, and electrocorundum) and they are not greatly inferior in polishing capacity to synthetic diamond for polishing hardened steels.

2. From the large quantity of experimental data obtained, it was established that during polishing micropowders of quasi-metallic compounds are ground more rapidly than are those of the well-known abrasives green silicon carbide, boron, carbide, and synthetic diamond. Consequently, the new materials have a wider field of application; moreover they permit the range of pastes differing in grain-size number to be greatly reduced and, other things being equal, they enable a higher class of finish to be obtained on the polished surface.

3. It is shown that during polishing micropowders of the new and existing abrasive materials are ground in almost the same way and that they have identical geometry of the grains.

Antiabrasion Material Resistant to
Erosive and Aggressive Media

V. V. Karlin, A. P. Garshin, and A. A. Novikov

Self-bonded silicon carbide is suggested as a material suitable for
making components which operate in acid media and under strong
erosive or abrasive conditions. The microstructure of finished components, after siliconizing under optimum conditions and compacted
at various specific pressures, has been examined.

In the production of double superphosphate a vertical drying
chamber is used to treat a pulp consisting of fine crystals of mono-
calcium phosphate, a small amount of the phosphorus salts of aluminum, iron, and calcium, and 40% of liquid phase (water and phosphoric acid). The concentration of phosphoric acid in the pulp is
20% (20 g of 100% phosphoric acid per 100 g of pulp). This pulp
is sprayed in the drying chamber by means of a disk having four
detachable nozzles which are mounted in the spray head of the
machine rotating at a speed of 9000 rpm.

The choice of a material for making the disks and nozzles
operating under the above conditions is a very complex task owing
to erosion of the nozzles and the action of the phosphoric acid on
them. As a result, disks and nozzles made of the following materials have been investigated: chromium electrocorundum, tungsten carbide, alloy steels, and silicon carbide. The test results
are given in Tables 1 and 2. The most suitable material was one
based on silicon carbide and made in the Federal German Republic. However, the use of imported materials is expensive, the

TABLE 1. Comparative Resistance of Disks Made of Various Materials

Material	Operating time, h	Amount of pulp passing through the spray head, tons	Nature of wear of disks
Chromium corundum produced in Federal German Republic	5	40	Formation of a circular groove to a depth of 1.2 mm around the periphery of the disk (see Fig. 2)
Tungsten carbide produced in Federal German Republic	14	112	The same, to a depth of 2 mm
Alloy steel produced in Federal German Republic	50	400	The same, to a depth of 2.5 mm
	127	1016	Complete wear (see Fig. 3)
Chromium cast iron produced in USSR	300	2400	Formation of a circular groove to a depth of 2-3 mm around the periphery of the disk
Silicon carbide produced in Federal German Republic	600—700	2000—7000	—
Self-bonded silicon carbide produced in the USSR (VNIIASh)	980—1000	9800—10 000	—

TABLE 2. Resistance of Nozzles Made of Various Materials

Material	Operating time, h	Operating time, h	Amount of pulp passing through the nozzles, tons	Nature of wear
Chromium corundum produced in Federal German Republic	8	5.0	40	Negligible wear; surface became rougher
	12	5.5	44	The same
	4	79.7	638	Groove 1.5 mm deep appeared in direction of nozzle
Silicon carbide produced in Federal German Republic	4	600—700	6000—7000	The same to a depth of 1.3 mm
Tungsten carbide produced in Federal German Rep.	8	14	112	Groove 2 mm deep appeared in direction of rotation of nozzle
Self-bonded silicon carbide produced in USSR (VNIIASh)	4	980—1000	9800—10000	The same, to a depth 1.2 mm

annual cost of sets of disks and nozzles for two plants (approximately 200 nozzles and 50 disks) being about one million crowns or about 130,000 rubles.

Self-bonded silicon carbide has been suggested by VNIIASh as a suitable material for the manufacture of disks and nozzles. Investigations have been carried out in USA [4, 7], Federal German Republic [6, 8, 9, 12], USSR [1, 2, 3, 5], and other countries into the production, properties, and fields of application of a similar material.

The components (disks and nozzles) are compacted, using an organic bonding agent, from charges consisting of SiC grains of various sizes and a carbonaceous material (petroleum coke, soot) in a predetermined ratio and siliconizing them in molten silicon or silicon vapor. As a result, secondary SiC is formed by reaction between the carbon and silicon, and in addition pores existing in the components are filled with silicon, which is present in excess during siliconizing. A bonded and fairly strong component is obtained.

A 200-kW resistance furnace was used for siliconizing the components. Heating was effected in the furnace by means of a core lined with flaky graphite and current-conducting electrodes

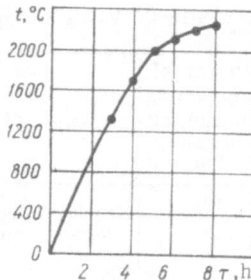

Fig. 1. Optimum heating
conditions for siliconizing.

400×400 mm in size. The components were placed in a special
space prepared from graphitized electrodes of various shapes
depending on the shape of the components and installed in the core
of the furnace.

Several sets of heating conditions were tried out in developing
the siliconizing process, the most effective results being obtained
under the conditions shown in Fig. 1.

The microstructure of the finished components, siliconized
under optimum conditions, were examined (Figs. 2 and 3), and it

Fig. 2. Microstructure of material treated under optimum conditions.
(Specific compaction pressure = 600 kg/cm^2; amount of carbon in the
original mass of the specimen before siliconizing = 30%).

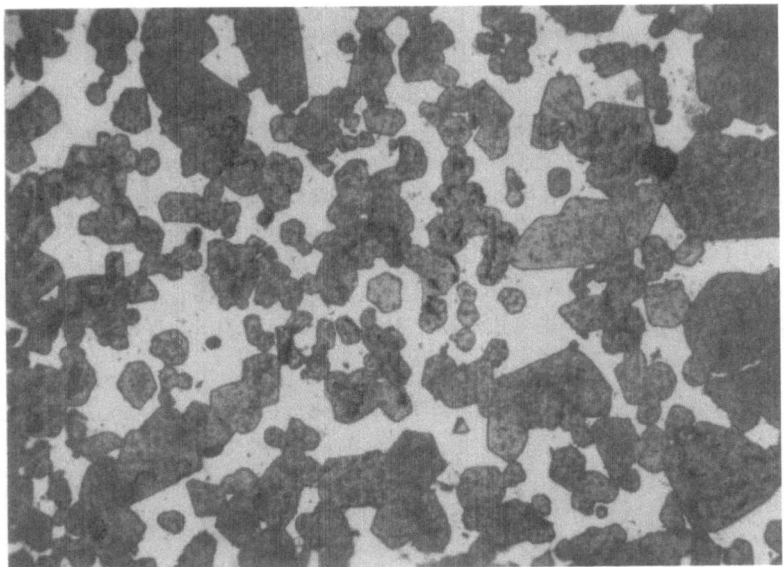

Fig. 3. Microstructure of specimen compacted under a specific pressure of 200 kg/cm² and siliconized under optimum conditions (the amount of carbon in the mass of the specimen before siliconizing = 30%).

can be seen that there are differences in the microstructures of specimens compacted under different pressures. With a large specific compaction pressure (see Fig. 2), the SiC grains are assembled into dense aggregates between which there are large areas filled with silicon.* For specimens compacted at lower pressures, the microstructure consists of SiC grains and aggregates of such grains (Fig. 3) "embedded" in a field of silicon. The grains of SiC and silicon are more uniformly distributed in the material. Hence it is possible to speak in the first case about a self-bonded SiC and in the second case about a silicon-bonded SiC.

X-ray structural analysis carried out on specimens containing 10-50% of carbon before siliconizing revealed that the SiC phase is present as the hexagonal modification found in commercial SiC, the main polytype being SiC-11 (56-73%).† The heat-treatment conditions are evidently such that the secondary silicon carbide

* The petrographic studies were carried out by T. P. Nikitina.
† The x-ray structural investigations were carried out in the x-ray laboratory of VNIIASh.

that is formed as β-SiC recrystallizes in the course of the process into α-SiC.

New technological procedures have been developed at VNIIASh in the course of their experience, and these enable the properties of SiC-base components to be controlled in accordance with their application. Components of the following composition have been produced by varying the percentage content of carbon in the mixture to be compacted, the size of the SiC grain, the specific compaction pressure, the form of the bonding material, and the siliconizing conditions: 60-96% SiC, 40-45% Si, 0.5-5% C_{free}, and 1.5-6% SiO_2.

Below are given some physical properties of self-bonded SiC

Specific weight	2.85-3.1 g/cm^3
Thermal coefficient of linear expansion (up to 1000°C)	$(4.1-5.5) \times 10^{-6} deg^{-1}$
Thermal resistance (cooling from 1600°C into water at 0°C)	7-10 thermal cycles
Modulus of elasticity (up to 1000°C)	$(26-33) \times 10^5$ kg/cm^2
Ultimate strength in compression (tests carried out on cylindrical specimens 20 mm in diameter and height)	1800-5000 kg/cm^2
Ultimate strength in fracture (tests carried out on octagonal specimens with a cross section at the fracture neck of 5 cm^2)	300-450 kg/cm^2
Ultimate strength in bending (tests carried out on specimens 20×20 mm in cross section and 120 mm long	950-1100 kg/cm^2

Specimens of the material described have been tested by VNIIASh in various fields of present-day technology. Figure 4 shows some types of components made of self-bonded SiC.

The valuable properties of components based on SiC have aroused interest in this material.

As test results have shown (see Tables 1 and 2), the high resistance to erosive and aggressive media is responsible for

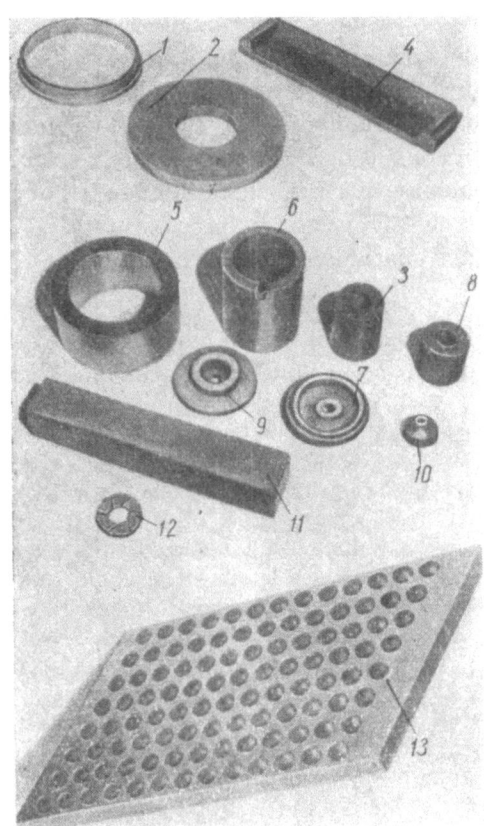

Fig. 4. Some examples of siliconized components based on SiC: 1) packing ring for the shaft of a mixer operating in an acid medium; 2, 3) disk and nozzle of spray head of drier used in the production of double superphosphate (unpolished components); 4) metallizator working in molten aluminum at 850°C (unpolished components); 5, 6) packing rings working under conditions of abrasive wear and in acid media;7, 8) jets working under conditions of abrasive wear and in acid media (unpolished components); 9, 10) pair of bearings (unpolished) working in an acid medium; 11) bar for lining a pipe-line through which abrasive material is transported; 12) mounting for a plasma torch; 13) grid for the suction boxes of paper-making machines.

the operation of disks and nozzles made at VNIIASh for 980-1000 h (up to 13,000 tons of pulp).

By appropriate special mechanical treatment of components made of the material described, it is possible to achieve a high grade of surface finish (higher than ∇10), and this opens up great possibilities for the production of friction pairs operating in severe corrosive conditions.

Conclusions

1. Optimum conditions have been determined for the production of components based on self-bonded SiC, with properties controlled to a known extent.

2. Components made of self-bonded SiC have been found to be capable of operating under erosive conditions and the action of phosphoric acid (in machines used for the production of double superphosphates) for 1000 h and more, i.e., the service life is increased by 200 and by 71 times as compared with that of components made of chromium electrocorundum and tungsten carbide respectively and by approximately 1.5 times compared with SiC-base material of foreign origin.

3. Self-bonded SiC can be recommended for the manufacture of components which have to work in acid media and also under conditions of severe erosive and abrasive action.

Literature Cited

1.   I. S. Kainarskiǐ and E. V. Degtyareva, Carborundum Refractories [in Russian] (1963).
2.   V. V. Karlin, V. A. Alferov, and V. E. Kopyt, Abrazivy i Almazy, No. 2, 33 (1965).
3.   A. N. Novikov, Ogneupory, No. 12, 557 (1957).
4.   U. S. Patent No. 2,784,112 (1957).
5.   I. N. Frantsevich et al., Tsvet. Metally, No. 6, 49 (1965).
6.   E. A. Bloch, Z. Erzbergbau u. Metallhuttenwesen, 14:400 (1961).
7.   U. S. Patent No. 3,100,688 (1963).
8.   P. Popper, Ber. Deut. Keram. Ges., 37:297 (1960).
9.   P. Popper, Special Ceramics, Heywood, London (1960), p. 208.
10. K. M. Taylor, Materials and Methods, 44:92 (1956).
11. K. M. Taylor, Ingenieriacind, 25:91 (1957).
12. M. L. Wrotten, Raketentechn. u. Raumfahrtforsch., 2:21 (1958).

Reduction of the Erosive Wear of Refractory Compounds in Plasma Streams by Means of Thermochemical Protection

V. M. Sleptsov, G. M. Shchegolev, Yu. P. Kukota, and E. M. Prshedromirskaya

The efficacy of using thermochemical protection to reduce the corrosive-erosive wear of refractory compounds in high-temperature streams is discussed. A method of producing porous materials is described. An apparatus for determining the resistance of materials in plasma flows has been developed. Concentration profiles at the porous wall of a channel when CO_2 is blown into a stream of air are given.

The growth of the new technology presents metallurgists with the problem of developing materials capable of operating for long periods in contact with aggressive high-temperature plasma streams, the main factors involved being high-temperature oxidation, erosion of the material, and thermal failure.

Refractory quasimetallic compounds of the transition metals with carbon, boron, nitrogen, and silicon, and above all the carbides and borides, possess properties that make them promising materials for use in particular parts of high-temperature apparatus. Because of their high melting point, hardness, electrical conductivity, strength at elevated temperatures, and low work function in thermionic emission, carbides and borides can be used as electrodes in magnetohydrodynamics — as generators and accelerators, in torches for the gas-plasma cutting of metals, and in illuminating arc technique.

Their use is restricted, however, by the necessity of having a neutral atmosphere. The low resistance of scaling in air greatly limits the temperature level at which carbides can be used in plasma streams containing O_2, CO, CO_2, H_2O, and aggressive gases. Great difficulties are also encountered in combating the erosive action of rapid streams at the hot surface of the material. Here, besides improving the structure and quality of the material, special measures have to be taken for the thermal and chemical protection of the surface.

One of the most effective means of thermal protection is porous cooling. When protective gases are used as the coolant, a film of gas is formed close to the wall, which provides chemical protection of the material against oxidation.

The use of thermochemical protection of the material exposed to the plasma stream makes it possible to alleviate conditions at the walls in the following ways:

1. By reducing the tangential stresses at the wall which are set up by friction between the stream and the material, and this is reflected in a sharp reduction in the erosive wear.

2. By preventing the diffusion of aggressive constituents – particularly oxygen – from the center of the stream to the layer close to the wall, where oxidation of the material could occur. This makes it possible to use carbides and other refractory compounds at wall temperatures exceeding the limit of their corrosion resistance in the given medium.

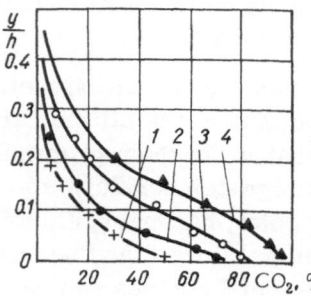

Fig. 1. Concentration profiles at the porous wall of a channel at the following blow-in ratios of CO_2 introduced into a stream of air: 1) 0.0157, 2) 0.0216, 3) 0.0288, 4) 0.0370. Reynolds criterion $Re_d = 3.7 \times 10^4$.

As can be seen from Fig. 1, the boundary layer becomes thicker with increase in the amount of gas blown in, and the concentration of the gas thus introduced approaches 100% at the wall making it difficult for the air to reach the protected surface.

3. Blowing in of the gas markedly reduces the convective heat flow to the wall, and in addition it cools the material by passing into the pores. By a suitable choice of composition, specific heat, and temperature of the coolant, practically any wall temperature can be attained within a given flow-rate range which is set by the need to achieve maximum effect of the erosive-corrosive protection.

The technology of producing porous refractory compounds is based on the preliminary granulation of finely divided powders.

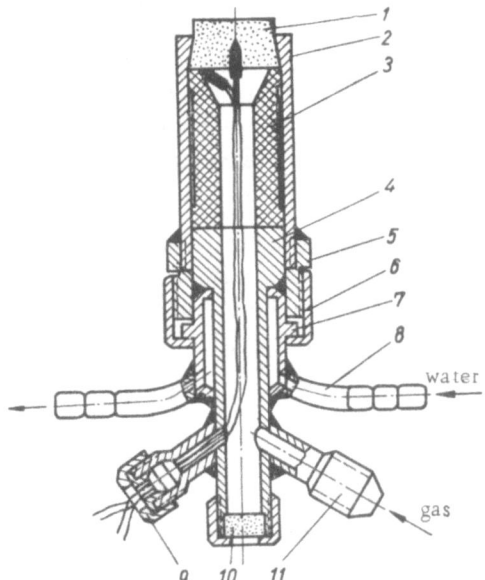

Fig. 2. Device for testing porous refractory materials in a plasma stream: 1) porous metalloceramic insert; 2) holder (porous tungsten impregnated with copper); 3) graphite sleeve; 4) body of the device; 5) intermediate connector; 6) clamping nut; 7) water-cooling jacket; 8) connecting pipe; 9) thermocouple leads; 10) quartz window; 11) inlet for protective gas.

It was found that the gas permeability and uniformity of pore distribution are greater in materials prepared from spheroidized powders. Spheroidization was effected by rolling the granules or by passing them through a plasma jet. Porous components made of refractory-metal carbides were produced by the compaction of spheroidized powders, followed by sintering.

Figure 2 depicts a section of a device which was developed in order to test porous materials in high-temperature streams of combustion products. The porous test specimen 1 in the form of a cylinder 25 mm in diameter with a conical shaft was gripped in the holder 2. The whole device was mounted in the nozzle of a combustion chamber operating on natural gas and oxygen. The velocity of the combustion products was 530-560 m/sec and their temperature 2750°K. The surface temperature of the porous insert reached 2100°K, which is higher than the scaling-resistance temperature of the material. Cinematography revealed that solid or liquid particles carried out from the combustion chamber by the gas stream are diverted from the surface of the porous material in the region where the protective gas is introduced, and it is this which is responsible for the reduction in erosive wear of the material.

It proved impossible to determine any loss in weight of the specimens after the tests, which lasted in the majority of cases for 5 min; this indicates a considerable increase in the erosion-corrosion resistance of the material.

Conclusions

The possibility of reducing the erosive-corrosive wear of refractory compounds in high-temperature streams by thermochemical protection is demonstrated. A device for testing the resistance of materials in plasma streams has been developed. The orders of magnitude of the proportion of protective gas introduced into the stream to reduce erosive wear have been determined.

Indentors for Measuring the Hardness of Materials at High Temperatures

V. V. Dzhemelinskii, M. S. Koval'chenko, V. A. Borisenko, and G. N. Makarenko

The possibility of using hot-compacted specimens of boron carbide and titanium diboride as materials for the indentors for measuring the hardness of tungsten carbide at high temperatures has been explored. It is shown that a titanium diboride indentor flattens at 1770°K owing to a fall in the hardness of the material at this temperature. A boron carbide indentor can be used repeatedly for measuring the hardness of tungsten carbide at temperatures up to 2170°K without any trace of chemical reaction between the material of the specimen and that of the indentor and also without failure of the latter.

Their hardness at elevated temperatures is an important property of refractory materials. The most widely used method of hardness measurement is the static penetration method based on the penetration of an indentor into the test specimen after heating to high temperatures [3, 5].

Diamond is used for measuring the hardness of refractory materials at comparatively low temperatures. The great hardness of diamond is connected with the localization of the valence electrons at the atom nuclei with the formation of very stable sp^3 configurations, which in turn determine the rigidity and directionality of the chemical bonds. These positive properties enable diamond crystals to be used as a material for indentors for measuring the

hardness of refractory compounds and materials based on them
at temperatures up to 1100°K. The diamond tips, which possess
great hardness at low temperatures, rapidly become blunt and less
stable under high-temperature conditions. It has been found that,
beginning at a temperature of 1123°K, hardness measurements
result in rapid wear of the diamond pyramids, while at 1370-1470°K
they are no longer fit for use after a single penetration. On being
held for long periods at high temperatures, the diamond tip undergoes graphitization, accompanied by a marked loss of strength
and softening. Diamond is transformed into graphite at temperatures above 1100-1150°K.

Indentor tips made of synthetic corundum (artificial sapphire),
which have been successfully employed for measuring the hardness of metals up to 2030°K [1, 6], cannot be used for measuring
the hardness of refractory compounds and materials based on
them at temperatures above 1270°K, since at high temperatures
its hardness is practically no different from that of the materials
being tested. Figure 1 presents the results of M. G. Lozinskii
and V. S. Mirotvorskii for the temperature dependence of the
change in hardness of artificial sapphire and titanium carbide.
An essential condition for carrying out hardness tests by the
penetration method is a substantial difference in hardness between the indentor material and the material being tested. The
hardness of the former should be at least 2-2.5 times that of the
latter. Thus, owing to the absence of reliable indentor materials,
it is not possible to determine the hardness of refractory materials
in the range 1270-2200°K by the static indentation method, using
a four-sided pyramid with a tip angle of 136°.

It has been shown in [7] that certain refractory compounds
can be used as indentors for measuring the hardness of carbides

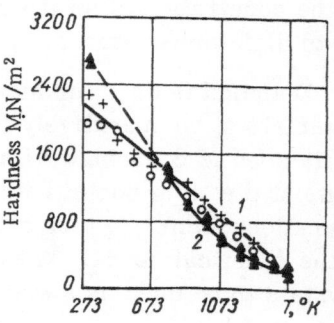

Fig. 1. Temperature dependence of the
change in hardness of: 1) artificial sapphire (synthetic corundum) according to
V. S. Mirotvorskii; 2) titanium carbide
according to M. G. Lozinskii (load on indentor = 1 kg).

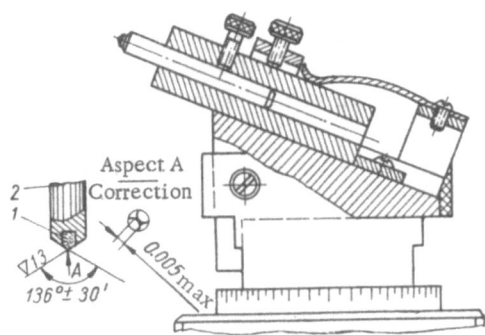

Fig. 2. Device for making indentors of the standard four-sided pyramid type with an angle of 136° between the opposite sides.

and borides. Saunders and Probst [8] employed boron carbide (B_4C) to determine the hot hardness of some bodies at 1898°K.

The present research was concerned with exploring the possibility of using indentors of boron carbide and titanium diboride to determine the hardness of tungsten carbide at high temperatures in vacuo.

The raw materials used for preparing the specimens consisted of: amorphous boron (of 99.5% purity), lampblack with an ash content of 0.2%, and finely divided boron carbide powder containing 76.8% B and 21.9% C and titanium diboride powder containing 69.3% Ti and 30.4% B.

Specimens of B_4C were prepared from the original powders by hot compaction in an argon atmosphere at 2270°K for 15 min under a pressure of 1800 MN/m². Titanium diboride specimens were prepared by hot compaction at 2600°K for 5 min under a pressure of 1200 MN/m².

The device illustrated in Fig. 2 was developed for making indentors of the standard four-sided pyramid type with an angle of 136° between opposite sides (as specified in GOST 2999-59). The indentor 1 is gripped in the holder 2. Holders for high-temperature indentors were made from molybdenum, which has a high melting point (2895°K) and a coefficient of linear expansion (5.5 × 10^{-6} deg^{-1}) which is close to that of the refractory compounds being used.

The faces of the indentor were first worked on a 3A64M universal lathe, using AChK diamond disks with an organic bond B1, grain size AC0 20, 100% concentration. Lapping (finishing) was carried out by means of diamond pastes of various grain sizes on lapping disks made of cast iron and fiber. The four-sided pyramid was produced by rotating the square stand with the indentor through 90° after one of the surfaces had been prepared. Quality control of the standard four-sided pyramid was achieved by means of instruments and metallographic microscopes. It was found from the results of these checks that the indentors made in the way described satisfy the requirements of the GOST standard.

The hardness tests were carried out in a vacuum of 10^{-5} mm Hg on a UTV-2 machine developed at the Institute of Problems of Strength of the Academy of Sciences of the Ukrainian SSR [6, 2], with which it is possible to determine the long- and short-term hardness of materials in vacuo or in inert media over a wide temperature range from 290 to 3270°K.

The indentors were tested by determining the hardness of hot-compacted tungsten carbide specimens. Before the impressions were made, the surfaces of the specimens were prepared in the same way as for metallographic examination. The tests were carried out in stages of 100-200°K with a constant time of load application of 1 min. Diamond was used as the indentor material at temperatures from 290 to 1100°K, while from 1300 to 2170°K boron carbide and titanium diboride were used.

The impressions made with a boron carbide pyramid on the surface of a tungsten carbide speciment at various temperatures have a good quality with no trace of chemical reaction between the material of the indentor and that of the specimen. Figure 3 shows the change in hardness of tungsten carbide as a function of temperature. The temperature-dependence curve of tungsten carbide

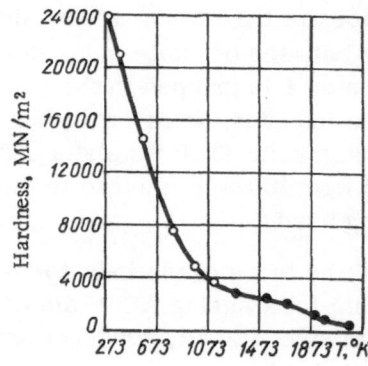

Fig. 3. Temperature dependence of the change in hardness of tungsten carbide (load on indentor 1 kg; load maintained for 1 min): O) impressions obtained by penetration of a diamond indentor; ●) impressions obtained by penetration of a boron carbide indentor.

measured by means of a boron carbide indentor is a continuation of the curve obtained at lower temperatures by using a diamond pyramid.

The tip of the titanium diboride indentor collapsed at 1770°K owing to a fall in the hardness of TiB_2 at this temperature.

Conclusions

1. The suitability of indentors made of hot-compacted boron carbide and titanium diboride for measuring the hardness of tungsten carbide at elevated temperatures has been investigated, and it was found that boron carbide indentors can be used for this purpose up to 2170°K.

2. It is shown that the temperature-dependence curve of the hardness of tungsten carbide measured by means of a boron carbide indentor is a continuation of the curve obtained at lower temperatures by means of a diamond pyramid.

3. It was found that when boron carbide is used for the indentor no chemical reaction takes place between the indentor and the tungsten carbide being tested. After penetration, sharp impressions, of good quality, are obtained, which provide fairly accurate hardness measurements.

Literature Cited

1. V. A. Borisenko, Trudy Nauchno-tekhnicheskogo Soveshchaniya, ITI, Kiev 230 (1961).
2. V. A. Borisenko and G. S. Pisarenko, Poroshkovaya Met., No. 5 (1961).
3. M. G. Lozinskii, The Structure and Properties of Metals and Alloys at High Temperatures [in Russian], Metallurgizdat, Moscow (1963).
4. M. G. Lozinskii and M. B. Guterman, Zavod. Lab., 22:1358 (1956).
5. I. L. Mirkin and D. E. Lifshits, Zavod. Lab., 15:1080 (1949).
6. G. S. Pisarenko et al., Trans. Seventh All-Union Conference on Powder Metallurgy, Erevan (1964), p. 50.
7. R. D. Koester and D. P. Moak, J. Less-Common Metals, 13:249 (1967).
8. W. A. Saunders and H. B. Probst, J. Am. Ceram. Soc., 49:231 (1966).

A Study of the External Friction of Refractory Carbides at High Temperatures in Vacuo

G. V. Samsonov and Yu. G. Tkachenko

The laws governing the external friction of TiC, ZrC, HfC, NbC, and TaC have been investigated from room temperature to 1800°C in a vacuum of 10^{-5} mm Hg. The coefficient of friction at first decreases and then increases sharply, beginning at a certain temperature in the range 950-1300°C which is characteristic of each carbide. The results obtained are interpreted in terms of both macro- and micro-factors. It is found that the higher the statistical weight of stable configurations (as regards energy) of the atoms of the components of the refractory carbide, the lower is its coefficient of friction.

Materials suitable for use in friction components working at high temperatures in vacuo should possess great hardness and thermal conductivity of heating, a high activation energy of diffusion, a small vapor pressure, and a low evaporation rate. These requirements are in the main satisfactorily met by the refractory carbides of transition metals of groups IV to VI.

Information on the laws governing the friction and adhesion reaction of refractory carbides is very limited [10-13], and in certain cases (ZrC and HfC) it is totally lacking.

In the present research the laws governing the external friction of TiC, ZrC, NbC, HfC, and TaC have been investigated in a vacuum of 10^{-5} mm Hg from room temperature to 1800°C, using a load of 5 kg/cm² and a rate of slip of 0.43 m/min. The tests involved determining the end friction of specimens in the form of hollow cylinders (external diameter 15 mm, internal diameter 8 mm, and height 12 mm). The tests were carried out on an apparatus specially devised and constructed for the purpose [1].

In preparing the specimens for testing the friction surfaces were polished with diamond disks and then finished with pastes. The mean square deviations of the surface irregularities did not exceed $0.20\ \mu$. The prepared test specimens were degassed by heating them in vacuo to a temperature 100 deg C above the maximum test temperature. The coefficient of friction was determined during subsequent cooling and reheating in stages of 50-100 deg C, the specimens being held for 12-15 min at each temperature. Experiment showed that after this time the coefficient of friction had become stable and underwent scarcely any further change.

Specimens of TiC, ZrC, and HfC were prepared by hot compaction [7], and those of NbC and TaC by cold compaction followed by sintering in a graphite-tube resistance furnace. The specimens were sintered in a hydrogen atmosphere in crucibles made of NbC. The sintering temperatures for NbC and TaC were 2400 and 2500°C, respectively, the holding time being 45 min. The chemical composition after sintering and the residual porosity of the test specimens are given in Table 1.

Chemical analysis of the specimens after testing revealed that the composition of the carbides was practically unchanged except for some loss of iron.

TABLE 1. Chemical Composition of Specimens

Carbide	Chemical composition of specimens, wt.%				Hot compaction	Method of production
	Me	c_{comb}	c_{free}	Impurities		
TiC	80.4	19.3	0.3	Fe — 0.2	2—3	Hot compaction
ZrC	87.4	12.1	0.5	Fe — 0.3	2—3	The same
HfC	93.2	6.7	0.3	—	1—2	The same
NbC	88.1	11.2	Not found	Fe — 0.5	12—14	Cold compaction and sintering
TaC	93.3	6.6	0.5	—	12—14	The same

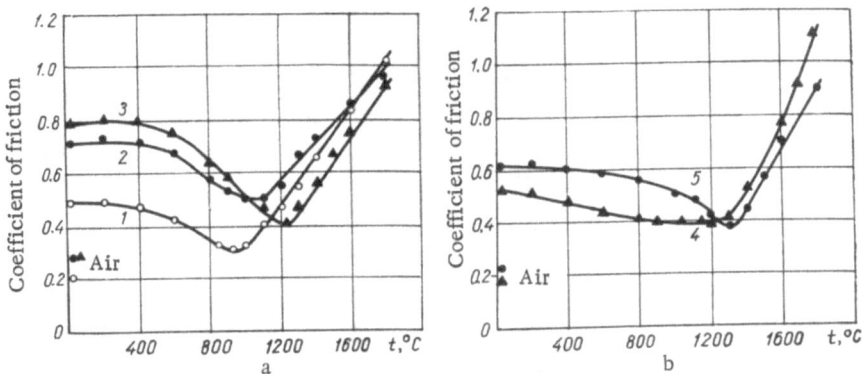

Fig. 1. Temperature dependence of the coefficient of friction during friction of identical specimens in vacuo: a) TiC (1), ZrC (2), HfC (3) (porosity 2-3%); b) NbC (4) and TaC (5) (porosity 12-14%).

The experiments revealed a general tendency for the coefficient of friction to change with rise in temperature: the coefficient first decreases and then rises sharply, beginning at a certain temperature in the range 950-1300°C which is characteristic of each carbide. The coefficient of friction in vacuum after degassing is 2.5-3 times greater than the value in air. Then the coefficient falls with rise in temperature, the minimum value occurring at 950°C for TiC, 1100°C for ZrC, 1250°C for HfC, 1200°C for NbC, and 1300°C for TaC. Further rise in temperature results in a sharp increase in the coefficient of friction (Fig. 1).

The relationships shown in the figure were obtained during heating of the specimens. In tests carried out under cooling conditions, the decrease and subsequent increase in the coefficient of friction with fall in temperature takes place with a certain delay so that a hysteresis loop is formed.

The reduction in the coefficient of friction with rise in temperature may be due to a weakening of the intergranular bond and to a change in the electron exchange between the surfaces.

Existing theories of friction relate the formation of frictional forces with different kinds of deformation of the projections on the surface and with molecular interaction between the bodies in contact [2-4]. It may be concluded that the nature of the friction between clean crystalline surfaces in vacuo, when the action of adsorbed surface films is reduced to a minimum after heating, depends on the deformation properties of the substances, their mutual solubility, and structural peculiarities. As the principal mechanical and physical properties of a substance are determined by its electron structure [8, 9], the external friction processes can be considered in the light of the electron structure, which, other things being equal (chemical composition, porosity, grain size, presence of impurities, etc.), is the determining factor in the friction of solids.

The mechanism of electron exchange in the friction of transition-metal carbides can be represented in the following way. Part of the localized valence electrons of Ti, Zr, Hf, Nb, and Ta form a wide spectrum of electron configurations varying in stability from d^0 to d^5, of which d^5 is the most stable from the energy point of view. Electron exchange is inhibited at the contact between two solids having a high statistical weight of stable electron configurations, and a lower coefficient of friction is therefore to be expected. With rise in temperature (to 950–1300°C), the less-stable configurations (d^1, d^2, d^3, etc.) break down and the more stable d^5 configurations are formed, which also results in a reduction in the coefficient of friction.

During the formation of refractory carbides, some redistribution takes place in the electron density of the metal and carbon. As a result of this, the properties of the carbides depend mainly on the extent of the breakdown or stabilization of the most stable sp^3 electron configuration of carbon in the carbide. Thus in TiC additional stabilization of the sp^3 configuration occurs at the expense of the unlocalized electrons of titanium. It may be assumed that with the reduction of the donor properties of group IV metals from titanium to hafnium, the degree of stabilization of the electron configuration of carbon decreases, and so does the statistical weight of the stable electron configurations as a whole. As the method of production, degree of deviation from stoichiometry, and porosity of the test specimens were practically the same in all cases, it may be concluded that the increase in the coefficient of friction in the series TiC → ZrC → HfC is due to a reduction in the statistical weight of the stable electron configurations of these substances.

In the formation of the carbides, niobium and tantalum atoms act as acceptors of electrons and to a certain extent destroy the sp³ configurations of carbon. The extent of the breakdown of the most stable sp³ configuration in the carbide depends on the acceptor capacity of the metals. A greater breakdown of the electron configuration of carbon is to be expected in TaC, the electrons of carbon being unable to pass over completely to the d shell of the metal, since it is already fairly full. This leads to an increase in the number of unlocalized electrons and consequently to a higher coefficient of friction than that for NbC.

Further softening of the carbides with rise in temperature (above 950-1300°C) is responsible for the plastic character of the deformation of the friction surfaces and the development of diffusion processes in the surface layers. All this leads to a closer approach of the surfaces and to the development of adhesion-reaction processes between the bodies in contact [5, 6], and thus causes a rapid increase in the coefficient of friction.

Conclusions

1. The coefficient of friction of similar specimens of TiC, ZrC, HfC, NbC, and TaC first decreases with rise in temperature and then rapidly increases. The minimum occurs at 950°C for TiC, 1100°C for ZrC, 1250°C for HfC, 1200°C for NbC, and 1300°C for TaC.

2. The higher the statistical weight of the stable (in terms of energy) configurations of the atoms of the components of the refractory carbide, the lower is its coefficient of friction.

Literature Cited

1. A. Ya. Artamonov and Yu. G. Tkachenko, Zavod. Lab., 34:609 (1968).
2. F. P. Bowden and D. Tabor, Friction and Lubrication [Russian translation], Mashgiz, Moscow (1960).
3. B. V. Deryagin, What is Friction? [in Russian], Izd. Akad. Nauk SSSR, Moscow (1963).
4. I. V. Kragel'skii, Friction and Wear [in Russian], Mashgiz, Moscow (1962).
5. N. N. Rykalin, M. Kh. Shorshorov, and Yu. L. Krasulin, in: The Nature of Metallic Phases and the Character of the Chemical Bond in Them [in Russian], IMET im. A. A. Baikova, Moscow (1965), p. 37.
6. G. V. Samsonov, A. L. Burykina, and O. V. Evtushenko, Avtomat. Svarka, No. 10, 30 (1966).
7. G. V. Samsonov and M. S. Koval'chenko, Hot Compaction [in Russian], Gostekhizdat, Kiev (1962).

8. G. V. Samsonov, Planseeber. Pulvermet., 15(1):3 (1967).
9. G. V. Samsonov, Poroshkovaya Met., No. 12, 40 (1966).
10. A. P. Semenov and V. V. Pozdnyakov, Mashinovedenie, No. 1, 116 (1967).
11. A. P. Semenov and V. V. Pozdnyakov, Dokl. Akad. Nauk SSSR, 160:811 (1965).
12. C. A. Brookes, Wear, 9:103 (1966).
13. B. L. Mordike, Wear, 3:374 (1960).

VIII. CHEMICAL AND REFRACTORY PROPERTIES OF CARBIDES

The Chemical Bond in Carbides of
the Rare-Earth Metals

T. Ya. Kosolapova, G. N. Makarenko,
and L. T. Domasevich

The results are given of a study of the gaseous products liberated during the decomposition of the carbide phases of the rare-earth metals by water. These products consist mainly of acetylene, ethylene, methane, ethane, and also hydrogen. From an analysis of the composition of the gaseous products in conjunction with an analysis of data on some of their physical properties, it is shown that with increase in the C/Me ratio of the carbides of the rare-earth metals, the nature of the chemical bond changes from being predominantly ionic in MeC phases to being predominantly covalent-metallic in MeC_2 phases.

Carbides of the rare-earth metals occupy a position intermediate between the carbides of the transition metals of groups IV to VI and the carbides of the alkaline earth metals. Phases of higher carbon content (MeC_2), besides having a high melting point, metallic-type conductivity and moderate values of the coefficient of thermal expansion [3], which make them similar to the transition-metal carbides, possess, in contrast to these carbides, a low chemical resistance, and they are easily decomposed not only by water but even by moist air.

Phases of lower carbon content (Me_2C_3 and MeC) have a low melting point and hardness, small thermodynamic stability, and a high electrical resistance which is typical in some cases of semiconducting materials [3]. Like the dicarbides, they are readi-

TABLE 1. Composition of Carbides Investigated

Carbide	Calculated composition, %		Composition of carbides produced, %		
	Me	C	Me	C_{comb}	C_{free}
YC_2	78.80	21.20	78.8	20.8	1.2
LaC_2	85.26	14.74	85.5	14.5	0.6
CeC_2	85.37	14.63	84.8	14.6	0.6
PrC_2	85.44	14.56	85.2	14.6	0.5
NdC_2	85.73	14.27	85.3	14.3	0.4
GdC_2	86.73	13.27	86.7	13.3	Not found
TbC_2	86.90	13.10	86.8	13.0	0.2
DyC_2	87.13	12.87	87.1	12.5	0.3
ErC_2	87.45	12.55	87.3	12.2	0.4
Y_2C_3	83.14	16.86	84.0	15.8	Not found
La_2C_3	88.53	14.47	88.4	11.5	» »
Ce_2C_3	88.61	11.39	88.3	11.4	» »
Pr_2C_3	88.67	11.33	89.0	11.2	» »
Nd_2C_3	88.90	11.10	99.0	10.8	» »
Gd_2C_3	89.71	10.29	89.0	10.2	» »
YC	88.10	11.90	85.3	12.0	» »
ScC	79.00	21.00	77.6	21.1	0.1

ly decomposed by water, but in this case the composition of the gases liberated is different.

Such combinations of properties in the carbides of the rare-earth elements indicate the complex nature of the chemical bond in them, and enable us to classify them as belonging to the salt-like covalent—metallic type [4].

The results of numerous investigations into the decomposition of carbides of the rare-earth elements and actinoids by water [5-8] are conflicting, but this is evidently due to differences in the composition of the carbides studied, which the authors often fail to report.

However, the determination of the composition of the gaseous products liberated during the decomposition of the rare-earth-metal carbides by water, in conjunction with an analysis of data on some of their physical properties, enables us to elucidate the nature of the chemical bond in the compounds concerned.

In the present research we have investigated the composition of the gaseous products liberated during the decomposition of the carbide phases of the rare-earth elements by water. The composition of the carbides studied is given in Table 1.

Decomposition of the carbides was carried out by the method described in [2]. The experimental results showed that (Table 2) the gaseous decomposition products consist of a mixture of hydrocarbons, containing mainly acetylene, ethylene, methane, and ethane, and also hydrogen.

It may be assumed that the evolution mainly of acetylene indicates the relative proportion of the covalent bond, which is stronger between the carbon atoms and weaker between the metal and carbon atoms, breakdown of which takes place on decomposition.

The acetylene content of the gaseous decomposition products increases as we pass from lanthanum dicarbide to the dicarbides of cerium, praseodymium, and neodymium; while on passing from gadolinium dicarbide to the dicarbides of terbium, dysprosium, and erbium, the acetylene content falls monotonically.

This is evidently due to the fact that on passing from lanthanum to gadolinium the possibility of electrons being transferred from the f to the d state diminishes [1]; at the same time the number of electrons participating in the $Me-C$ bonds decreases, these bonds become weaker, and the amount of acetylene evolved as a result of the action of water increases.

TABLE 2. Composition of the Gaseous Products of Decomposition of Rare-Earth-Metal Carbides by Water

Carbide	C_2H_2	H_2	C_2H_4	C_2H_6	CO	CH_4
YC_2	58.61	5.03	10.98	8.04	1.35	15.0
LaC_2	53.00	15.60	11.20	20.20	Not found	Not found
CeC_2	58.23	15.92	10.49	15.36	" "	" "
PrC_2	63.92	7.92	11.90	16.26	" "	" "
NdC_2	67.49	4.70	12.50	15.31	" "	" "
GdC_2	52.80	17.60	14.80	13.25	1.5	" "
TbC_2	51.10	6.60	24.60	14.40	Not found	3.3
DyC_3	50.10	9.70	14.90	18.40	Traces	6.9
ErC_2	45.90	8.60	20.40	13.30	Not found	11.8
Y_2C_3	15.8	4.4	7.8	8.9	" "	64.0
La_2C_3	Traces	36.1	2.7	Traces	" "	61.1
Ce_2C_3	"	23.5	2.0	»	" "	74.5
Pr_2C_3	6.4	6.4	16.1	16.6	" "	54.8
ScC	Not found		Not found	Not found	" "	
		4.8			" "	90.8
YC	" "	2.2	" "	" "	" "	97.7

As we pass from gadolinium to erbium, the probability of electrons being transferred from the f to the d state increases; the number of electrons taking part in the Me−C bonds rises, these bonds are strengthened, and the amount of acetylene evolved diminishes.

The reduction in the acetylene content of the decomposition products of gadolinium dicarbide. as compared with neodymium carbide, is due to the presence in gadolinium, as in lanthanum, of a single electron in the d state, which strengthens the Me−C bond.

The liberation of acetylene and ethylene as the principal gaseous products of decomposition of the dicarbides of the rare-earth elements by water is in good agreement with the results of an investigation of their electrophysical properties [3]. From a calculation of the current-carrier concentration, using a single-band model, it may be concluded that of four C_2 groups three have a triple bond $C \equiv C^{2-}$ and are responsible for the evolution of acetylene, and one has a double bond $C = C^{4-}$ which is responsible for the evolution of ethylene.

The appearance of methane in the gaseous products points to the presence of the C^{4-} group; in this the four carbon bonds are symmetrical, identical, and form with the metal atoms strongly polarized bonds with a highly asymmetric electron-density distribution, i.e., they characterize the existence of ionic bonding. Whereas carbon in acetylene and ethylene is typified by the hybrid groups sp and sp^2, respectively, carbon in methane is typified by the most stable configuration sp^3. The appearance of methane in the decomposition products of terbium dicarbide and the increase in methane content that occurs on passing from terbium carbide to thulium carbide indicate the existence of ionic bonding and an increase in the proportion of it during the passage. Such a phenomenon occurs as a result of the increasing possibility $f \rightarrow$ d electron transfers taking place on passing from terbium to thulium. Owing to the tendency of the metal atoms to increase the statistical weight of the stable electron configurations, these electrons tend to stabilize the sp^3 state of carbon.

A study of the physical properties indicates that as we pass from lanthanum dicarbide to thulium dicarbide the specific electrical resistivity rises and the melting point falls, confirming the increase in the proportion of ionic bonding [1].

According to [8], the evolution of hydrogen during the decomposition of dicarbides of the rare-earth-metal elements by water is proportional to the number of valence electrons per metal atom in the conduction band and characterizes the proportion of metallic bond; this is a maximum in carbides of metals having one electron in the d state, and it increases somewhat with rise in the atomic number of the metal present in the dicarbide.

Thus, the results we have obtained indicate a combination of covalent, metallic, and ionic bonds in the dicarbides of the rare-earth metals, with a preponderance of the covalent bond; however, the proportion of this bond diminishes as we pass from terbium dicarbide to thulium dicarbide as a result of the increase in the relative proportion of metallic and ionic bonding.

The liberation during decomposition, in addition to the above components, of ethane and also propane, propylene, n-butane, isobutane, and butadiene, which were found in the case of the decomposition of praseodymium dicarbide to amount in all to 2.8%, is evidently the result of catalytic action of the lower oxides of the rare-earth elements, the evolution of the gases resulting from reactions which can be represented as follows:

$$MeC_2 + H_2O \rightarrow MeO + C_2H_2,$$
$$MeO + H_2O \rightarrow Me(OH)_3 + H_2,$$
$$C_2H_2 + 2H_2 \rightarrow C_2H_6 \text{ and products.}$$

The data on the composition of the gaseous products of decomposition of the sesquicarbides of the rare-earth elements and the monocarbides of scandium and yttrium (see Table 2) indicate that the acetylene content of the gaseous products decreases in the decomposition of the sesquicarbides as compared with its content in the decomposition products of dicarbides, while the methane content rises sharply. This points to a large proportion of ionic bonding in the compounds concerned, as is confirmed by their lower melting point and higher specific electrical resistivity. However, the positive coefficient of the electrical resistivity indicates a significant proportion of metallic bonding too.

The principal gaseous product of the decomposition of scandium and yttrium monocarbides is methane, indicating that bonding in them is predominantly ionic in character. This conclusion is supported by the results of a study of the physical properties: yttrium monocarbide has an electrical resistivity typical of semiconductor materials, together with a low melting point and hardness.

Thus, with increase in the C/Me ratio in the rare-earth-metal carbides, the character of the chemical bond changes from being predominantly ionic in the MeC phases to being predominantly covalent—metallic in the MeC_2 phases.

Literature Cited

1.   D. Jost, G. Russell, and S. Garner, Rare-Metal Elements and Their Compounds [Russian translation], IL, Moscow (1949).
2.   T. Ya. Kosolapova et al., Zh. Neorg. Khim., 10:2453 (1965).
3.   G. N. Makarenko, Author's Summary of Candidate's Dissertation, IPM, Kiev (1965).
4.   G. V. Samsonov, Poroshkovaya Met., No. 1, 98 (1965).
5.   A. Damiens, Compt. rend., 157:214 (1913).
6.   N. Greenwood and A. J. Osborn, J. Chem. Soc., No. 4, 1775 (1961).
7.   G. J. Polenik and J. G. Waff, J. Inorg. Chem., 1:345 (1962).
8.   T. H. Speddin, R. Gschneidner, and A. H. Daane, J. Am. Chem. Soc., 80:3399 (1958).

A Study of the Reaction of Zirconium Dioxide with the Carbides of Group VI Metals

V. B. Fedorus, T. Ya. Kosolapova, Yu. B. Kuz'ma, and L. N. Kugai

The character of the reaction between zirconium dioxide and the carbides of group VI metals Cr_3C_2, Mo_2C, and WC has been investigated by means of x-ray, chemical, and metallographic analyses. The nature of the intermediate and end products of the reactions has been determined, and the dependence of the phase composition of the reaction products on the sintering temperature has been established. A study has been made of the resistance of the carbides of molybdenum and tungsten and zirconium dioxide to acids and to mixtures of acids with oxidizing and complex-forming agents. A method is proposed for the chemical phase separation of the compounds concerned.

There are few data available in the literature on the reaction of the carbides of transition metals of group VI, Cr_3C_2, Mo_2C, and WC, with refractory oxides. A detailed study has been made only of the system $Al_2O_3 - Mo_2C$ [21], materials based on this system being widely used as tools for cutting metals on account of their great hardness, wear resistance, and thermal conductivity [1, 3, 14]. The carbide−oxide mixture $Al_2O_3 - WC$ is also used for making cutting tools [20].

The kinetics of the reduction of the oxides of the transition metals Ti, Zr, V, Nb, Ta, and Cr by tungsten carbide has been

TABLE 1. Character-
istics of the Initial
Carbides

Carbide	Content, %		
	Me	C_{total}	C_{free}
Cr_3C_2	86.4	13.4	0.2
Mo_2C	94.1	6.00	Not found
WC	93.8	6.2	0.1

studied in [12, 18]. The authors of [15, 16, 19] report the powerful
reducing properties of WC. According to [17], WC reacts with
Fe_2O_4 to form the double carbide $(Fe,W)_{23}C_6$.

In the present research we have investigated the nature of
the reaction between zirconium dioxide and the carbides of Cr, Mo,
and W in vacuo.

The materials used consisted of the group VI metal carbides
Cr_3C_2, Mo_2C, and WC, the characteristics of which are given in
Table 1, and the monoclinic modification of ZrO_2 of analytical
purity. The experiments were carried out over the range 1000–
2000°C in a laboratory vacuum furnace with a graphite heater.

TABLE 2. Results of the Reaction of Chromium Carbide with
Zirconium Dioxide in Vacuo (5×10^{-3} mm Hg)

T, °C	Content, %					Phase composition from the analytical data
	Cr	Zr	C_{comb}^*	Σ, $Cr+$ $+Zr+C$	$O=$ $100-\Sigma$	
Charge	51.5	30.1	7.8		10.6	ZrO_2, Cr_3C_2
1000	51.4	30.2	7.8	89.4	10.6	ZrO_2, Cr_3C_2
1100	51.5	30.1	7.7	89.3	10.7	ZrO_2, Cr_3C_2
1200	50.5	32.9	7.5	89.9	10.1	ZrO_2, Cr_3C_2
1300	49.5	33.7	7.2	90.4	9.6	ZrO_2, Cr_3C_2, Cr_7C_3
1400	49.0	35.2	6.8	91.0	9.0	ZrO_2, Cr_3C_2, Cr_7C_3
1500	48.8	37.4	5.8	92.0	8.0	ZrO_2, Cr_3C_2, Cr_7C_3, ZrC_x ($a_{ZrC_x} = 4.667$Å)
1600	45.0	40.3	7.1	92.4	7.6	ZrO_2, Cr_3C_2, Cr_7C_3, ZrC_x ($a_{ZrC_x} = 4.670$A)
1700	43.8	43.1	9.2	96.1	3.9	Cr_7C_3, ZrC ($a_{ZrC} = 4.675$ Å)
1800	42.5	46.1	11.1	99.7	0.3	Cr_3C_2, ZrC ($a_{ZrC} = 4.687$ Å)

*No free carbon was found at the temperatures used.

To avoid contact reaction with a graphite substrate, the briquets were put on a plate made of a previously sintered charge. The experimental results were assessed on the basis of x-ray, chemical, and metallographic analyses.

The results of the reaction of zirconium dioxide with chromium carbide are given in Table 2. Beginning at 1300°C, the reaction is accompanied by a change over from the zirconium dioxide lattice to the zirconium carbide lattice. Lines from zirconium carbide and Cr_7C_3 are present on the x-ray diffraction diagrams obtained from specimens at 1300°C and higher. At the same time a silvery deposit is observed on the walls and cover of the container; in all probability this consists of metallic chromium which has been formed in the course of the reaction and which, as a result of the high vapor pressure, has sublimated in vacuum at the temperature of the experiment. This conclusion is supported by the results of chemical analyses of the reaction products. The chromium content of the test specimens falls sharply with rise in temperature. According to the results of x-ray analysis, at 1600°C and above the specimens contain only Cr_7C_3 and zirconium carbide. The lattice parameters of the chromium carbide are $a = 14.00$ Å, $c = 4.55$ Å, in close agreement with data in the literature $(a = 14.01$ Å, $c = 4.55$ Å, $c/a = 0.322)$.

The lattice parameter of zirconium carbide increases with rise in temperature, indicating an increase in carbon content; however, it does not reach the value given in tables $(a = 4.695$ Å) within the temperature range investigated. This may be due to the solubility of chromium carbide in the ZrC_x lattice. According to [10], the solubility of chromium carbides at 1300°C is 6 at.% Cr, which leads to some contraction of the zirconium carbide lattice.

In addition, the reaction probably takes place through the formation of the lower chromium carbide $Cr_{23}C_6$, although lines from this carbide are absent from the x-ray diffraction diagrams of specimens obtained over the whole temperature range, owing to the metastability of the lower chromium carbide at the temperatures concerned.

The reaction of ZrO_2 with the carbides of molybdenum and tungsten was investigated over the range 1000-2000°C.

First a study was made of the resistance of these three compounds to a mixture of acids with oxidizing agents, a mixture of

TABLE 3. The Results of the Reaction of Mo_2C with ZrO_2 in Vacuo (5×10^{-3} mm Hg)

T, °C	General chemical analysis, %					Content of insoluble residue, %†			Formula of insoluble residue	Soluble portion, %	Content of soluble portion, %			Phase composition based on the results of x-ray structural analysis
	Zr	Mo	C_{comb}	$\Sigma(Zr+Mo+C)$	O*	H/O	Zr	O*			Zr	Mo	C	
Charge	17.1	72.3	4.5	93.9	6.1	23.2								
1000	17.0	72.0	5.0	94.0	6.0	23.1	73.9	26.1	ZrO_2	76.9	Not found	94.0	6.0	ZrO_2, Mo_2C
1100	17.0	72.0	4.9	93.9	6.1	23.2	73.8	26.2	ZrO_2	76.9	»	93.6	6.4	ZrO_2, Mo_2C
1200	17.2	72.3	4.6	94.1	5.9	22.9	74.5	25.5	$ZrO_{1.95}$	77.1	» »	93.8	5.98	ZrO_2, Mo_2C
1300	17.3	72.1	4.6	94.0	6.0	21.1	75.3	24.7	$ZrO_{1.92}$	78.9	1.77	91.4	5.84	ZrO_2, Mo_2C
1400	17.8	72.9	4.3	95.0	5.0	20.2	74.4	25.6	$ZrO_{1.96}$	79.8	3.5	91.4	5.4	ZrO_2, Mo_2C
1500	18.1	73.0	4.2	95.3	4.7	17.3	74.6	25.4	$ZrO_{1.94}$	82.7	6.4	88.3	5.1	ZrO_2 (traces), Mo_2C, Mo. ZrC (a_{ZrC} = 4670Å)
1600	18.3	74.1	3.6	96.0	4.0	14.4	75.9	24.1	$ZrO_{1.82}$	85.5	8.54	86.7	4.21	ZrO_2 (traces), Mo_2C, Mo, ZrC (a = 4.672 Å)
1700	18.0	74.0	3.3	95.3	4.7	11.3	72.5	27.5	$ZrO_{2.15}$	88.7	11.05	83.4	3.72	Mo_2C (traces), Mo, ZrC
1800	18.7	75.9	2.5	97.1	2.9	7.6	75.0	25.0	$ZrO_{1.9}$	92.4	14.1	82.1	2.7	Mo, ZrC, a_{ZrC} = 4.676Å
1900	18.9	76.1	1.8	96.8	3.2	7.1	—	—	—	92.9	15.0	82.0	1.94	Mo, ZrC, a_{ZrC} = 4.668 Å
2000	18.9	76.9	1.8	97.6	2.4	3.6	72.5	27.5	$ZrO_{2.15}$	96.4	16.9	79.8	1.87	Mo, ZrC, a_{ZrC} = 4.667 Å

* The oxygen content was calculated from the difference 100% − Σ(Zr + Mo + C).

† Molybdenum and carbon were not found in the insoluble residue.

acids with oxidizing and complex-forming agents, and a mixture
of complex-forming and oxidizing agents. It was found that both
molybdenum carbide and tungsten carbide dissolves completely in
a 30% solution of hydrogen peroxide, in a mixture of sulfuric acid
(1:3) and hydrogen peroxide, and in a mixture of oxalic acid and
hydrogen peroxide on heating for 1 h. Zirconium dioxide has a
high chemical resistance and is practically insoluble under the
above conditions. The possibility of separating zirconium dioxide
from the carbides of molybdenum and tungsten was verified on artificial mixtures. The best results were obtained by using as sol-
vent a mixture of sulfuric acid (1:3) and a 30% solution of hydrogen
peroxide [4, 6]. The method is as follows. A weighed sample of
the finely ground powder (40 μ) is heated for 1 h in a mixture of
50 ml of sulfuric acid (1:3) and 50 ml of a 30% solution of hydrogen peroxide. During dissolution 2-3 additions of 10 ml of hydrogen peroxide are made at intervals. The solution is cooled, the
precipitate is allowed to settle completely, and the clear solution
filtered through a Schott No. 4 crucible. The precipitate is again
covered with the mixture and heated for 30-40 min, filtered through
the same crucible, washed first with dilute sulfuric acid and then
with water, dried, and weighed. The repeated treatment results in
complete separation of the components.

The contents of zirconium, carbon, and molybdenum or tungsten are determined in the residue. To determine the zirconium
content, the residue is filtered off through an ashless filter, washed,
ignited in a platinum crucible, and fused with potassium pyrosul-
fate. The fused mass is leached with a 2 N solution of sulfuric
acid, and the zirconium content is determined by complexonometric
titration, using xylyl orange as indicator. The zirconium and molybdenum or tungsten contents are determined in the solution.

Analysis of the insoluble residues from the systems ZrO_2–
Mo_2C and ZrO_2–WC showed that they contain no carbon, molybdenum, or tungsten, but that the zirconium content corresponds to
the stoichiometric content for ZrO_2. At 1500°C and higher temperatures the amount of insoluble residue (ZrO_2) diminishes, while
the zirconium content of the solution increases.

According to the results of chemical analysis, reaction in the
ZrO_2–Mo_2C system begins at 1300°C, but the amount of reaction
products is small and is not detected by x-ray methods (Table 3).
Only at 1500°C does x-ray structural analysis of the specimens

reveal lines from two new phases — ZrC and Mo — which become stronger with rise in temperature, and at 1900-2000°C the x-ray diffraction diagrams contain only lines from these phases, the microhardness values for which are 2350-2400 and 350 kg/mm^2, respectively. The second phase revealed by x-ray analysis is ZrC deficient in carbon; its lattice parameter a increases with rise in temperature from 4.670 Å (at 1500°C to 4.676 Å (at 1800°C) as a result of the increase in complexity of the metalloid sublattice ZrC_x, and, according to the parameter, at 1800°C it corresponds to the formula $ZrC_{0.6}$.

The reduction in the lattice parameter of the ZrC_x phase to 4.668-4.667 Å that occurs with further rise in temperature is probably due to the solubility of molybdenum in zirconium carbide at these temperatures. According to [2, 10, 11], the solubility of molybdenum in ZrC_x is considerable: this is due to the reasonable closeness of the atomic radii of zirconium (r_{Zr} = 1.60 Å) and molybdenum (r_{Mo} = 1.39 Å) and of the chemical factor (the location of both metals in the same period of the periodic system). The ZrC_x lattice contracts when molybdenum dissolves in it, and its parameter [2] corresponds to that of a solid solution (Zr,Mo)C containing 1.2 at.% Mo. Samsonov et al. [8] have also reported the reaction of ZrC with molybdenum in the given temperature range with the formation of a phase having a microhardness (H_μ = 2599 kg/mm^2) similar to that of the (Zr,Mo)C phase which we have found. In all probability the reaction in the system $ZrO_2 - Mo_2C$ proceeds thus:

$$ZrO_2 + Mo_2C \rightarrow Mo + ZrC + CO$$

followed by partial solution of the reaction products.

Investigation of the reaction in the system $ZrO_2 - WC$ showed that the reaction takes place to a significant extent at 1500°C. Below this temperature no reaction is observed, and the reaction products consist, according to the results of chemical and x-ray structural phase analyses, of the original components (Table 4). Raising the temperature leads to rapid reaction, and at 1800°C all the carbon in the tungsten carbide is used up in reducing the zirconium dioxide. According to the results of x-ray structural analysis, specimens produced at 2000°C consist of two phases: tungsten and a solid solution based on ZrC (a = 4.648 Å).

TABLE 4. The Results of the Reaction of WC with ZrO_2 in Vacuo $(5 \times 10^{-3}$ mm Hg)

T, °C	General chemical analysis, %				Content of insoluble residue, %†			Formula of insoluble residue	Soluble portion, %	Content of soluble portion, %				Phase composition based on the results of x-ray structural analysis
	Zr	W	C_{comb}	O^*	C	Zr	O^*			Zr	W	C	O^*	
Charge	17.72	71.43	4.64	6.21										
1000	17.8	71.6	4.7	5.9	23.8	74.4	25.6	$ZrO_{1.96}$	76.2	Not found	93.9	6.16	—	ZrO_2, WC
1100	17.7	71.7	4.8	5.6	23.6	74.2	25.8	$ZrO_{1.99}$	76.4	" "	93.8	6.28	—	ZrO_2, WC
1200	17.3	71.7	4.7	6.3	23.3	73.4	26.6	$ZrO_{2.1}$	76.7	"	93.4	6.13	0.47	ZrO_2, WC
1300	17.5	71.8	4.7	6.0	22.2	77.4	22.6	$ZrO_{1.87}$	77.8	"	92.3	6.04	1.7	ZrO_2, WC
1400	17.7	72.1	4.68	5.5	21.6	75.9	24.1	$ZrO_{1.82}$	78.4	0.64	92.0	5.97	1.4	ZrO_2, WC
1500	17.8	75.8	3.5	2.9	15.1	74.2	25.8	$ZrO_{1.99}$	84.9	7.0	89.3	4.10	—	W, ZrC(traces), W_2C (traces), ZrO_2
1600	17.8	77.0	2.8	2.9	7.8	74.3	25.7	$ZrO_{1.97}$	92.2	13.0	83.5	2.5	1.0	W, ZrC, $a_{ZrC} = 4.670$ Å, ZrO_2
1700	17.9	77.7	2.2	2.2	6.75	74.0	26.0	ZrO_2	93.25	13.1	83.3	2.4	1.2	W, ZrC, $a_{ZrC} = 4.664$ Å
1800	18.1	78.1	1.9	1.9	5.4	74.0	26.0	ZrO_2	94.6	13.9	82.6	2.0	1.5	W, ZrC, $a_{ZrC} = 4.661$ Å
1900	18.3	78.8	1.6	1.3	4.4	75.0	25.0	$ZrO_{1.9}$	95.6	15.7	82.4	1.7	0.2	W, ZrC, $a_{ZrC} = 4.656$ Å
2000	18.4	79.0	1.5	1.1	2.4	—	—		97.6	16.4	80.9	1.54	1.1	W, ZrC, $a_{ZrC} = 4.648$ Å

*Oxygen content by difference.
†Tungsten and carbon were not found in the insoluble residue.

A certain reduction observed in the lattice parameter of ZrC is due to solution of tungsten. According to [7], this parameter corresponds to a (Zr,W)C solution containing 3.8 at.% W.

The microhardness of the (Zr,W)C solid solution is 2680-2700 kg/mm^2, and that of tungsten 550-600 kg/mm^2.

It is interesting to compare the present results with those which we have already published [4, 5] on the reaction of ZrO$_2$ with carbides of the metals of groups IV and V. The products of the reaction of ZrO$_2$ with the carbides of titanium and zirconium in the range 1000-2000°C are complex oxycarbide phases. In systems involving group V metal carbides the reaction leads to the formation of a phase based on ZrC and metal is precipitated, i.e., the reaction proceeds in a way similar to that of the reaction between ZrO$_2$ and carbides of group VI metals, although the temperature at which the reaction begins is higher.

It can be seen that in the reaction of carbides of metals belonging to groups IV to VI with ZrO$_2$ complex oxidation−reduction processes occur which, in the temperature range investigated, can be represented in the following way if certain assumptions are made:

$$ZrO_2 + Me^{IV}C \rightarrow (Zr_y Me_{1-y}^{IV})(C_x O_{1-x}) + CO,$$
$$ZrO_2 + Me^{V}C \rightarrow Me^{V} + ZrC_x + CO,$$
$$ZrO_2 + Me^{VI}C \rightarrow Me^{VI} + ZrC_x + CO.$$

Thus, with rise in temperature the metal carbides are gradually impoverished in carbon as a result of the reaction with ZrO$_2$ while the latter is reduced to ZrC$_x$ ($x \rightarrow 1$). The impoverishment of the carbide reducing agents in carbon is accompanied by a decrease in the lattice parameter with temperature, by the formation of the lower carbides (V$_2$C, Cr$_7$C$_3$, W$_2$C), and subsequently by the appearance of the metals (V, Nb, Cr, Mo, W). The fall in the temperature at which the reaction begins in the systems MeC − ZrO$_2$ as we pass from ZrO$_2$−Me^{IV}C to ZrO$_2$−Me^{V}C and ZrO$_2$−Me^{VI}C is due to a weakening of the Me − C bond in the carbides and to a reduction in the thermodynamic stability of the carbides in the order Me^{IV}C → Me^{V}C → Me^{VI}C [9].

Literature Cited

1.   F. Eisenkolb, Powder Metallurgy [Russian translation], Metallurgizdat, Moscow (1959).

2. Investigations in the Field of the Chemistry of Silicates [in Russian], Nauka, Moscow and Leningrad (1965), p. 220.

3. Cermets [Russian translation], IL, Moscow (1962).

4. T. Ya. Kosolapova, V. B. Fedorus, Yu. B. Kuz'ma, and E. E. Kotlyar, Izv. Akad. Nauk SSSR, Neorg. Mat., 2:1521 (1966).

5. T. Ya. Kosolapova, V. B. Fedorus, and Yu. B. Kuz'ma, Izv. Akad. Nauk SSSR, Neorg. Mat., 2:1516 (1966).

6. E. E. Kotylar and T. N. Nazarchuk, Izv. Akad. Nauk SSSR, Neorg. Mat., 2:1782 (1966).

7. Yu. B. Kuz'ma, T. F. Fedorov, and E. A. Shvets, Poroshkovaya Met., No. 2, 22 (1965).

8. G. V. Samsonov, L. V. Strashinskaya, and E. A. Shiller, Izv. Akad. Nauk SSSR, Otd. Tekh. Nauk, Met. i Toplivo, No. 5, 167 (1962).

9. G. V. Samsonov, Ukrainsk. Khim. Zh., 31:1233 (1965).

10. T. F. Fedorov and Yu. B. Kuz'ma, Poroshkovaya Met., No. 3, 75 (1965).

11. T. F. Fedorov, Yu. B. Kuz'ma, and L. V. Gorshkova, Poroshkovaya Met., No. 3, 69 (1965).

12. G. P. Shveikin, Poroshkovaya Met., No. 6, 70 (1962).

13. A. M. Yakimova, in: The Structure and Properties of High-Temperature Metallic Materials [in Russian], Nauka, Moscow (1967), p. 54.

14. C. Agte, R. Kohlermann, and E. Heymel, Schneidkeramik, Academie Verlag, Berlin (1959), p. 46.

15. Freundlich and Josein, Bull. Soc. Chim. France, 557 (1957).

16. Freundlich, Josien, and Erb, Bull. Soc. Chim. France, 281 (1960).

17. Josien and Renaud, Rev. Chim. Min., 1, 415 (1964).

18. Josien and Renaud, Compt. rend., 260:2239 (1965).

19. Josien, Bull. Soc. Chim. France, 480 (1956).

20. F. Kolbl, Maschinenwelt u. Elektrotechn., 13:7 (1958).

21. W. Seith and H. Schmecken, Heraeus Festschrift, Hanau (1950), p. 218.

The Reaction of Refractory Carbides with Molten Steels and Cast Irons

G. V. Samsonov, G. K. Kozina, A. D. Panasyuk, and S. N. Bondarchuk

A study has been made of the reaction between the refractory carbides of metals of groups IV to VI and molten carbon and alloy steels and cast irons containing various forms of graphite. The reaction was determined from the nature and degree of wetting of carbide plates by the molten alloys. The sessile-drop method was used in the experiments. It was found that the carbides of group-IV metals are the most resistant to molten steels and cast irons, the resistance decreasing on passing to the carbides of metals of groups V and VI. The contact angle of wetting of the carbides decreases with rise in the carbon content of commercially pure alloys, beginning at a carbon content of 0.8%.

Although refractory materials are finding ever-increasing application in new fields of science and technology, cast iron and steel remain the most important industrial alloys. Hence a problem of primary importance is their compatibility with various working media (ceramics, melts, and metal vapors). It is particularly important to know the nature and rate of their physicochemical reaction with the materials of ingot molds, crucibles, pump components, pipelines, and the protective casings of measuring and regulating apparatus.

As oxygen-free refractory compounds have been finding increasingly wide application in recent years as high-temperature materials, it is of interest to investigate their resistance to molten media and in particular to molten steels and cast irons.

433

TABLE 1. Carbon Content and Porosity of Carbides Investigated

Carbide	C_{total}, %	C_{free}, %	Porosity, %
TiC	19.1	Traces	4—8
ZrC	11.2	0,5	3—8
HfC	6.23	Not found	3—6
NbC	10.8	" "	12—16
TaC	6.3	" "	25
Cr_3C_2	13.3	" "	3—8
Mo_2C	5.3	" "	9—12
WC	6.1	" "	3—8

TABLE 2. Chemical Composition of the Steels and Cast Irons Used for Wetting the Carbides

Type of alloy	C	Mn	Si	Cr	S	P
Steel 20	0.17	0.42	0.06	—	0,03	0,03
Steel 45	0.48	0.56	0.21	—	0,03	0.03
U8	0.75	0.20	0.16	—	0.025	0.03
U10	0,98	0,23	0.15	—	0.023	0.03
U12	1.20	0.21	0.18	—	0.024	0.03
ShKh15	1,02	0.4	0.3	1.4	—	—
ShKh15SG	0,95	1.1	0.6	1.35	—	—
White iron	2.56	0.98	1.0	0.27	0.02	0.03
KCh30-6	3.1	0.44	1.55	—	0.018	—
CKh50-4	2.65	0.42	1,24	0.05	0.16	0.05
VCh50-1.5	4,0	1.4	2.5		0,1	
DI.-1*	0.15	0.5	0.5	14.5	0.025	0.032

* Ni — 2.7%; W — 0.3%; Mo — 0.3%.

TABLE 3. Contact Angles of Wetting of Carbides of Metals of Groups IV to VI by Steels and Cast Irons *

Carbide	Steel U8	White iron	KCh-30-6	KCh-50-4	VCh50-1.5	ShKh15	ShKs 15SG	DI-1
TiC	90	82	98	108	118	102	100	58
ZrC	131	132	100	122	125	122	145	102
HfC	148	140	122	145	132	132	145	128

*The carbides NbC, TaC, Cr_3C_2, Mo_2C, and WC are completely wetted ($\theta = 0°$).

There is scarcely any information in the literature on the reaction of carbides with steels and cast irons, only a few scraps of information being given in isolated articles [5, 9, 12]. It has been shown in [5, 12] that TiC is destroyed in molten steels and cast irons, while it is reported in [9] that SiC reacts well with steel, the reaction increasing with rise in the carbon content of the steel.

In the present research we have investigated the reaction of the refractory carbides of metals belonging to groups IV to VI of the periodic system with the molten steels 20, 45, U8, U12, Kh15, ShKh15SG, and DI-1 and with the molten ions: white iron, wrought ferritic KCh 30-6, wrought pearlitic KCh 50-4, and high-strength magnesium cast iron VCh-50-1.5.

As the majority of the carbides investigated consist of a phase of variable composition having a wide homogeneity range, carbides with a composition close to the stoichiometric one were chosen for investigation.

The test specimens took the form of disks 8-14 mm in diameter and 3-4 mm high. Specimens of the carbides TiC, ZrC, HfC, Cr_3C_2, Mo_2C, and WC were prepared by hot compaction in graphite dies, in which the processes of compaction and sintering went on simultaneously. Specimens of NbC and TaC were prepared by the compaction of samples in metallic dies, followed by sintering. The amount of carbon in the initial mix and the porosity of the sintered specimens are given in Table 1. The chemical composition of the steels and cast irons used in the experiments is given in Table 2.

The reaction was determined from the nature and extent of the wetting of the carbide specimens by the molten steels and cast irons. The sessile-drop method was used in the experiments [2, 20], the contact angles being measured by means of an instrument similar to that described in [4] in an atmosphere of purified argon at the melting point and on subsequent superheating by 100-150 deg C. To reduce the effect of surface relief, all the carbide specimens were subjected to the same treatment before the tests, viz. polishing on a series of progressively finer grades of emery paper, followed by careful degreasing. The metal specimens were also carefully degreased. Table 3 gives the contact angles of wetting of the carbides by the various kinds of steel and cast iron.

In this research we have investigated the effect of certain impurities and alloying additions on the contact angle, in particular,

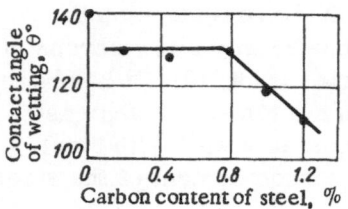

Fig. 1. The effect of carbon content on
the contact angle of wetting of ZrC by
the steels: steel 20, steel 45, U8, U10,
and U12. Holding time = 15 min.

additions of sulfur and carbon. To study the effect of sulfur, specimens of steel 45 with the following sulfur contents were prepared:
0.036, 0.072, 0.11, and 0.132 wt.%. To study the effect of carbon,
we chose specimens of the commercial alloys whose compositions
are given in Table 2. The results obtained are presented in Figs.
1 and 2. The effect of porosity of the carbides was investigated
for the case of ZrC wetted by steel U8 with the following results:

Porosity, %	4	8	10	16	21	30	34
Contact angle of wetting, $\theta°$	125	123	126	125	128	123	125

We also studied the effect on the reaction between the carbides and the molten steels of such important practical factors
as the temperature and holding time (Fig. 3). The experimental
results show that the carbides of group IV metals are the most resistant to liquid steels and cast irons, and this was confirmed by
the existence of a strong bond between the metal and carbon atoms
in these carbides.

It has previously been shown that [10] the wetting of refractory compounds by molten metals is a function of the electron exchange between the compounds and the metal. From this it may be
assumed that the reaction between carbides and molten simple sub-

Fig. 2. The effect of sulfur content on
the contact angle of wetting of ZrC by
steel 45 (holding time = 15 min).

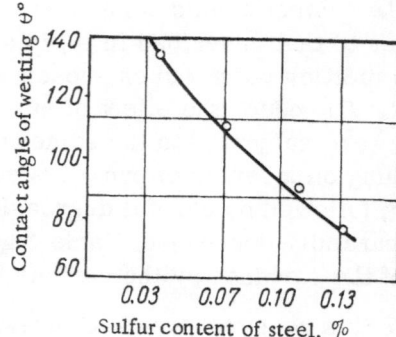

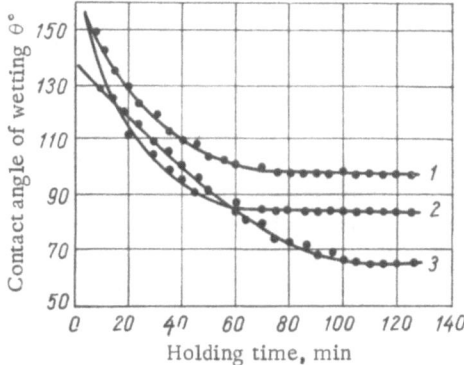

Fig. 3. The effect of holding time on the contact angle of wetting of group IV metal carbides by magnesium cast iron VCh50-1.5 at the melting point: 1) HfC, 2) ZrC, 3) TiC.

stances is determined by the donor—acceptor properties of the partners.

Carbides of group IV metals are the most stable in their properties because the metal atoms in them act as donors, giving up their unlocalized electrons to the carbon atoms. This conclusion emerges from work on the determination of the charges on titanium and carbon atoms in TiC [1], and also from the determination of the charge on carbon atoms in solid solution in titanium and other transition metals by the electrical-transfer method [6]. On the formation of carbides of transition metals of group IV, the metal atoms surrender their unlocalized electrons to the atoms of carbon, thereby stabilizing its sp^3 electron configurations [11]. Consequently, the proportion of unlocalized electrons taking part in the bonds with the wetting metal is small, and the contact angles of wetting of the group IV metal carbides by steels and cast irons are large. It also emerges from [6] that the metal atoms in carbides of metals belonging to groups V and VI act as acceptors and destroy the sp^3 electron configurations of carbon. The latter attempt to reestablish themselves by reaction with the wetting metal, and so the contact angles of wetting of carbides of metals of groups V and VI by steels and cast irons are virtually zero.

Information on the effect of carbon, sulfur, and alloying elements on the surface tension of pure iron and its alloys is to be found in the literature. It has been established [8, 15-17, 19] that the surface tension of an alloy decreases with rise in its carbon content, the reduction being more marked in commercially pure alloys than in pure iron—carbon alloys. And as the contact angle of wetting is directly related to the surface tension [$w_A = \sigma_{liq} (1 + \cos \Theta)$], it is natural that the angle should also decrease with the reduction in surface tension, and this is in fact observed in the references cited above.

Pitak and P'yanykh [9] have determined the direct effect of carbon in steel on the contact angle of wetting of SiC by the steel, and have found that the angle decreases with rise in carbon content of the steel. The present research has confirmed that the contact angle of wetting of carbides, in particular ZrC, by commercially pure alloys decreases with rise in carbon content of the alloys. Up to 0.8% C (corresponding to pure eutectoid steel) this effect is absent or is so small that it lies within the limits of experimental error; beginning at 0.8% C the contact angle of wetting of ZrC by steel decreases, possibly as a result of the formation of carbide solid solutions.

It has been shown in [17, 3] that the more rapid fall in the surface tension of commercial alloys as compared with pure iron−carbon alloys is due to the presence in them of a higher sulfur content and to the fact that adhesion between the melt and the solid phase increases with rise in sulfur content. Our experiments have confirmed that sulfur is a surface-active element and that it sharply reduces the contact angle of wetting of carbides, in particular ZrC, by steel. This effect of sulfur is evidently connected with the high acceptor capacity of its atoms, the configuration of which (s^2p^4) attempts to form the stable s^2p^6 configuration by attracting unlocalized electrons from the wetted carbide.

The effect of Cr, Mn, Si, Ni, Ti, V and other transition metals on the surface properties of pure iron and its commercial alloys has been studied in [7, 13, 14, 18, 21, 8], and it has been shown that alloying with small amounts of these elements has practically no effect on the surface properties. This has been confirmed in the present work in the case of the wetting of carbides by the steels ShKh15 and ShKh15SG, and it can be explained by the completion of unfilled d shells of the atoms of these metals at the expense of the valence electrons of the iron in the steel (the iron atoms, which have d^6s^2 configurations, acquire stable d^5 configurations by the transfer of ds^2 electrons from the atoms of the transition metals introduced into the steel).

Regarding the wetting of carbides by cast irons containing graphite in various forms it should be noted that the most inert toward the carbides is the high-strength, magnesium-modified iron containing spheroidized graphite. This is due to the fact that in this type of iron the atoms at the surface of the globules of graphite acquire sp^3 configurations at the expenses of the acceptance by the atoms of ordinary graphite, which have sp^2 configurations, of valence electrons from the magnesium atoms. Electron exchange with these globules is suppressed, leading to large contact angles of wetting of carbides by magnesium cast iron. Data on the effect of porosity on the contact angle of wetting have been given above. As the scatter of the results lies within the limits of experimental error, it may be concluded that porosity in the carbides has no effect on the contact angle of wetting of the carbides by steels, at any rate when $\Theta > 90°$.

The nature of the effect on the contact angle of such an important practical factor as temperature can hardly be determined on the basis of a theoretical analysis, since three values of the surface tension enter into the equation for $\cos \Theta = (\sigma_{sol} - \sigma_{liq})/\sigma_{liq}$, and each of them varies in its own way with temperature, while data for σ_{sol} and σ_{liq} are practically nonexistent. Obviously, a study of the temperature dependence, as well as the time dependence, is necessary in each particular case.

Thus, by comparing the experimental results we have obtained for the reaction between refractory-metal carbides and molten steels and cast irons with their electron structure, we can account for the laws governing the contact reaction between the carbides and the molten metals and can also explain the effect of alloying additions on the process concerned.

Literature Cited

1. M. P. Arbuzov and B. V. Khaenko, Poroshkovaya Met., No. 4, 74 (1966).
2. A. I. Belyaev, Physicochemical Processes in the Electrolysis of Aluminum [in Russian], Metallurgizdat, Moscow (1947).
3. Wang Chi-T'ang, R. A. Karasev, and A. M. Samarin, Izv. Akad. Nauk SSSR, Otd. Tekh. Nauk, Met. i Toplivo, No. 2, 49 (1960).
4. V. N. Eremenko and Yu. V. Naidich, The Wetting of the Surface of Refractory Plates by Rare Metals [in Ukrainian], Izd. Akad. Nauk UkrSSR, Kiev (1958).

5. M. S. Koval'chenko and N. N. Sereda, Poroshkovaya Met., No. 1, 17 (1968).
6. I. I. Kovenskii, Fiz. Tverd. Tela, 5:1423 (1963).
7. T. T. Kolesnikova and A. M. Samarin, Izv. Akad. Nauk SSSR, Otd. Tekh. Nauk, No. 5, 63 (1956).
8. A. M. Levin, Electrometallurgy [in Russian], Nauchnye Trudy DMETI, No. 27, 105 (1952).
9. N. V. Pitak and N. L. P'yanykh, Ogneupory, No. 5, 31 (1965).
10. G. V. Samsonov, in: Surface Phenomena in Melts and in Powder-Metallurgy Processes [in Russian], Izd. Akad. Nauk UkrSSR, Kiev (1960), p. 90.
11. G. V. Samsonov, Poroshkovaya Met., No. 12, 49 (1966).
12. G. V. Samsonov, G. A. Yasinskaya, and T'ai Shou-Wei, Ogneupory, No. 1, 35 (1960).
13. L. A. Smirnov, S. I. Popel, and A. M. Pastukhov, Izv. VUZ. Chernaya Met., No. 4, 13 (1965).
14. L. A. Smirnov, S. I. Popel, and B. V. Tsarevskii, Izv. VUZ. Chernaya Met., No. 3, 10 (1965).
15. B. V. Stark and S. I. Filippov, Izv. Akad. Nauk SSSR, Otd. Tekh. Nauk, No. 3, 413 (1949).
16. S. I. Filippov, Theory of the Process of Decarburization of Steel [in Russian], Metallurgizdat, Moscow (1956).
17. B. V. Tsarevskii and S. I. Popel, Izv. VUZ. Chernaya Met., No. 8, 15 (1960).
18. B. V. Tsarevskii and S. I. Popel, Izv. VUZ. Chernaya Met., No. 12, 12 (1960).
19. F. A. Halden and W. D. Kingery, J. Phys. Chem., 59:577 (1955).
20. W. D. Kingery and M. Humenik, J. Phys. Chem., 57:350 (1953).
21. P. Kozakevitch and G. Urbain, J. Iron Steel Inst., 186:467 (1957).

A Study of the Corrosion of Nonmetallic Materials in Molten Borax

M. B. Gutman, L. A. Mikhailov, V. G. Kaufman, T. V. Dubovik, and V. K. Kazakov

Experiments were carried out in molten borax with and without electrolysis. Corrosion relationships were calculated from the results obtained for the groups of materials tested. Oxide-base materials were
found to have a resistance considerably lower than that of materials
based on the nitrides of boron and silicon. It was found that in each
group materials containing silicon have the least resistance.

In order to increase the service life of metal components
that are in frictional contact, their surface layer is subjected to a
thermochemical treatment, one of the most promising of such
treatments being boronizing. Of the methods of boronizing at
present known, the most efficient is an electrolytic method carried
out in molten borax $Na_2B_4O_7$, or a mixture of borax with other salts
at a temperature of the order of 950°C. The boronized component
acts as the cathode and a graphite electrode, the anode. The process takes place in an electric bath with a crucible made of cast
Ni–Cr steel. In its molten state borax reacts strongly with the
crucible materials and with oxides on its surface. The service
life of the crucible is short, depending to a large extent on the
quality of the crucible casting. A more reliable apparatus for
electrolytic boronizing is one with a crucible made of a nonmetallic

material that is resistant to molten borax. The life of such an apparatus is determined only by the rate of attack of the crucible material by borax.

To construct such an apparatus, it is essential to discover a nonmetallic material which is capable of resisting molten borax over a prolonged period (several thousand hours). In the electrolysis of borax, reduction of sodium takes place at the cathode:

$$Na^+ + e = Na^0.$$

Since the temperature at which the process is carried out is higher than the boiling point of sodium (890°C), the sodium evaporates and burns away. Several reactions occur at the anode

$$B_4O_7^{2-} \rightarrow 2B_2O_3 + \frac{1}{2} O_2 + 2e.$$

The oxygen evolved reacts with the graphite of the anode to form CO_2:

$$O_2 + C = CO_2.$$

The B_2O_3 formed at the anode may react with the graphite anode:

$$2B_2O_3 + 4C = B_4C + 3CO_2.$$

Part of the sodium formed at the cathode may react with the $B_4O_7^{2-}$ ion:

$$B_4O_7^{2-} + 12Na \rightarrow 4B + 6Na_2O + O^{2-}.$$

The elementary boron formed impregnates the surface layer of the component being treated. In the general case, attack on the materials by molten borax may depend on a variety of processes, but it is determined to the largest extent by the nature of the chemical reaction of the material with borax.

All the materials based on metal oxides react with molten borax to form a salt of metaboric acid:

$$Na_2B_4O_7 + FeO = 2NaBO_2 + Fe(BO_2)_2;$$
$$3Na_2B_4O_7 + Al_2O_3 + 6NaBO_2 + 2Al(BO_2)_3.$$

A complex aluminoborate compound is formed in the attack on aluminum oxide (Al_2O_3). In the attack on silicon (SiO_2), B_2O_3 may be displaced from the salt and a complex borosilicosodium glass may be formed. If there is more silicon oxide than boron oxide, a silicon-based glass is formed. From the point of view of resistance to attack by molten borax, the most outstanding materials are those based on nitrides and carbides. Such materials have been developed at the Institute for Problems in Materials Science of the Academy of Sciences of the Ukrainian SSR, where experiments have been made to determine the rate of attack of specimens of various materials in molten borax. The process of electrolytic boronizing takes place for part of the time under conditions of salt electrolysis and for part of the time without electrolysis. Consequently the experimental investigation of the corrosion of nonmetallic materials in molten borax was conducted both under conditions of electrolysis and in their absence.

The apparatus used consisted of a crucible electric furnace. To carry out electrolysis in the crucible the anode and cathode were suspended on brackets, between which the test specimens were arranged on supports. Specimens were tested for corrosion in molten borax without electrolysis in the same apparatus, but without the electrodes.

Each specimen was weighed on an analytical balance to an accuracy of 0.01 g, and its geometrical dimensions were measured to an accuracy of 0.1 mm. Several specimens of the same composition were tested at the same time in order to eliminate the effect of elements contained in a single specimen. The specimens prepared for testing were suspended over an open pool of molten borax and heated in order to avoid the possibility of thermal shock. Then the support with the specimens attached to it was slowly lowered, and the specimens were immersed in the crucible containing the molten borax. At predetermined intervals the specimens were removed one by one from the melt, cooled, and the borax remaining on them washed off with hot water. Then the specimens were weighed on the analytical balance and their geometrical dimensions measured. After examination, the specimens were photographed and then took no part in further tests. The borax in the crucible was replaced at the end of the tests on each group of specimens. We chose as the principal criterion determining the corrosion of the materials in molten borax the change in thickness of the

vertical sides of the specimens. We did this because the crucible
of the apparatus also has extended vertical walls which are attacked
too by the borax. The attack on the vertical surfaces of the specimens was determined from the formula

$$\delta = \frac{\delta_i \pm \delta_f}{2} ,$$

where δ_i is the initial thickness of the specimen (mm), δ_f the final
thickness of the specimen (mm), and 2 is a factor to take account
of the fact that in the tests corrosion takes place from two sides of
the specimen.

Then graphical relationships $\delta = f(\tau)$ were constructed, and
a calculated relationship of the form $\delta = k\tau$ was determined, where
k is an empirical coefficient and τ is the time of operation of the
apparatus, (h).

All the materials tested can be divided into the following
main groups:

Oxide-base materials, including those based on the oxides of
   aluminum, silicon, cesium, zirconium, and magnesium;
Various grades of graphites;

Materials based on the nitrides of boron and silicon developed
   in the Institute for Problems in Materials Science of the
   Academy of Sciences of the Ukrainian SSR.

From the experimental results obtained, calculated corrosion
relationships $[\delta = f(\tau)]$ (Fig. 1) were determined for the groups of
materials tested, on the following basis:

SiO$_2$	Cs$_2$O	MgO	ZrO$_2$	Al$_2$O$_3$
0.169τ	0.07τ	0.064τ	0.06τ	0.057τ
graphite	Si$_3$N$_4$ — SiC	Si$_3$N$_4$	Si$_3$N$_4$ — BN	BNC
0.0036τ	0.024τ	0.022τ	0.017τ	0.005τ

The longest duration of the tests was 1000 h. As the results
show, the resistance of oxide-base materials is considerably lower
than that of materials based on the nitrides of boron and silicon.
The lowest resistance in each group related to those materials which
contain silicon. In the group of oxide-base materials, those based
on silica have the least resistance, while in the group based on the
nitrides of boron and silicon the resistance of the materials increases as the silicon content falls. Least resistant are the ma-

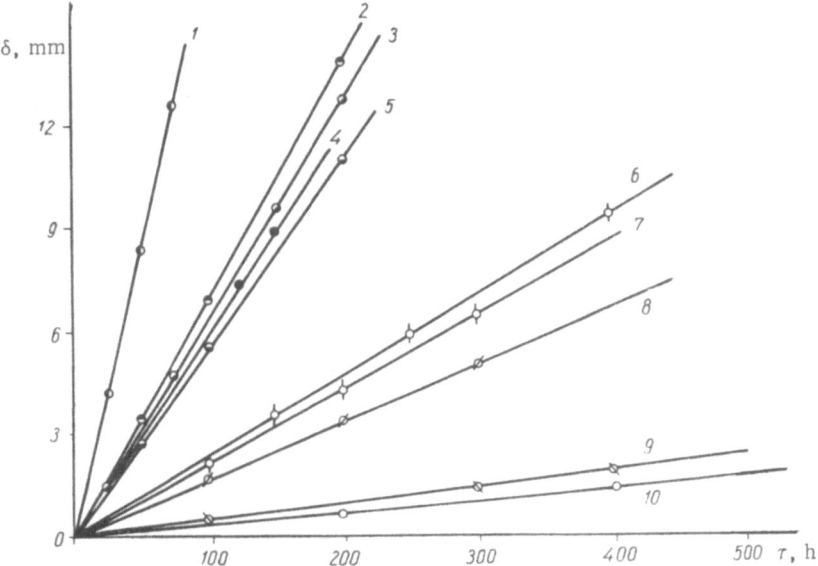

Fig. 1. The corrosion of various materials in molten borax: 1) SiO_2; 2) Cs_2O; 3) MgO; 4) ZrO_2; 5) Al_2O_3; 6) Si_3N_4–SiC; 7) Si_3N_4; 8) Si_3N_4–BN; 9) BNC; 10) graphite.

terials based on Si_3N_4–SiC, then those based on Si_3N_4 and Si_3N_4–BN. One reason for this may be the tendency of silicon to form glass with boron. Boron carbonitride (BNC) and graphite have the highest resistance. However, graphite cannot be used for the crucible of the apparatus owing to the technological conditions of the process. Oxide-base materials have been found to have the same rate of attack by molten borax both with and without electrolysis. On the other hand, the rate of attack of materials based on the nitrides of boron and silicon is considerably less under conditions of electrolysis than it is in molten borax without electrolysis. Calculated corrosion relationships [$\delta = f(\tau)$] (Fig. 2) have been determined from the experimental results obtained for materials based on the nitrides of boron and silicon in molten borax under conditions of electrolysis.

The longest duration of the tests was 1000 h.

$$
\begin{aligned}
Si_3N_4\text{–SiC} &\quad - \quad 0{,}000625\,\tau \\
Si_3N_4 &\quad - \quad 0{,}000328\,\tau \\
Si_3N_4\text{–BN} &\quad - \quad 0{,}00032\,\tau \\
BNC &\quad - \quad 0{,}0003\,\tau
\end{aligned}
$$

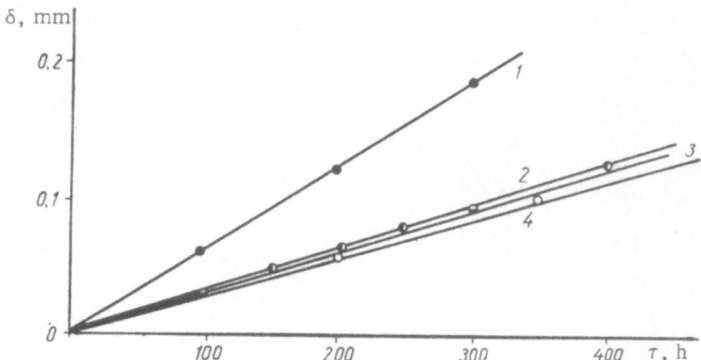

Fig. 2.   Corrosion of materials based on the nitrides of boron and
silicon in molten borax under conditions of electrolysis: 1) Si_3N_4–
SiC; 2) Si_3N_4; 3) Si_3N_4–BN; 4) BNC.

As can be seen from the above results, the rates of corrosion
of materials based on the nitrides of boron and silicon are of the
same order except in the case of materials based on Si_3N_4–SiC.
However, the rate of corrosion of BNC materials in molten borax
in the absence of electrolysis is lower by an order of magnitude
than it is in the most resistant materials. Under technological
conditions electrolytic boronization occupies approximately three
quarters of the total time of the process. Hence for an approximate
service life of the crucible of 10,000 h, the thickness of the boron

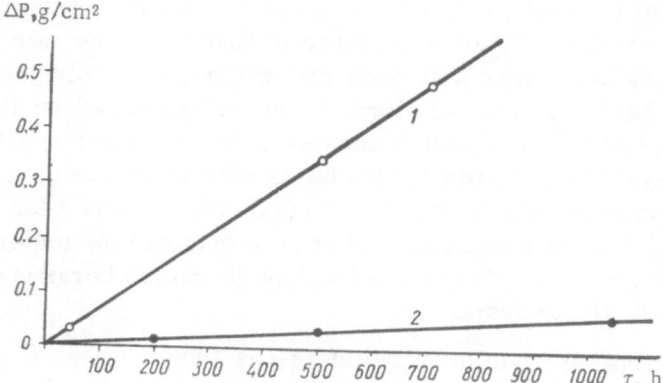

Fig. 3.   Change in weight of boron carbonitride specimens in molten borax: 1) without electrolysis; 2) under conditions of electrolysis.

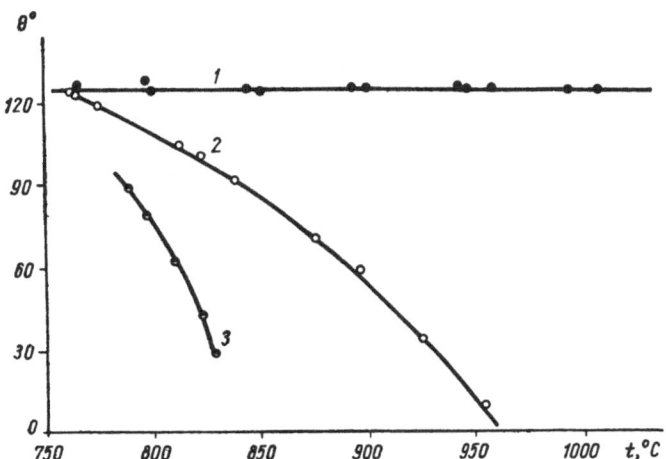

Fig. 4. Temperature dependence of the angle of wetting of materials by molten borax: 1) boron carbonitride; 2) graphite; 3) Nichrome.

carbonitride corrosion attack is $0.0003 \times 7500 + 0.005 \times 2500 =$ 14.75 mm. On the other hand, the most resistant oxide-base material (Al_2O_3) with a thickness of 15 mm gives a crucible life of only 250 h. As can be seen from Fig. 3, the losses in weight of the specimens on electrolysis are an order of magnitude lower. The wettability of the materials by molten borax has been investigated, and the temperature dependence of the angle of wetting is shown in Fig. 4 for boron carbonitride, graphite, and Nichrome. As can be seen from these results, borax hardly wets boron carbonitride, it begins to wet graphite at 840°C, and Nichrome is wetted at temperatures of the order of 780°C. On the basis of the results obtained in the present investigation, boron carbonitride is recommended as a suitable material for the crucible of the apparatus used for electrolytic boronizing. Equipment for this process built at VNIIETO has confirmed the very good working properties of the boron carbonitride crucible. This year VNIIETO, in conjunction with the Zaporozhe Branch of the OKB of the Institute for Problems in Materials Science of the Academy of Sciences of the Ukrainian SSR has built at one of their plants the first industrial-experimental equipment for electrolytic boronizing, using a boron carbonitride crucible.

The Reaction of Highly Refractory Oxides, Refractory Carbides, and Boron Carbonitride with Molybdenum and Tungsten under the Operating Conditions in High-Temperature Vacuum Electric Furnaces

I. A. Etinger, L. F. Mal'tseva, L. A. Savranskaya, E. N. Marmer, V. S. Kindysheva, V. A. Nikolaeva, and T. V. Dubovik

A study has been made of the contact reaction of the most promising materials for electric-furnace construction (Al_2O_3, ZrO_2, NbC, and boron carbonitride) with molybdenum and tungsten under conditions approximating to the operating conditions in industrial vacuum electric furnaces in the temperature range 1200-2400°C and pressures of 10^{-3} - 10^{-4} mm Hg. The temperatures at which reaction begins have been determined, and the reaction of all the materials has been found to proceed more rapidly via the gas phase. The reaction between boron carbonitride and molybdenum and tungsten is shown to be accompanied by the formation of new phases on the metal surface.

The compatibility of the materials employed is one of the principal requirements that have to be met in the construction of electric furnaces operating at temperatures above 1800°C and pressures of 10^{-3}-10^{-4} mm Hg.

In building vacuum and gas-filled electric furnaces of various types for heating molybdenum and tungsten billets and forgings for rolling and extrusion, and also for the heat treatment of re-fractory-metal components, it is rather difficult to chose materials for the lining, supports for the billets, dies, etc., which will not contaminate the heated material by reacting with it. Of the high-temperature materials, those that satisfy this requirement most fully are the oxides Al_2O_3, ZrO_2, BeO, ThO_2, the refractory com-pounds NbC, ZrC, TaC, HfC, and boron carbonitride. The results of investigations reported in [1-5, 10-12] are conflicting owing to differences in the method of production, chemical composition, and structure of the materials, and also the test conditions.

The object of the present research was to investigate the contact reaction of the most widely used and promising materials — Al_2O_3, ZrO_2, NbC, ZrC, and boron carbonitirde — with molybdenum and tungsten under conditions approximating to the operating con-ditions in industrial vacuum electric furnaces.

The highly refractory oxide materials investigated consisted of commercial specimens [9] and also of specimens prepared under laboratory conditions. The chemical composition of the latter is given in Table 1.

The carbides of niobium and zirconium were obtained by various powder-metallurgy methods at IPM AN UkrSSR and at NIIVT, while the boron carbonitirde was prepared at IPM AN UkrSSR and its Zaporozhe branch.

The chemical composition of these materials was as follows: MbC contained 82.2-89.8% Nb, 8.6-13.0% C_{total}, 0.1-1.4% C_{free}, 0.2-2.3% N_2, and 0.1-0.4% O_2: ZrC contained 86.8% Zr, 11.4% C_{total}, 1.2-1.4% C_{free}; and boron carbonitride contained 5-10% C, 34-37% N_2, 51-55% B, and 2.5% B_2O_3.

The molybdenum and tungsten specimens were prepared from commercial rods.

The specimens under investigation were brought into contact with one another by means of tungsten clamps; in the case of the oxides, metal extrusions were held inside the ceramic specimens.

TABLE 1. Chemical Composition of Ceramic
Specimens Prepared in IPM AN UkrSSR and
NIIVT

Material	Chemical composition, %					
	ZrO_2	Al_2O_3	SiO_2	CaO	MgO	Fe_2O_3
Aluminum oxide	—	99.20	0.60	0.10	Traces	0.10
Aluminum oxide containing 9%SiO_2	—	89.29	9.37	0.33	0.03	0.10
Aluminum oxide containing 18% SiO_2	—	80.83	18.4	0.39	0,06	0.12
Aluminum oxide containing 55% SiO_2	—	44.27	55.00	0.35	0,06	0,13
Commercial zirconium dioxide stabilized with CaO	90.62	0.96	1.65	5.34	0,95	0.48
Pure zirconium dioxide stabilized with CaO	94.27	0.26	0.72	4.45	0.10	0.21
Pure zirconium dioxide stabilized with MgO	92,11	0.18	0,61	0.24	6.58	0,28

The specimens prepared in this way were heated in vacuum high-temperature electric furnaces of two types: with thermal insulation in the form of screens or with a highly refractory lining. The test materials were placed in a container made of tungsten sheet in order to avoid the possibility of oil vapors, evaporation products, and gases evolved by the elements of the heating chamber coming in contact with them.

After the tests microsections were prepared from the specimens for metallographic examination. The microhardness was measured under a load of 50 g on a PMT-3 machine.

The reaction of molybdenum and tungsten with highly refractory oxide ceramics based on Al_2O_3 containing various amounts of SiO_2 was studied in the temperature range 1200-1950°C for a period of 2-75 h under a pressure of about 10^{-4} mm Hg. The onset of the reaction was determined from the change in structure of the metal. Microstructural analysis of the metals and microhardness measurements showed that reaction between the oxide and the metal is accompanied by oxidation of the latter. At an exposed surface of the metal oxidation takes place via the gaseous phase and leads to the appearance of cavities in the metal owing to the evaporation of the readily volatilized oxides of molybdenum and

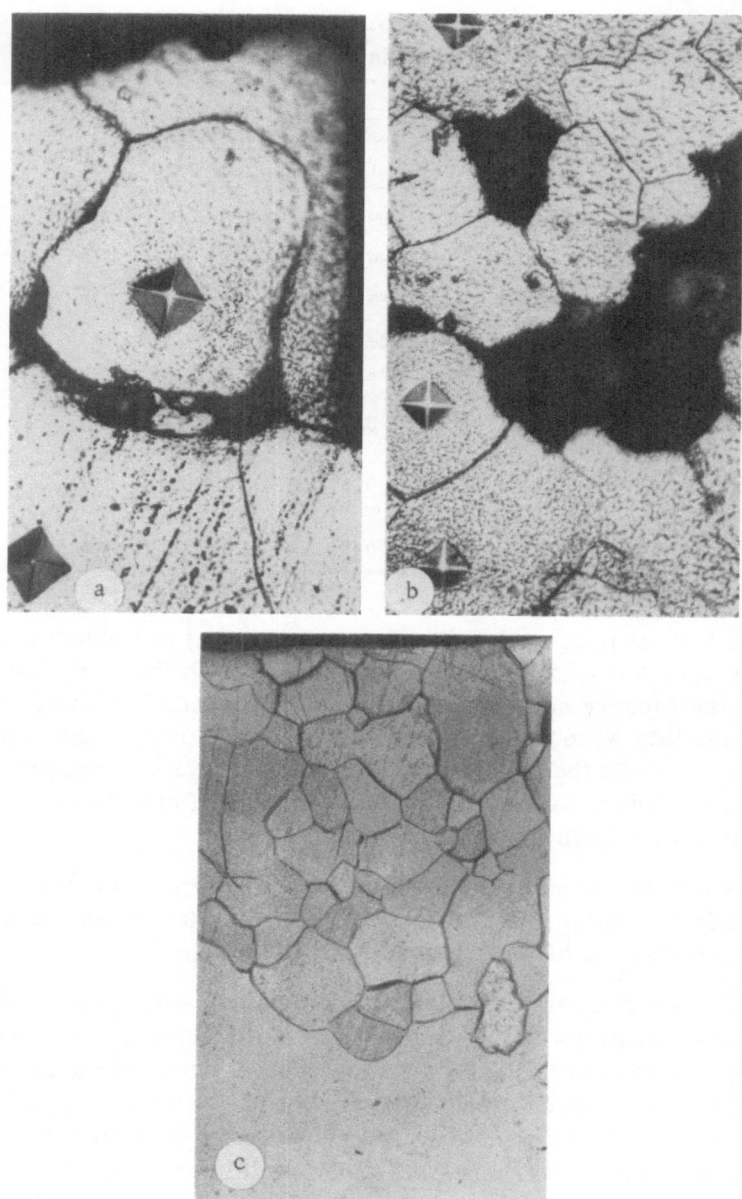

Fig. 1. Microstructure of molybdenum and tungsten specimens after reaction with ceramic based on Al_2O_3 at a pressure of about 10^{-4} mm Hg: a) molybdenum (1900°C, 5 h, reaction via the gaseous phase) × 500; b) tungsten (1900°C, 5 h, reaction via the gaseous phase) × 500; c) tungsten (1950°C, 5 h, reaction by direct contact) × 200.

tungsten that are formed. The porosity of a metal which is in direct contact with the oxide is considerably less marked. In some cases we observed in the structure of the metal the formation of fine grains at the edge and coarse grains at the center, the result, no doubt, of saturation of the surface layers of the metal with oxygen. Microstructures of molybdenum and tungsten after the reaction with Al_2O_3 are shown in Fig. 1.

Aluminum oxide begins to react with molybdenum and tungsten at 1900°C; aluminum oxide containing additions of 9, 18, and 55% SiO_2 begins to react at 1800, 1700, and 1450°C, respectively. The temperatures at which the reaction of molybdenum and tungsten with Al_2O_3 begins are given in Table 2.

The reaction of specimens of zirconium dioxide of varying degrees of purity with molybdenum and tungsten was investigated at a pressure of about 10^{-4} mm Hg in the temperature range 1450–2300°C for periods from 2 to 50 h.

TABLE 2. Temperature of the Beginning of the Reaction between Specimens of Highly Refractory Oxide Ceramics Based on Al_2O_3 and ZrO_2 and Molybdenum and Tungsten at Pressures of 10^{-3}–10^{-4} mm Hg

Composition of ceramic	Temperature of beginning of reaction, °C		Holding time, h
	with molybdenum	with tungsten	
Aluminum oxide	1900	1900	To 75
Aluminum oxide with 9% SiO_2	1800	1800	" "
Aluminum oxide with 18% SiO_2	1700	1700	" "
Aluminum oxide with 55% SiO_2	1450	1450	" "
Commercial zirconium dioxide stabilized with CaO	1900	1900	" "
Pure zirconium dioxide stabilized with CaO	1900	1900	" "
Pure zirconium dioxide stabilized with MgO	1900	1900	" "
Aluminum oxide	No reaction at 1750	Above 1800	From 100 to 1000
Aluminum oxide with 12% SiO_2	—	1750	The same
Aluminum oxide with 30% SiO_2	1450	—	The same
Aluminum oxide with 45% SiO_2	1400	—	The same
Commercial zirconium dioxide stabilized with CaO	No reaction at 1650	1900	The same
Zirconium carbide (ZrC)	200	2200	2.0
Niobium carbide (NbC) with atomic ratios of 1.0	1800	2300	2.0
0.89	—	2200	2.0
0.81	—	2200	2.0
0.73	—	2200	2.0

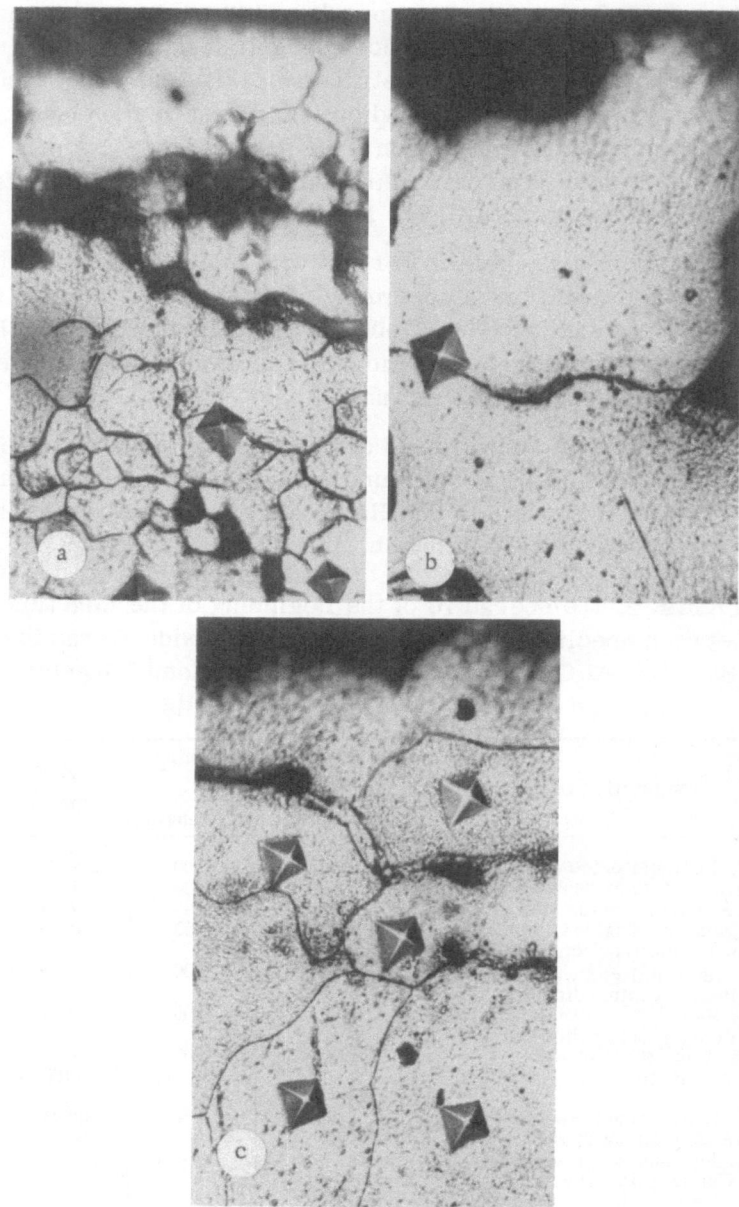

Fig. 2. Microstructures of molybdenum and tungsten specimens after reaction through the gas phase with ceramics based on ZrO_2 of varying degrees of purity under a pressure of about 10^{-4} mm Hg (1900°C, 5 h)
× 500: a) tungsten (pure ZrO_2 stabilized with MgO); b) molybdenum
(pure ZrO_2 stabilized with MgO); c) molybdenum (commercial ZrO_2
stabilized with CaO).

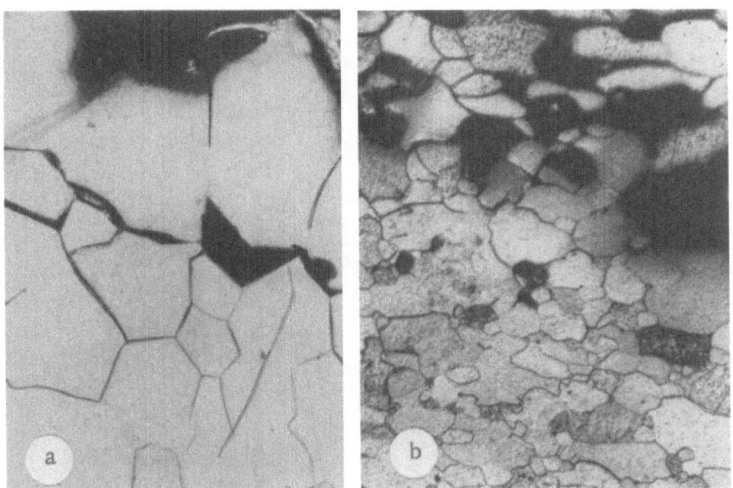

Fig. 3. Microstructures of molybdenum and tungsten heating elements
after service in a vacuum electric furnace with a lining based on Al_2O_3
and SiO_2 at a pressure of about 10^{-4} mm Hg: a) molybdenum (1450°C,
50 h, lining made of Al_2O_3 containing 45% SiO_2) × 300; b) tungsten
(1750°C, 200 h, lining made of Al_2O_3 containing 12% SiO_2) × 100.

The onset of the reaction was indicated by the appearance
of considerable porosity at the edge of the metal resulting from
its oxidation and the subsequent evaporation of the reaction products.

Metallographic studies revealed that zirconium dioxide of
varying degrees of purity began to react with molybdenum and
tungsten at 1900°C under a pressure of about 10^{-4} mm Hg after a
period of heating of 2-5 h. The microstructures of molybdenum
and tungsten obtained as a result of reaction with ZrO_2 are shown
in Fig. 2. The temperatures at which the reaction between ZrO
and molybdenum and tungsten began are given in Table 2.

Examination of microsections of material based on Al_2O_3
and ZrO_2 showed that in specimens with extruded molybdenum and
tungsten rods inside, the metal penetrated into the depth of the
ceramic. This can be attributed to the fact that the readily volatile
metal oxides formed as a result of the reaction had diffused into
the ceramic specimens, where they were again reduced to the pure
metal in consequence of disproportionation reactions taking place.

In addition to the tests on oxide specimens, we also investigated
the reaction of refractory linings based on Al_2O_3 and ZrO_3 with

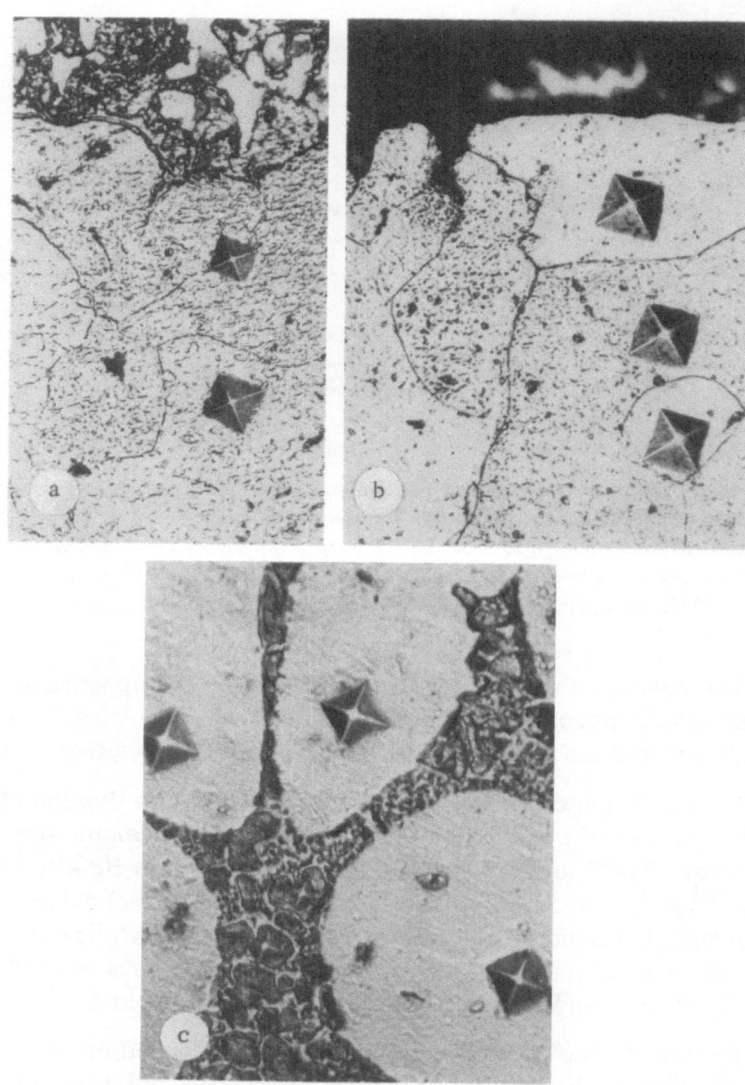

Fig. 4. Microstructures of molybdenum and tungsten specimens after reacting with NbC specimens of various compositions within the homogeneity range on heating for 2 h ($\times$ 500): a) tungsten (2200°C, $NbC_{0.81}$); b) molybdenum (1800°C, $NbC_{1.0}$); c) molybdenum (2200°C, $NbC_{1.0}$).

molybdenum and tungsten used as heating elements suspended from hooks along the lining walls so as to simulate the actual operating conditions of industrial vacuum electric furnaces. The method has been described in [6-8]. The tests were carried out at a pressure of about 10^{-4} mm Hg in the temperature range 1300-1900°C for periods from 100 to 1000 h. The start of the reaction was determined from the change in electrical resistance of the heating element over a long period and from the reduction in its diameter. The microstructure of the heating elements was also examined. The temperatures at which reaction began in these materials are given in Table 2. They are in satisfactory agreement with the results obtained on the test specimens. Photographs of the microstructure of the molybdenum and tungsten heating elements are shown in Fig. 3.

An investigation of the resistance of NbC and ZrC to molybdenum and tungsten was made at a pressure of about 10^{-4} mm Hg in the temperature range 1800-2400°C for a period of 2 h.

It was found from microstructural analysis of the contact area that a new phase appeared in disseminated form or as a frontal

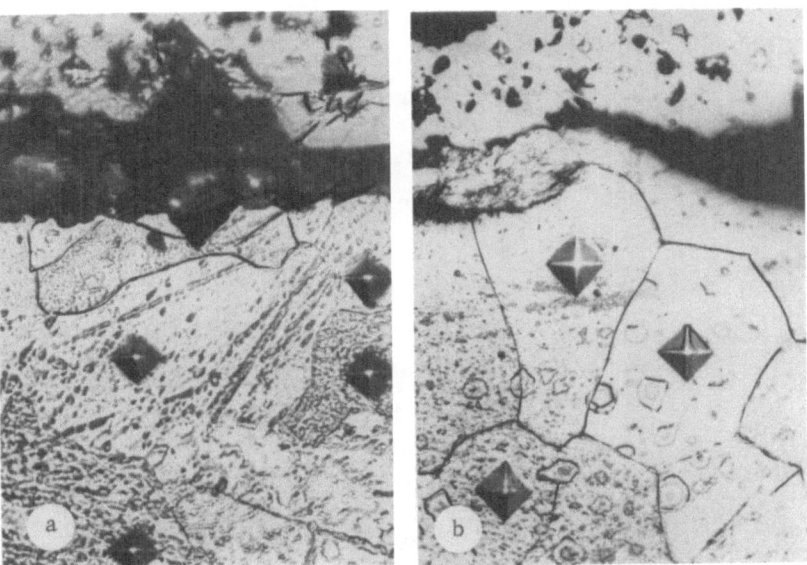

Fig. 5. Microstructures of molybdenum (a) and tungsten (b) specimens after reacting with ZrC for 2 h at 2200°C under a pressure of about 10^{-4} mm Hg. × 500.

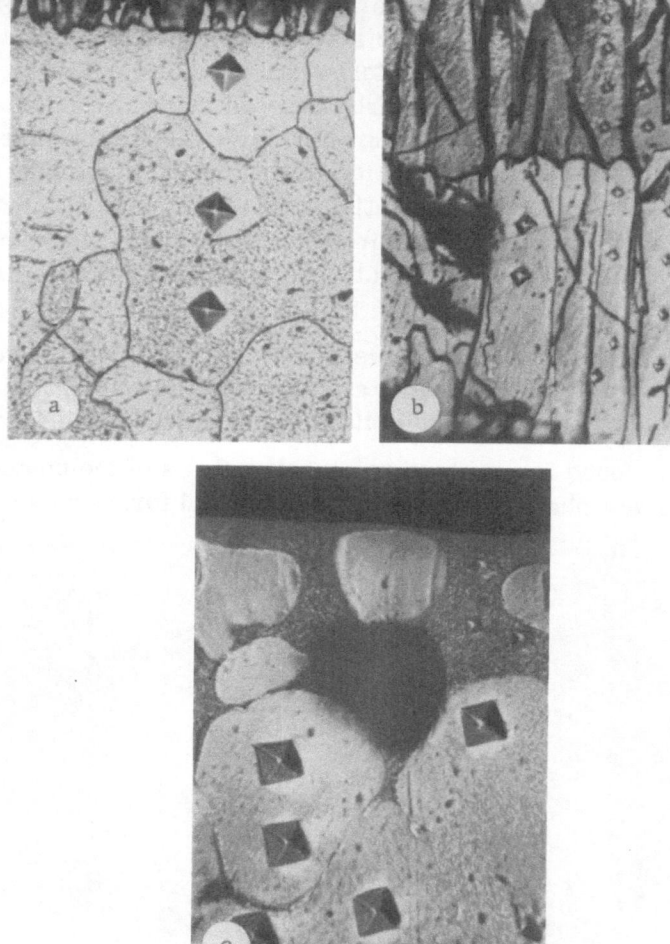

Fig. 6. Microstructures of molybdenum and tungsten specimens after reacting with boron carbonitride at a pressure of 10^{-3}-10^{-4} mm Hg (× 500): a) tungsten (1700°C, 0.3 h); b) tungsten (2000°C, 2 h); c) molybdenum (1800°C, 0.3 h).

TABLE 3. Microhardness of Various Parts of Molybdenum and Tungsten Specimens after Heating Them in Boron Carbonitride Containers at a Pressure of 10^{-3}-10^{-4} mm Hg

Temperature, °C	Time, h	Microhardness Hμ, kg/mm²		Depth of reaction zone, mm	Nature of contact	Manufacturer of boron carbonitride
		Reaction zone	Principal phase (metal)			
			Molybdenum			
1650	0.5	600—750	220	—	—	Zaporozhe Branch of IPM AN UKrSSR
1700	0.3	2300—2850	250	0.070	—	The same
1800	0.3	350—2600	220	0.012	—	The same
1800	1.3	2100—2850	230	0.096—0.126	—	The same
1900	0.1	1500—2500	230	0.130	—	The same
1800	2	1600—2500	190	0.072—0.117	—	IPM AN UKrSSR
1800	2	1350—2100	250	0.045—0.144	—	Boron carbonitride with 10% Mo.
1800	2	1650—2300	190	0.018—0.290	—	IPM AN UKrSSR chemically pure boron carbonitride
			Tungsten			
1850	2	2850—3200	500	0.207—0.225	Direct contact	Zaporoshe Branch of IPM ANUkrSSR
		2300—2500	400	0.297—0.315	The same	The same
		2500—2850	500	0,207—0.261	Via gas phase	The same
		2300—2850—3200	450	0.297—0.324	The same	The same
		2850—3200	500	0.100	The same	The same
		2500—2850—3200	500	0.160—0.320	The same	The same
2000	2	2100—2300—2500	500	0.126—0.135	Direct contact	The same
		2100—2300—2500	500	0.432—0.459	The same	The same
		2500—2850—3200	450	0.500—0.640	The same	The same
		1600—2300—2500—2900	450	0.792—1.035	Via gas phase	The same
		2300—2600—2300—3200	450	0.756—0,765	The same	The same
		2100—2300—2500	450	0.594—0.837	The same	The same
		2100—2300—2500—2850	500	0.702—0.927	The same	The same
			400	0.009—0,018	Direct contact	IPM AN UkrSSR
		1900—2100—2900	400	0.054—0,127	Via gas phase	The same

layer at the metal boundary at a temperature which was characteristic of each metal—carbide pair. This temperature was taken
as the temperature at which the reaction started. The thickness
of the new phase layer increased with rise in temperature. Measurement of the microhardness of the phases formed and a comparison of the results with the values for the carbides and their
solid solutions made it possible to determine the nature of the reaction process between the materials. The microhardness values
of the new phases lay in the range 500-2800 kg/mm^2.

The temperatures at which reaction began between molybdenum and tungsten and NbC and ZrC of various compositions
within the homogeneity range are given in Table 2, and photographs
of the microstructures are shown in Figs. 4 and 5.

The reaction of boron carbonitride with molybdenum and
tungsten was studied in the temperature range 1650-2000°C at a
pressure of 10^{-3}-10^{-4} mm Hg for periods from 0.1 to 2 h. Besides
making tests on specimens in direct contact, we also investigated
the reaction via the gaseous phase. In these tests, the tungsten
specimens were loaded into a boron carbonitride container.

The results of the metallographic analysis (Fig. 6) and the
microhardness values, given in Table 3, showed that the reaction
between boron carbonitride and molybdenum and tungsten begins
at 1650-1700°C. It is accompanied by the formation of new phases
on the metal surface in the form of a frontal layer, the thickness
of which increases with rise in temperature and time. Thus, for
example, at 1800°C the thickness of the reaction zone between
boron carbonitirde and molybdenum increased from 0.012 to 0.126 mm
with change of holding time from 0.3 to 1.3 h. It must be noted
that the reaction between boron carbonitride and tungsten takes
place more rapidly via the gaseous phase. Thus, for example, at
2000°C the depth of the reaction zone with direct contact is approximately 0.135 mm, whereas via the gaseous phase it is 0.837 mm.

The extent of the reaction between boron carbonitride and
molybdenum and tungsten has been found to be independent of the
composition of the boron carbonitride and of the way in which it
was produced. From the change in microhardness in the contact
zone it may be concluded that the reaction is accompanied by the
formation of several phases. The composition of the compounds
formed cannot be determined from the microhardness values owing
to the wide scatter of the results obtained.

Irrespective of their original composition, the boron carbonitride specimens changed color during heating from white to black; their porosity increased at the same time, indicating decomposition of the boron carbonitride in vacuum at temperatures above 1650-1700°C.

Conclusions

1. A study has been made of the reaction of high-temperature materials based on Al_2O_3, ZrO_2, ZrC, NbC, and boron carbonitride with molybdenum and tungsten under the conditions prevailing in vacuum electric furnaces in the temperature range 1200-2400°C and in the pressure range $10^{-3}-10^{-4}$ mm Hg.

2. The temperatures at which the reaction begins have been determined from the results of metallographic analysis.

3. In all materials the reaction proceeds more rapidly via the gaseous phase. In the case of oxides the reaction is accompanied by the appearance of a porous structure, whereas in the case of the carbides and boron carbonitride a new phase is formed.

Literature Cited

1. V. I. Ivanov et al., Ogneupory, No. 7 (1963).
2. Ogneupory, No. 9, 118 (1964).
3. G. V. Samsonov, Ogneupory, No. 3, 122-134 (1956).
4. G. V. Samsonov, G. A. Yasinskaya, and E. A. Shiller, Ogneupory, No. 7 (1961).
5. V. Espe, The Technology of Electrovacuum Materials [in Russian], Gosenergoizdat, Moscow and Leningrad (1962).
6. I. A. Etinger, O. S. Gurvich, and E. N. Marmer, Electrotermiya, No. 5, 23-26 (1962).
7. I. A. Etinger et al., Elektrotermiya, No. 6, 12-14 (1963).
8. I. A. Etinger, O. S. Gurvich, and E. N. Marmer, Elektrotermiya, 31, 21-23 (1964).
9. Collection of the Actual Interrepublic Technological Conditions in the Output of the Refractory Industry, Approved by the State Metallurgical Committee in 1963-64 [in Russian], State Committee for Ferrous and Nonferrous Metallurgy in the State Plan of the USSR (1964).
10. P. D. Johnson, J. Am. Ceram. Soc., 33:168 (1950).
11. G. Economos and W. D. Kingery, J. Am. Ceram. Soc., 36:403 (1953).
12. R. Kieffer and I. Benesovsky, Metallurgia, 58:119-124 (1958).

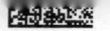